AQUATIC EFFECTS
OF
ACIDIC DEPOSITION

AQUATIC EFFECTS
OF
ACIDIC DEPOSITION

Timothy J. Sullivan

WITHDRAWN

LEWIS PUBLISHERS
Boca Raton London New York Washington, D.C.

Library of Congress Cataloging-in-Publication Data

Sullivan, Timothy, J., 1950-
    Aquatic effects of acidic deposition / Timothy J. Sullivan.
        p.   cm.
    Includes bibliographical references and index.
    ISBN 1-56670-416-2 (alk. paper)
        1. Acid deposition--Enviromental aspects. 2. Acid pollution of rivers, lakes,
etc.--United States. I. Title.

TD427.A27 S85 2000
628.1'683—dc21   99-058887
                                                                                    CIP

© 2000 by CRC Press LLC
Lewis Publishers is an imprint of CRC Press LLC

No claim to original U.S. Government works
International Standard Book Number 1-56670-416-2
Library of Congress Card Number 99-058887
Printed in the United States of America 1 2 3 4 5 6 7 8 9 0
Printed on acid-free paper

To Debbie, Laura, and Jenna

Acronyms

ABW	Absaroka-Beartooth Wilderness
AERP	Aquatic Effects Research Program
ALSC	Adirondack Lakes Survey Corporation
ALTM	Adirondack Long-Term Monitoring Program
ANC	Acid neutralizing capacity
ANC_G	ANC as measured by Gran titration in the laboratory
AQRV	Air quality related values
ASI	Acidic stress index
BMW	Bob Marshall Wilderness
CAAA	Clean Air Act Amendments
CALK	Calculated ANC
CCA	Canonical correspondence analysis
CEC	Cation exchange capacity
CLIMEX	Climate Change Experiment
DDF	Dry deposition factor
DDRP	Direct Delayed Response Project
DIC	Dissolved inorganic carbon
DOC	Dissolved organic carbon
ELS	Eastern Lake Survey
ELS-I	Phase I of the Eastern Lake Survey
ELS-II	Phase II of the Eastern Lake Survey
EPA	Environmental Protection Agency
ERP	Episodic Response Program
EXMAN	Experimental Manipulation of Forest Ecosystems in Europe Program
FADS	Florida Acid Deposition Study
FISH	Fish in Sensitive Habitats Project
FLAG	Federal Land Managers AQRV Group
FLM	Federal Land Manager
GLAC	Glacier National Park
GLEES	Glacier Lakes Ecosystem Experiment Site

GRTE	Grand Teton National Park
HBEF	Hubbard Brook Experimental Forest, NH
HUMEX	Humic Lake Acidification Experiment
IA	Integrated Assessment
IAG	Internal Alkalinity Generation Model
IAM	Integrated Assessment Model
ILWAS	Integrated Lake–Watershed Acidification Study
IWS	Integrated Watershed Study
LAC	Limits of acceptable change
LTM	Long-term monitoring program
LTRAP	Long Range Transboundary Pollution Program
MAGIC	Model of Acidification of Groundwater in Catchments
MAGIC-WAND	Model of Acidification of Groundwater in Catchments with Aggregated Nitrogen Dynamics
MERLIN	Model of Ecosystem Retention and Loss of Inorganic Nitrogen
MPCA	Minnesota Pollution Control Agency
NADP	National Atmospheric Deposition Program
NAPAP	National Acid Precipitation Assessment Program
NDDN	National Dry Deposition Network
NIICCE	Nitrogen Isotopes and Carbon Cycling in Coniferous Ecosystems Model
NITREX	Nitrogen Saturation Experiments Program
NIVA	Norwegian Institute for Water Research
NLS	National Lake Survey
NPP	Net primary production
NPS	National Park Service
NSS	National Stream Survey
NSWS	National Surface Water Survey
NTN	National Trends Network
NuCM	Nutrient Cycling Model
OTA	Office of Technology Assessment
OWLS	Object Watershed Link System
PIRLA	Paleoecological Investigation of Recent Lake Acidification

PIRLA-II	Continuation of Paleoecological Investigation of Recent Lake Acidification
PRL	Proton reference level
PSD	Prevention of significant deterioration
QA/QC	Quality assurance/quality control
RADM	Regional Acid Deposition Model
RAIN	Reversing Acidification in Norway Project
RIA	Randomized intervention analysis
RILWAS	Regionalized Integrated Lake Watershed Acidification Study
RMSE	Root mean square error
ROMO	Rocky Mountain National Park
SAMI	Southern Appalachian Mountains Initiative
SBW	Selway-Bitterroot Wilderness
SNSF	Sur Nedbørs Virkning på Skog og Fisk (Acid Precipitation—Effects on Forest and Fish)—acidic deposition research program in Norway
SOS/T	State-of-Science/Technology Report
SWAP	Surface Water Acidification Program
TAF	Tracking and Analysis Framework
TOC	Total organic carbon
UN/ECE	United Nations/Economic Commission for Europe
USGS	U.S. Geological Survey
VTSSS	Virginia Trout Stream Sensitivity Study
VWM	Volume weighted mean
WACALIB	Weighted-averaging calibration
WLS	Western Lake Survey
WMP	Watershed Manipulation Project
WY	Water year
YBP	Years before present
YELL	Yellowstone National Park

Acknowledgments

Aquatic Effects of Acidic Deposition covers a wide range of topics and scientific disciplines in which I have been involved in my own research over the last two decades. My interest in these areas of research has been stimulated by a large number of colleagues, many of whom are specialists in the various elements and fields of study covered in this book. This state-of-the-science summary was made possible by the concerted efforts of a great many scientists who provided the research foundation that I have summarized and by my interactions with them to enhance my understanding of the key elements that are discussed.

I am very grateful to Jayne Charles who did all of the word processing and worked extensively with the reference list. Susan Binder kindly assisted with graphics and many other aspects of document production. Many of the figures presented in the book were prepared by Susan, Kellie Vaché, and Joe Bischoff.

Several colleagues kindly provided helpful comments on an earlier draft of the book. These include Jack Cosby, Steve Kahl, and Kathy Tonnessen.

Funding to support preparation of this book was provided by the National Oceanic and Atmospheric Administration, through the National Acid Precipitation Assessment Program; by the National Park Service, Air Resources Division in Denver, through a grant to the University of Virginia; and the U.S. Environmental Protection Agency, through a grant to Rensselaer Polytechnic Institute. Much of my recent research that is summarized herein was supported by the U.S. Department of Energy, the National Park Service, the U.S. Environmental Protection Agency, and the USDA Forest Service. None of the previously mentioned agencies have cleared this book through their peer or administrative review processes, and no official endorsement is implied.

I am very grateful to Mike Uhart, Kathy Tonnessen, Jack Cosby, John Karish, and Jon Zehr for assistance with funding arrangements, as well as support and encouragement.

Preface

In 1990, the U.S. National Acid Precipitation Assessment Program (NAPAP) completed the initial phase of what was then the largest environmental research program ever conducted. NAPAP research investigated the causes and effects of, and mitigation strategies for, acidic atmospheric deposition throughout the U.S. This massive environmental research and assessment effort took 10 years to complete, involved hundreds of scientists, engineers, and economists, and cost in excess of $500 million. The scientific culmination of this research was embodied in a series of 27 State of Science and Technology (SOS/T) Reports that were published in 1990. In addition, a policy report was published in 1991 as the NAPAP Integrated Assessment. After 1990, research funding for work on acidic deposition effects decreased suddenly and substantially. However, many significant research programs were still in progress and/or not yet published, especially in the areas of aquatic effects of nitrogen and sulfur deposition. Results from these efforts appeared in the scientific literature during the early 1990s. In addition, a suite of research projects was initiated post-1990, albeit with somewhat lower funding levels than were common during the heyday of the NAPAP research program. These latter projects took advantage of the significant knowledge gains of the 1980s and, therefore, tended to be more focused and productive than earlier research efforts. Many addressed significant knowledge gaps that had been identified in the SOS/T reports, particularly regarding the interactions between acidic deposition and other sources of natural and anthropogenic acidity. As a consequence, a large body of scientific information related to aquatic effects of atmospheric nitrogen and sulfur deposition has been produced since publication of the SOS/T series of reports. New findings have added support to the state of scientific understanding in some areas, modified it in others, and led to the development of new paradigms and perspectives in still other areas of research.

The primary aim of this book is to summarize and synthesize major advancements since 1990 in the state of scientific understanding of the aquatic effects of atmospheric deposition of nitrogen and sulfur. It is intended to emphasize advancements in those aspects of aquatic effects research that are of direct policy relevance. Thus, topics concerning quantification of the magnitude of effects and recent developments in the area of predictive modeling capabilities are deemed to be of great importance, for the purposes of this book. Special attention is given to those aspects of aquatic effects research that had either been poorly studied pre-1990 or for which major research efforts have been completed in recent years. Topics of special interest include virtually all aspects of nitrogen effects research, as well as the importance of natural sources of acidity, the influence of land use and landscape change on the chemistry of drainage waters, and the role of short-term episodic events.

This book is intended as a teaching resource and reference source. It provides a comprehensive update on the state of scientific understanding regarding an important environmental topic. It also illustrates the progression and refinement of the scientific knowledge base as research in this field has evolved from general basic research to more narrowly focused efforts aimed at answering specific questions. The target audience includes advanced students of environmental science and engineering and applied environmental practitioners. The latter group includes federal and state land managers and environmental stewards, many of whom are tasked with protecting sensitive natural resources from air pollution degradation and overseeing and prioritizing efforts to mitigate past damage. The wealth of recent knowledge gains summarized here will assist environmental professionals in making informed judgments regarding air pollution sensitivities, effects, and remediation.

The effects of atmospheric nitrogen and sulfur inputs to watershed systems, and the interactions between such inputs and other natural and anthropogenic features and stressors, provide an ideal framework for the study of upland hydrobiogeochemistry. Such study requires understanding of the key aspects of myriad disciplines and ecosystem compartments. Major components include mass balance input–output calculations, the study of hydrological flowpaths as water moves through the watershed system, and a wide range of interactions between drainage water and soils, geological substrates, and both terrestrial and aquatic biota. Understanding how hydrobiogeochemical processes and cycles govern the response of the entire watershed to atmospheric inputs and the associated interactions with natural features of the landscape, climate, and human disturbance aids our understanding of global ecosystems and the influence of human activities on ecosystem function and integrity.

Timothy J. Sullivan

Contents

1

Introduction

1.1   1990 NAPAP Reports and Integrated Assessment

It is well known that emissions of sulfur and nitrogen from power plants, industrial facilities, and motor vehicles are linked to acidic deposition in many parts of the world. The potential effects of acidic deposition on human health and the environment became a major concern in the U.S. during the 1970s. Published reports linked acidic deposition with surface water acidification, fish kills, damage to crops and materials, and adverse effects on human health. The U.S. Congress considered actions to limit the emissions into the atmosphere of acid-forming precursor materials, including sulfur and nitrogen oxides, but information was limited regarding key processes and cause/effect relationships.

In order to provide sufficient scientific information with which to make the policy and regulatory decisions thought to be necessary for the protection of the environment and public welfare, Congress mandated a 10-year research program. The study was initiated under the Acid Precipitation Act of 1980 (PL 96-294, Title VII). An Interagency Task Force was established and the National Acid Precipitation Assessment Program (NAPAP) was started. NAPAP set out to answer important questions regarding the distribution of acid-sensitive natural resources and their degree of sensitivity, emissions and deposition of sulfur and nitrogen, source–receptor relationships and processes, the probable extent of damage and potential need for mitigation, and the availability of emission control technologies and mitigation options. Over the next 10 years, NAPAP was to spend over $500 million and became, at the time, what was the largest environmental research program ever conducted.

NAPAP developed a complex research and assessment process that involved many hundreds of scientists throughout the U.S., Canada, and Europe. Major objectives were to identify the causes and extent and magnitude of the effects of acidic deposition. The culmination of the NAPAP process in 1990 included two major elements. A series of 27 State of Science and Technology (SOS/T) Reports was prepared to summarize NAPAP's technical findings. The SOS/T reports addressed the full spectrum of acidic deposition

1

issues, from emissions (SOS/T 1) through valuation (SOS/T 27). Of these reports, six covered aspects of acidic deposition aquatic effects (SOS/T 9 through 15) that were thought to comprise the most significant components of environmental impacts. The second major element that culminated the NAPAP effort was the Integrated Assessment (IA), a policy report to Congress that was published in 1991.

In 1990, as the NAPAP research program was winding down, Congress passed the Federal Acid Deposition Control Program as Title IV of the Clean Air Act Amendments. The objective of Title IV was to reduce the adverse effects of acidic deposition by reducing the emissions of sulfur dioxide in particular and to a lesser extent nitrogen oxides. Title IV stated that "reduction of total atmospheric loading of sulfur dioxide and nitrogen oxides will enhance protection of the public health and the environment." An annual 10 million ton reduction in SO_2 emissions below 1980 levels was mandated and targeted to electric utilities. In addition, a reduction in the emissions of nitrogen oxides of about 2 million tons from 1980 levels was specified. Upon full implementation of the control program, reductions of about 40% in SO_2 emissions and 10% in NO_x emissions from the 1980 base year are anticipated. There will be a national cap on the utility and industrial emissions of SO_2 at this level, but a national cap on the emissions of NO_x is not legislated, and NO_x emissions are projected to rise in the 21st century (NAPAP 1992). The Federal Acid Deposition Control Program includes an innovative market-based approach for achieving emissions reductions. Electric utilities have been given considerable flexibility in achieving reductions in efficient, cost-effective ways. Emissions allowances were issued to affected utility units based on their historic fuel consumption and rate of SO_2 emissions. Each allowance entitles a unit to emit 1 ton of SO_2 during or after the year specified on the allowance. Once allocated, allowances are marketable, allowing affected utilities to buy, sell, or bank allowances for future use (NAPAP 1992).

The first phase of implementation of the control program began January 1, 1995. About 260 units at the 110 highest emitting electric-utility plants in the eastern U.S. were allocated emissions allowances. Phase II began January 1, 2000, and will affect about 2300 electric generating units throughout the U.S., serving generators with capacities of 25 megawatts or greater. When the program is fully implemented in 2010, the annual allocation of emissions allowances will result in a national emissions cap of 8.9 million tons of SO_2 from utility units. Nonutility SO_2 emissions will be capped at 5.6 million tons per year.

NAPAP expected a number of environmental benefits in response to implementation of Title IV. Lakes and streams acidified by acidic deposition were expected to recover and support fish life. The risks of long-term soil degradation, ecosystem change, and loss of biological diversity were expected to be reduced. Average visibility was expected to improve, allowing for increased enjoyment of scenic vistas throughout the nation. Stresses on forest health were expected to decrease, particularly in red spruce forests along the ridges of the Appalachian Mountains (NAPAP 1991).

It is too early to judge the extent to which reductions in acid deposition in response to implementation of Title IV have or have not affected aquatic chemistry or biology. Chemical effects owing to changes in atmospheric deposition exhibit lag times of one to many years. Lags in measurable effects on aquatic biota can be longer. Continued monitoring of water quality for several years or more will be required to assess potential improvements that may occur as a consequence of emissions reductions already realized. It is clear that the concentrations of sulfate have decreased substantially in surface waters in many areas of the eastern U.S. It is expected that the concentrations of sulfate in surface waters will continue to decline in many areas, especially in the Northeast. However, the extent to which surface water acidity may be reduced in response to the expected continued decreases in sulfate concentrations as well as the extent to which biological recovery may be realized remains uncertain.

NAPAP was reauthorized under the 1990 Clean Air Act Amendments (PL 101-549, Title IX). Although the research activities of NAPAP have been completed, additional assessment efforts have continued (e.g., NAPAP 1992, 1998). Since 1990, aquatic effects research has also continued, albeit at a much lower level of effort than was seen during the 1980s. The recent research has been more heavily focused, however, and has benefitted from the substantial progress in understanding that was made during the heyday of acid deposition research funding of the previous decade. In many ways, the post-1990 research has been more efficient, and great strides have been made through relatively modest levels of research funding. Many of the knowledge gaps identified by NAPAP in 1990 have been filled in large part by a series of narrowly focused, carefully designed studies. This book attempts to bring together the key findings of these recent research efforts. By summarizing advancements in the state of the science of aquatics effects since 1990, this book contributes to NAPAP's ongoing assessment activities.

1.2   Scope

The scope of this book is limited mainly to the aquatic effects of acidic deposition. The NAPAP SOS/T reports included technical summaries for six major areas of aquatic effects research:

SOS/T 9   Current Status of Surface Water Acid-Base Chemistry (Baker et al., 1990a).

SOS/T 10   Watershed and Lake Processes Affecting Surface Water Acid-Base Chemistry (Turner et al., 1990).

SOS/T 11   Historical Changes in Surface Water Acid-Base Chemistry in Response to Acidic Deposition (Sullivan, 1990).

SOS/T 12   Episodic Acidification of Surface Waters Due to Acidic Deposition (Wigington et al., 1990).

SOS/T 13   Biological Effects of Changes in Surface Water Acid-Base Chemistry (Baker et al., 1990c).

SOS/T 14   Methods for Projecting Future Changes in Surface Water Acid-Base Chemistry (Thornton et al., 1990).

SOS/T 15   Liming Acidic Surface Waters (Olem, 1990).

This book summarizes recent advancements in scientific understanding that pertain mainly to areas addressed in the SOS/T Reports 9, 11, 12, and 14, with lesser treatment of areas addressed in the SOS/T Reports 10 and 13. Liming issues (SOS/T 15) are not addressed. The greatest attention is focused on recent findings in the U.S. Research elsewhere, especially in Europe, is discussed to the extent that it has complemented research in this country. An effort is made to summarize what is new in the understanding of the science of aquatic effects of acidic deposition, particularly in those research areas that have direct bearing on policy-relevant assessment activities. Research completed and published between 1990 and 1998 receives the greatest attention.

1.3   Goals and Objectives

The major goal of this book is to summarize important scientific findings regarding the aquatic effects of acidic deposition subsequent to publication of the State of Science and Technology Reports by NAPAP in 1990. Because the focus is on advancements in the science that are of direct policy relevance, improved modeling capabilities and improved understanding of acidification responses are of greater interest than specific advancements with respect to the understanding of acidification processes.

Specific objectives of the analyses reported here are to

1. Clarify current understanding of the extent to which lake and stream systems in the U.S. have experienced chronic acidification owing to acidic deposition.

2. Quantify acidification dose–response relationships for sensitive surface waters and recent advancements regarding specification of the critical loads of acidifying compounds required to protect sensitive aquatic receptors from adverse effects.

3. Describe improvements in predictive capabilities for aquatic effects and the results of model testing efforts.

4. Clarify current understanding regarding the relative importance of various causes of surface water acidification.

5. Describe recent advancements in the understanding of episodic acidification of surface waters.

6. Describe the results of ecosystem manipulation experiments that have involved short-term increases or decreases in the levels of acidic deposition to forested plots or small catchments.

1.4   Outline of State of Science Update

Chapter 2 provides background material on acidic deposition, the major response variables of concern, and methods for evaluating acidification. Chapters 3 through 9 comprise a state of science update on advancements in acidic deposition aquatic effects research since 1990. The focus is on aspects of aquatic effects that have direct relevance to public policy. Special emphasis is given to advancements in the science that enhance our predictive capabilities and shed light on dose–response relationships and the establishment of critical loads of S and N for the protection of sensitive aquatic resources. Chapters 10 and 11 comprise case studies of two important acid-sensitive regions, one heavily impacted (Adirondack Mountains) and one highly sensitive but only minimally impacted to date (high-elevation portions of the Rocky Mountains and Sierra Nevada).

Chapter 3 covers advancements in the understanding of chronic acidification. The characteristics of sensitive systems are described. Causes of chronic acidification are reviewed and summarized, including an assessment of the relative importance of each.

Current understanding of the extent and magnitude of chronic surface water acidification is reviewed in Chapter 4 for the major acid-sensitive regions of the U.S. An assessment is provided of the sensitivity of aquatic resources to acidification from acidic deposition in each of the regions. Quantitative data are presented regarding the extent of acidification to date.

Chronic acidification chemical dose–response relationships are summarized in Chapter 5 for those sites for which such data are available. Recent steps toward the establishment of deposition standards or critical loads are described.

Chapter 6 addresses issues related to episodic acidification, short-term (hours to weeks) decreases in pH and ANC of surface waters in response to increased hydrologic discharge associated with snowmelt or rain events. Although less is known about episodic acidification than chronic acidification, it is thought that the biological impacts of acidification are most often first manifested as episodic, rather than chronic, processes. The characteristics of aquatic systems that are sensitive to episodic acidification are presented. The extent and magnitude of episodic acidification are summarized to the extent that they are known. Finally, the major causes of episodic acidification and their relative importance are described.

Chapter 7 provides a summary of recent advancements in the scientific understanding of N dynamics and the aquatic effects associated with elevated N deposition. The N cycle is described, with particular emphasis on recent results from experimental field studies conducted during the past decade in Europe.

Chapter 8 summarizes the results of recent selected examples of whole-ecosystem experimental acidification and de-acidification studies. Many such experiments have been conducted in Europe and a few in the U.S. during the past decade. The results of these experimental manipulation studies provide invaluable quantitative dose–response data as well as a basis for extensive testing of predictive models.

Chapter 9 covers aspects of acidification modeling of aquatic effects. Selected new models are described for N effects modeling. Results of model testing activities for S effects modeling are described, including recent modifications to the MAGIC model, the principal modeling tool used thus far in the U.S. and Europe to predict chronic acidification responses on a regional as well as site-specific basis.

In Chapters 10 and 11, detailed case studies are presented for Adirondack Park, NY, and for national parks and wilderness areas of the West, with particular emphasis on Sequoia National Park, CA, and Rocky Mountain National Park, CO. A great deal of aquatic effects research has been conducted within these three parks during the last one to two decades. Discussion of major findings in these areas provides an excellent overview of recent scientific developments in this research field as well as important insights into key acidification processes.

Finally, in Chapter 12, major conclusions of the book are highlighted and topics for future work are considered. Significant remaining knowledge gaps in the state of scientific understanding are underscored. Major research and assessment needs are described.

It is hoped that the recently completed and ongoing research highlighted in this book will aid in the preparation of much improved NAPAP assessments in the years 2000 and 2010. Acidic deposition research developments provide an excellent example of the interconnections between environmental research and public policy. Formulation of sound environmental policy requires the kind of iterative research and modeling program that has been implemented for acidic deposition.

2

Background and Approach

2.1 Overview

2.1.1 Atmospheric Inputs

The approach taken for this book has been to review and summarize important results of aquatic effects research efforts undertaken or completed since 1990. The major emphasis is on the results presented in peer-reviewed publications in the scientific literature, although results of some agency reports are also discussed. Conclusions are drawn on the basis of a variety of assessment tools, using a weight-of-evidence approach, as followed by Sullivan (1990) and NAPAP (1991). Emphasis is placed on studies conducted in regions that contain large numbers of acid-sensitive aquatic systems. Regions in which aquatic resources are either not very sensitive or are primarily influenced by environmental perturbations other than acidic deposition receive less coverage.

The natural cycling of S, N, and C has been fundamentally altered by human activities across large areas of the earth since the last century. Both S and N have the capacity to acidify soils and surface waters. Nitrogen can also lead to eutrophication of lakes, streams, estuaries, and near-coastal ocean ecosystems and can cause reduction in visibility. Disruptions of the carbon cycle have caused increasing concerns about global climate change. A need has therefore arisen to develop a more complete scientific understanding of key processes that regulate elemental transport of S, N, and C among the various environmental compartments: atmosphere, soils, water, and biomass.

The term acidic deposition refers to deposition from the atmosphere to a surface of the hydrosphere, lithosphere, or biosphere (i.e., any portion of a watershed) of one or more acid-forming precursors. The latter can include oxidized forms of S and oxidized or reduced forms of N. Such atmospheric deposition occurs in several forms, the best understood of which is wet deposition, or deposition as dissolved SO_4^{2-}, NO_3^-, and NH_4^+ in rain or snow. A sizable component of the acidic deposition to a watershed can also occur in dry form, when gaseous or particulate forms of S or N are removed from the atmosphere by contacting watershed features, especially vegetative surfaces. In some environments, particularly at high elevation, a substantial component of the total deposition of S and N occurs as cloudwater intercepts exposed watershed surfaces. Thus, the total deposition of S and N to a watershed includes wet, dry, and cloudwater (occult) deposition. The wet component is most easily measured of the three, and in most (but not all) cases it makes up the largest fraction of the total.

This chapter includes discussion of the primary chemical variables of concern in acidification research, historical water quality assessment techniques, and predictive models. It is important that each of these topics is understood in order to make sense of the state-of-the-science summary presented in Chapters 3 through 12.

We have a general idea of wet deposition levels of S and N throughout the U.S. on a regional basis, largely by virtue of the National Atmospheric Deposition Program/National Trends Network (NADP/NTN) of monitoring sites. However, few data are available from high-elevation sites where many of the most sensitive aquatic and terrestrial resources are located. In addition, knowledge is limited of the amounts of deposition other than wet deposition.

Some aspects of measuring air pollution and air pollution effects are evolving, and scientists remain divided with respect to appropriate assessment techniques. Among these topics is the measurement or estimation of atmospheric deposition in remote areas. The estimation of deposition of atmospheric pollutants in high-elevation areas is problematic, in part because all components of the deposition (e.g., rain, snow, cloudwater, dryfall, and gases) have seldom been measured concurrently. Even measurement of wet deposition remains a problem because of the logistical difficulties in operating a site at high elevation. Portions of the deposition have been measured by using snow cores (or snow pits), bulk deposition, and automated sampling devices such as those used at the NADP/NTN sites. All of these approaches suffer from limitations that cause problems with respect to developing annual deposition estimates. The snow sampling includes results for only a portion of the year and may seriously underestimate the load for that period if there is a major rain-on-snow event prior to sampling. Bulk deposition samplers are subject to contamination problems from birds and litterfall and automated samplers have insufficient capacity to measure snowfall events.

Cloudwater, dryfall, and gaseous deposition monitoring further complicate the difficult task of measuring total deposition. Cloudwater can be an important portion of the hydrologic budget in forests at some high-elevation sites, and failure to capture this portion of the deposition input could lead to substantial underestimation of total annual deposition. Furthermore, cloudwater chemistry has the potential to be much more acidic than rainfall. Dryfall from wind-borne soil can constitute a major input to the annual deposition load of some constituents, particularly in arid environments. Aeolian inputs can provide a major source of acid neutralization, not generally measured in other forms of deposition. Gaseous deposition is calculated from the product of ambient air concentrations and estimated deposition velocities. The derivation of deposition velocities is subject to considerable debate. In brief, there is great uncertainty regarding current deposition of atmospheric pollutants throughout much of the mountainous regions of the U.S.

Dry and/or occult (i.e., fog) deposition of major anions and cations can be extremely important components of the total atmospheric deposition to a watershed. At some locations, total deposition of S or N may be only slightly higher (e.g., less than 50%) than the measured wet deposition. This often seems to be the case in areas remote from major emission sources. Such a situation is not universally generalizable, however. The Bear Brook watershed in Maine provides a good example of particularly high levels of S deposition above what is recorded in precipitation. Rustad et al. (1995) calculated average water yields, after evapotranspiration, of 65 and 70%, respectively, for the East and West Bear Brook catchments. The volume-weighted average concentration of SO_4^{2-} in precipitation was about 26 µeq/L from 1987 to 1991, and this should account for about 39 µeq/L in runoff after adjusting for the water yield. However, the average SO_4^{2-} concentration in discharge actually measured 105 µeq/L in both streams prior to the chemical manipulation of the West Bear Brook watershed. Rustad et al. (1995), Norton et al. (1999), and Kahl et al. (in press) concluded that the additional SO_4^{2-} was not from weathering of S-bearing minerals because there were no identified sources of sulfide in the watershed and because the $^{34}S/^{32}S$ ratio in streamwater was approximately the same as in the incoming precipitation (Stam et al., 1992). Furthermore, the watershed soils appeared to be generally adsorbing, rather than desorbing, S. Thus, Norton et al. (1999) concluded that dry and occult deposition delivered at least an additional 150% S to the watershed. This conclusion was further supported by the chemistry of fog samples collected at the watershed summit, which averaged 127 to 160 µeq/L SO_4^{2-} during three years of study. Input/output data for other first order streams in Maine also suggested quite high levels of dry and occult deposition of S (Norton et al., 1988).

Dry and occult deposition of N are also undoubtedly high at the Bear Brook watershed. Norton et al. (1998) reported average fog concentrations of NO_3^- ranging from 56 to 64 µeq/L and average concentrations of NH_4^+ ranging from 28 to 53 µeq/L in 1989, 1990, and 1991. Mass balance calculations for N do not allow quantification of dry and occult inputs, however, because the forest canopy actively takes up deposited N.

Lovett (1994) summarized the current understanding of atmospheric deposition precesses, measurement methods, and patterns of deposition in North America. National monitoring networks for wet and dry deposition, such as NADP/NTN and CASTNET, provide data for regional assessment. Model formulations are available for estimating deposition at sites where direct measurements are not available. The reader is referred to the review of Lovett (1994) for further details.

2.1.2 Sensitivity to Acidification

Surface waters that are sensitive to acidification from acidic deposition of S or N typically exhibit a number of characteristics. Such characteristics either predispose the waters to acidification and/or correlate with other parameters that predispose the waters to acidification. Although precise guidelines are not widely accepted, general ranges of parameter values that reflect sensitivity are as follows (Peterson and Sullivan, 1998):

> *Dilute*–Waters have low concentrations of all major ions and, therefore, specific conductance is low (less than 25 µS/cm). In areas of the West that have not experienced substantial acidic deposition, highly sensitive lakes and streams are often ultradilute, with specific conductance less than 10 µS/cm.

> *Acid neutralizing capacity*–ANC is low. Acidification sensitivity has long been defined as ANC < 200 µeq/L, although more recent research has shown this criterion to be too inclusive (Sullivan, 1990). Waters sensitive to chronic acidification generally have ANC < 50 µeq/L, and waters sensitive to episodic acidification generally have ANC < 100 µeq/L. Throughout the acid-sensitive regions of the western U.S., where acidic deposition is generally low and not expected to increase dramatically, ANC values of 25 µeq/L and 50 µeq/L probably protect waters from any foreseeable chronic and episodic acidification, respectively.

> *Base cations*–Concentrations are low in non-acidified waters, but increase (often substantially) in response to acidic deposition. The amount of increase is dependent on the acid-sensitivity of the watershed. In relatively pristine areas, the concentration of (Ca^{2+} + Mg^{2+} + K^+ + Na^+) in sensitive waters will generally be less than about 50 to 100 µeq/L.

> *Organic acids*–Concentrations are low in waters sensitive to the effects of acidic deposition. Dissolved organic carbon (DOC) imparts substantial pH buffering and causes water to be naturally low in pH and ANC, or even to be acidic (ANC < 0). Waters sensitive to acidification from acidic deposition in the West generally have DOC less than about 3 to 5 mg/L.

> *pH*–pH is low, generally less than 6.0 to 6.5 in acid-sensitive waters. In areas that have received substantial acidic deposition, acidified lakes are generally those that had pre-industrial pH between 5 and 6.

> *Acid anions*–Sensitive waters generally do not have large contributions of mineral acid anions (e.g., SO_4^{2-}, NO_3^-, F^-, Cl^-) from geological or geothermal sources. In particular, the concentration of SO_4^{2-} in drainage waters would usually not be substantially higher than could be attributed reasonably to atmospheric inputs, after accounting for probable dry deposition and evapotranspiration.

Physical characteristics–Sensitive waters are usually found at moderate to high elevation, in areas of high relief, with flashy hydrology and minimal contact between drainage waters and soils or geologic material that may contribute weathering products to solution. Sensitive streams are generally low order. Sensitive lakes are generally small drainage systems. An additional lake type that is often sensitive to acidification is comprised of small seepage systems that derive much of their hydrologic input as direct precipitation to the lake surface.

2.2 Chemical Response Variables of Concern

An important objective of this book is to quantify change in the principal chemical constituents that respond to atmospheric deposition of S and N. In order to standardize the voluminous information available from a variety of sources (e.g., paleolimnology, historical data, measurements of recent trends, empirical distributions, modeling, surveys, manipulation experiments), changes are typically presented proportionally, on an equivalent basis (e.g., the equivalent change in ANC ÷ the equivalent change in SO_4^{2-}). Such an approach facilitates quantification and intercomparison.

Several watershed processes control the extent of ANC consumption and rate of cation leaching from soils to drainage waters as water moves through undisturbed terrestrial systems. Of particular importance is the concentration of anions in solution. Naturally-occurring organic acid anions, produced in upper soil horizons, normally precipitate out of solution as drainage water percolates through lower mineral soil horizons. Soil acidification processes reach an equilibrium with acid neutralization processes (e.g., weathering) at some depth in the mineral soil (Turner et al., 1990). Drainage waters below this depth generally have high ANC. The addition of strong acid anions from atmospheric deposition allows the natural soil acidification and cation leaching processes to occur at greater depths in the soil profile, thereby allowing water rich in mobile anions to emerge from mineral soil horizons. If these anions are charge balanced by hydrogen and/or aluminum cations, the water will have low pH and could be toxic to aquatic biota. Thus, the mobility of anions within the terrestrial system is a major factor controlling the extent of surface water acidification.

2.2.1 Sulfur

Sulfate has been the most important anion, on a quantitative basis, in acidic deposition in most parts of the U.S. Consequently, sulfate and the controls on its inputs and processing have received the greatest scientific and policy attention to date (Turner et al., 1990). Virtually all of NAPAP's major aquatic modeling and integration efforts leading up to the Integrated Assessment (NAPAP, 1991) focused predominantly on the potential effects of S deposition (e.g., Church et al., 1989; Turner et al., 1990; Baker et al., 1990a; Sullivan et al., 1990a). The response of S in watersheds, and to a lesser extent its chronic effects on surface water quality, are now reasonably well understood. This understanding has been developed largely through the efforts of three large multidisciplinary research efforts: the Norwegian SNSF program (Acid Precipitation Effects on Forests and Fish, 1972–1980), NAPAP (1980–1990), and the British-Scandinavian Surface Water Acidification Program (SWAP 1984–1990).

2.2.2   Nitrogen

The second important acid anion found in acidic deposition, in addition to sulfate, is nitrate. Nitrate (and also ammonium that can be converted to nitrate within the watershed) has the potential to acidify drainage waters and leach potentially toxic Al from watershed soils. In most watersheds, however, N is limiting for plant growth and, therefore, most N inputs are quickly incorporated into biomass as organic N with little leaching of NO_3^- into surface waters. A large amount of research has been conducted in recent years on N processing mechanisms and consequent forest effects, mainly in Europe (cf., Sullivan, 1993). In addition, a smaller N research effort has been directed at investigating effects of N deposition on aquatic ecosystems. For the most part, measurements of N in lakes and streams have been treated as outputs of terrestrial systems. However, concern has been expressed regarding the role of NO_3^- in acidification of surface waters, particularly during hydrologic episodes, the role of NO_3^- in the long-term acidification process, and the contribution of NH_4^+ from agricultural sources to surface water acidification (Sullivan and Eilers, 1994).

Until quite recently, atmospheric deposition of N has not been considered detrimental to either terrestrial or aquatic resources. Because most atmospherically deposited N is strongly retained within terrestrial systems, atmospheric inputs of N have been viewed as fertilizing agents, with little or no N moving from terrestrial compartments into drainage waters. More recently, however, N deposition has become quantitatively equivalent to S deposition in many areas owing to emissions controls on S, and biogeochemical N cycling has become the focus of numerous studies at the forest ecosystem level. It has become increasingly apparent that, under certain circumstances, atmospherically deposited N can exceed the capacity of forest and alpine ecosystems to take up N. This N saturation can lead to base cation depletion, soil acidification, and leaching of NO_3^- from soils to surface waters. Aber et al. (1989) provided a conceptual model of the changes that occur within the terrestrial system under increasing loads of atmospheric N. Stoddard (1994) described the aquatic equivalents of the stages identified by Aber et al. (1989), and outlined key characteristics of those stages as they influence seasonal and long-term aquatic N dynamics. The N-saturation conceptual model was further updated by Aber et al. (1998).

2.2.3 Acid Neutralizing Capacity

Acid neutralizing capacity (ANC) is the principal variable used to quantify the acid-base status of surface waters. Acidic waters are defined here as those with ANC less than or equal to zero. Acidification is often quantified by decreases in ANC, and susceptibility of surface waters to acidic deposition impacts is often evaluated on the basis of ANC (Altshuller and Linthurst, 1984; Schindler, 1988). In regional investigations of acid-base status, ANC has been the principal classification variable (Omernik and Powers, 1982). Acid neutralizing capacity is widely used by simulation models that predict the response of ecosystems to changing atmospheric deposition (Christophersen et al., 1982; Goldstein et al., 1984; Cosby et al., 1985a,b; Lin and Schnoor, 1986). Historical changes in surface water quality have been evaluated using measured (titration) changes in ANC (c.f., Smith et al., 1987; Driscoll and van Dreason, 1993; Newell, 1993) or estimated by inferring past and present pH and ANC from lake sediment diatom assemblages (Charles and Smol, 1988; Sullivan et al., 1990a; Davis et al., 1994).

ANC is a measure of titratable base in solution to a specified endpoint. It is measured by quantifying the amount of strong acid that must be added to a solution to neutralize this base. The end point of this strong-acid titration would be easily identified except for the presence of weak acids and the relatively small amounts of strong base present in low-ANC waters. Together, these factors obscure the end point. For such systems, the Gran procedure (Gran, 1952) is commonly used to determine the end point and thus the ANC. ANC measured by Gran titration is designated ANC_G.

ANC can be calculated by two distinct methods that have been shown to be mathematically equivalent, using the principles of conservation of charge and conservation of mass (Gherini et al., 1985). In one method (Stumm and Morgan, 1981), ANC is calculated as the difference between the sum of the proton (H^+-ion) acceptors and the sum of the proton donors, relative to a selected proton reference level:

$$ANC = [HCO_3^-] + 2[CO_3^{2-}] + [OH^-] + [\text{other proton acceptors}] - [H^+] \quad (2.1)$$

Here, brackets denote molar concentrations. The other method relates ANC to the total non-hydrogen cation concentrations, the individual uncomplexed cation charges (z_i) at the equivalence point (the point at which, during titration, the concentration of proton donors equals the concentrations of proton acceptors), the total strong-acid anion concentrations, and the individual uncomplexed anion charges (z_j), at the equivalence point (Gherini et al., 1985;

Church et al., 1984; Schofield et al., 1985). Using this approach, ANC is approximated with the following relation:

$$ANC = 2[Ca^{2+}] + 2[Mg^{2+}] + [K^+] + [Na^+] + [NH_4^+] + x[Al_T^{n+}]$$
$$- 2[SO_4^{2-}] - [NO_3^-]-[Cl^-]-[F^-]$$

(2.2)

where brackets indicate molar concentrations. The charges z_i and z_j, and thus the concentration multipliers in Eq. (2.2) are determined by the predominant charges of the uncomplexed constituents at the equivalence point.

For most of the species, there is little uncertainty as to the predominant uncomplexed charge at the equivalence point. For example, the charge of calcium is 2+, and thus the multiplier is 2 in Eq. (2.2). However, because of complexation with OH^-, F^-, and organic ligands, the charge of Al, shown as x in Eq. (2.2), is not always obvious. Designation of the charge, however, establishes the proton reference level (PRL). Two PRLs have frequently been used for aluminum, 3+ and 0 (Cosby et al., 1985c; Church et al., 1984; Schofield et al., 1985). These levels have different advantages; the former yields results that are closer to ANC_G values; the latter eliminates the need to include Al in ANC calculations.

Data collected during the Regional Integrated Lake–Watershed Acidification Study (RILWAS; Goldstein et al., 1987; Driscoll and Newton, 1985) from 25 lake–watershed systems in the Adirondack Mountains of New York were used by Sullivan et al. (1989) to estimate the Al PRL. The speciation of Al was calculated using the chemical equilibrium model ALCHEMI (Schecher and Driscoll, 1994), and the equivalent charge on the Al species was determined. The mean charge on Al increases with decreasing pH. However, over the pH range from 4.8 to 5.2 that corresponds to the equivalence point of dilute waters (Driscoll and Bisogni, 1984), an Al charge of 2+ appears more representative than 3+ or 0 (Sullivan et al., 1989). This is equivalent to a PRL species for Al of $Al(OH)^{2+}$ instead of Al^{3+} or $Al(OH)_3^{\circ}$.

The difference between calculated and measured ANC_G values increases as organic-acid concentration, reflected by DOC, increases. The discrepancy between Gran titration ANC and calculated ANC caused by organic acid influence and/or differences in defining the proton references for Al have major implications for aquatic effects assessment activities. Gran ANC is used primarily for classification, evaluation of current status, monitoring of temporal trends, and calibration of paleolimnological transfer functions. Calculated ANC is used (defined in different ways) for dynamic model predictions (see, e.g., Reuss et al., 1986) and for interpretation of trends data in some instances. Unfortunately, the differences between the various definitions of ANC are seldom considered. These differences can drastically affect interpretation of chemical change (Sullivan, 1990). Both Al and DOC become increasingly important at lower pH and ANC values. For the lakes and streams of greatest interest, the acidic and near acidic systems, the influence of Al and/or DOC on Gran titration results is often considerable.

2.2.4 pH

pH is one of the major controlling variables for chemical and biological response. Biota respond strongly to pH changes and to chemical variables affected by pH (Schindler, 1988). pH (or more appropriately H^+ activity) has a large influence on other important chemical reactions such as dissociation of organic acids (Oliver et al., 1983) and concentration and speciation of potentially toxic Al (Driscoll et al., 1980; Dickson, 1980; Schofield and Trojnar, 1980; Muniz and Leivestad, 1980; Baker and Schofield, 1982). Thus, pH is certainly one of the most important variables to consider in assessing temporal trends in surface water chemistry. A difficulty, however, is that as groundwater emerges to streams and lakes, it is typically oversaturated with respect to CO_2 that combines with water to form carbonic acid and depresses solution pH. As excess CO_2 degasses from solution, the pH rises. Because of this instability in surface water pH, and the strong pH buffering of carbonic acid, ANC is often used preferentially over pH for documenting temporal change.

The previous discussion of ANC and pH illustrates four points, which obfuscate efforts at quantification of historical acidification (Sullivan, 1990):

1. ANC is often the chemical variable of choice for quantification of acidification because pH measurements are sensitive to CO_2 effects (Stumm and Morgan, 1981) and because pH change is not a reliable indication of acidification in waters that have not lost most or all bicarbonate buffering (Schofield, 1982).

2. Gran ANC measurements are easily interpreted, except in dilute waters having elevated concentrations of Al and/or organic acids (Sullivan et al., 1989). Unfortunately, these are often the waters of primary interest with respect to surface water acidification.

3. Mobilization of inorganic monomeric Al (Al_i) from soil to surface waters in response to increased levels of mineral acidity *does not* result in decreased ANC_G, although Al_i is biologically deleterious.

4. Quantification of acidification is routinely accomplished using ANC_G, and/or a variety of definitions of ANC (based on charge balance). These different approaches can yield radically different estimations of acidification for systems having elevated Al and/or DOC.

2.2.5 Base Cations

The ANC (and to a large degree pH) of surface waters lacking high-DOC concentrations is determined primarily by differences between the concentration of base cations (Ca^{2+}, Mg^{2+}, K^+, Na^+) and mineral acid anions. The extent to which base cations are released from soils to drainage waters in response to increased mineral acid anion concentrations from acidic deposition is perhaps the most important factor in determining concomitant change in pH,

ANC, Al, and biota. Principal factors that determine the degree of base cation release include bedrock geology, soil characteristics, soil acidification, and hydrologic pathways. The importance of base cation concentrations in regulating surface water ANC is discussed in detail by Baker et al. (1990a, 1991a).

Base cation release from the watershed is not the only aspect of base cation dynamics that is important with respect to acidification from acidic deposition. Significant amounts of base cations also are contributed to the aquatic and terrestrial systems from the atmosphere. Driscoll et al. (1989a) suggested that atmospheric deposition of base cations can have a major effect on surface water response to changes in atmospheric inputs of SO_4^{2-}. They presented a 25-year continuous record of the chemistry of bulk precipitation and stream water at the Hubbard Brook Experimental Forest (HBEF) in New Hampshire. The decline in SO_2 emissions in the northeastern U.S. during that time period (National Research Council, 1986; Likens et al., 1984; Hedin et al., 1987; Husar et al., 1991) was reflected in a decrease in the volume-weighted concentration of SO_4^{2-} in wetfall. Stream-water SO_4^{2-} concentration also declined, but stream-water pH showed no consistent trend. On the basis of generally constant dissolved silica concentrations and net Ca^{2+} export (stream output less bulk precipitation and biomass storage), Driscoll et al. (1989a) concluded that changes in weathering rates were unlikely. The observed decline in atmospheric deposition of base cations explained most of the decline in the concentration of base cations in stream water. The processes responsible for the changes in base cation deposition were unclear, but the potential ramifications of these findings for acidification and recovery of surface waters are important.

Base cations are released from the bedrock in a watershed in amounts and proportions that are determined by the geologic make-up of the primary minerals available in the watershed for weathering. In the absence of acidic deposition or other significant disturbance, an equilibrium should exist between the weathering inputs and leaching outputs of base cations from the soil reservoir. Under conditions of acidic deposition, strong acid anions (e.g., SO_4^{2-}, NO_3^-) leach some of the accumulated base cation reserves from the soils into drainage waters. The rate of removal of base cations by leaching may accelerate to the point where it significantly exceeds the resupply via weathering. Thus, acid neutralization of acidic deposition via base cation release from soils should decline under long-term, high levels of acidic deposition. This has been demonstrated by the results of the experimental acidification of West Bear Brook (c.f., Kahl et al., in press).

Base cation depletion has been recognized as an important effect of acidic deposition on soils for many years and the issue was considered by the Integrated Assessment in 1990. However, scientific appreciation of the importance of this response has increased with the realization that watersheds are generally not exhibiting ANC and pH recovery in response to recent decreases in S deposition. The base cation response is quantitatively more important than was generally recognized in 1990.

As sulfate concentrations in lakes and streams have declined, so too have the concentrations of Ca^{2+} and other base cations. There are several reasons for this. First, the atmospheric deposition of base cations has decreased in some areas in recent decades (Hedin et al., 1994), likely owing to a combination of air pollution controls, changing agricultural practices, and the paving of roads (the latter two affect generation of dust that is rich in base cations). Second, decreased movement of SO_4^{2-} through watershed soils has caused reduced leaching of base cations from soil surfaces. Third, soils in some sensitive areas have experienced prolonged base cation leaching to such an extent that soils have been depleted of their base cation reserves. Such depletion greatly prolongs the acidification recovery time of watersheds and may adversely impact forest productivity (Kirchner and Lyderson, 1995; Likens et al., 1996).

2.2.6 Aluminum

Aluminum is an important parameter for evaluation of acidic deposition effects in drainage systems because of its influence on ANC, and also because of its toxicity to aquatic biota (Schofield and Trojnar, 1980; Muniz and Leivestad, 1980; Baker and Schofield, 1982; Driscoll et al., 1980). Inorganic Al is mobilized from soils to adjacent surface waters in response to increased levels of mineral acidity (Cronan and Schofield, 1979). Processes controlling Al mobilization, solubility, and speciation are not well understood (Sullivan, 1994). In general, inorganic monomeric Al (Al_i) concentrations in surface waters increase with increasing H^+ concentration (decreasing pH), and are present in appreciable concentrations (greater than 1 to 2 µM) in drainage lakes and streams having pH less than about 5.5. Short-term temporal variations in Al_i concentration and speciation are determined by hydrologic conditions. Partitioning of runoff water between organic and mineral soil horizons and possibly reaction kinetics appear to be the most important determinants of runoff Al_i concentrations (Cronan et al., 1986; Neal et al., 1986; Sullivan et al., 1986; Sullivan, 1994).

Al_i cannot be measured directly, but is estimated based on operationally defined labile (mainly inorganic) and nonlabile (mainly organic) fractions (Driscoll, 1984). One procedure involves measurement of total monomeric Al (Al_m) by complexation with either 8-hydroxyquinoline (Barnes, 1975) or pyrocatechol violet (Seip et al., 1984; Røgeberg and Henriksen, 1985), followed by colorimetric determination, or sometimes in the case of 8-hydroxyquinoline complexation, atomic absorption spectroscopy. Nonlabile monomeric Al (Al_o) is measured in a similar fashion using a sample aliquot that has passed through a cation exchange column. Al_i concentration is then obtained as the difference between the concentrations of Al_m and Al_o.

For drainage lakes in the Adirondack Mountains of New York, an area that has experienced considerable surface water acidification, the concentration of Al_i is highly correlated with H^+, as would be expected from solubility constraints. Based on analysis of data from Phase II of the Eastern Lake Survey (ELS-II, Herlihy et al., 1991), the relationship between Al_i and H^+ appears to vary seasonally, and Al_i is higher at a given H^+ concentration in the spring than it is during the fall. This is attributable to seasonal differences in hydrology (e.g., related to spring snowmelt) and contact time of solution in the various soil horizons. It illustrates the limitation of mineral solubility equations for predicting Al_i concentration (Hooper and Shoemaker, 1985; Sullivan et al., 1986). The fall ELS-II data yielded the following relationship (Sullivan et al., 1990a):

$$[Al_i] = 0.75(0.26) + 0.41(0.02) [H^+] \quad r^2 = 0.92, \quad n = 33 \qquad (2.3)$$

where brackets indicate concentrations, units are in μM, and standard errors of the parameter estimates are given in parentheses. During spring the relationship was equally significant ($p < 0.0001$, $r^2 = 0.94$), but the slope was 0.54 (SE = 0.05), considerably higher than that observed during fall.

Aluminum has also been implicated as a causal factor in forest damage from acidic deposition. The adverse, soil-mediated effects of acidic deposition are believed to result from increased toxic Al in soil solution and concomitant decreased Ca^{2+} or other base cation concentration (Ulrich, 1983; Sverdrup et al., 1992; Cronan and Grigal, 1995). Specifically, a reduction in the Ca/Al ratio in soil solution has been proposed as an indicator reflecting Al toxicity and nutrient imbalances in sensitive tree species. This topic was reviewed in detail by Cronan and Grigal (1995), who concluded that the Ca/Al molar ratio provides a valuable measurement endpoint for identification of approximate thresholds beyond which the risk of forest damage from Al stress and nutrient imbalances increases. Base cation removal in forest harvesting can have a similar effect and can exacerbate the adverse effects of acidic deposition. Based on a critical review of the literature, Cronan and Grigal (1995) estimated that there is a 50% risk of adverse impacts on tree growth or nutrition under the following conditions:

- Soil solution Ca/Al is less than or equal to 1.0.
- Fine root tissue Ca/Al is less than or equal to 0.2.
- Foliar tissue Ca/Al is less than or equal to 12.5.

Al toxicity to tree roots and associated nutrient deficiency problems are largely restricted to soils having low base saturation. The Ca/Al ratio indicator was recommended for assessment of forest health risks at sites or in geographic regions where the soil base saturation is less than 15%.

2.2.7   Biological Effects

Matzner and Murach (1995) summarized several of the current hypotheses regarding the impacts of S and N deposition on forest soils and the implications for forest health in central Europe. This region has experienced decades of extremely high levels of both S and N deposition, in many places three- to five-fold or more higher than deposition levels in the impacted areas of the U.S. Despite needle losses in some areas, there has been a significant increase in forest growth in other areas (c.f., Kauppi et al., 1992). No simple causality between forest damage and air pollution has been identified in areas without large local emission sources. Matzner and Murach (1995) contended that an integrating hypothesis of regional effects of air pollution on forests is almost untestable because of the long-time lags in forest response, large number of natural and anthropogenic stresses that interact with each other, and long history of local forest management. Based on a review of the literature, these authors postulated that:

1. Al stress and low Mg supply in some forests of central Europe cause tree root systems to become more shallow and root biomass to decline.
2. High N deposition reduces fine root biomass and root length.
3. Changes in tree root systems in response to increased soil acidity and N supply will increase drought susceptibility of trees and is a major reason for needle and leaf losses in some areas.

The occurrence of acid stress is restricted to areas where soils are strongly acidified by S and N deposition and where past forest management practices have contributed to base cation depletion. Thus, Matzner and Murach (1995) saw no contradiction between the proposed links between air pollution and forest damage and the finding of Kauppi et al. (1992) that N surplus has resulted in increased forest growth in many areas of Europe.

Concentrations of root-available Ca^{2+} (exchangeable and acid-extractable forms) in forest floor soils have declined in the northeastern U.S. during recent decades (Shortle and Bondietti, 1992; Johnson et al., 1994). Lawrence et al. (1995) proposed that Al, mobilized in the mineral soil by acidic deposition, is transported to the forest floor in a reactive form that reduces Ca^{2+} storage and, therefore, its availability for root uptake. They presented soil and soil solution data from 12 undisturbed red spruce stands and 1 stand that has received experimental treatments of $(NH_4)_2SO_4$ since 1989. The stands, located in New York, Vermont, New Hampshire, and Maine, were selected to represent the range of environmental conditions and stand health for red spruce in the northeastern U.S. The Ca/Al molar ratio in B-horizon soil solution ranged from about 1 to 0.06, and was strongly correlated ($r^2 = 0.73$, $p < 0.001$) with exchangeable Al concentrations in the forest floor. Increased Al will potentially slow growth and reduce the stress tolerance of trees by reducing the availability of Ca^{2+} in the primary rooting zone (Lawrence et al., 1995).

Many species of aquatic biota are sensitive to changes in pH and other aspects of surface water acid–base chemistry. Such biological effects occur at pH values as high as 6.0 and above, but become more pronounced at lower pH, especially below 5.0. Individual species and life forms differ markedly in their sensitivity to acidification (Table 2.1). Biological effects on fish are better understood than are effects on other aquatic life forms, but it is clear that virtually the entire aquatic ecosystem is affected when acidification is pronounced.

The most important chemical parameters that cause or contribute to the adverse effects of acidification on aquatic biota are decreased pH (increased H^+), increased inorganic Al, and decreased Ca^{2+} concentrations. Different species and stages in the life history of a given species differ in their tolerance to variations in these three critical parameters. For example, egg and larval stages are often more sensitive to H^+ and Al stress than are adult life stages. Both H^+ and inorganic Al are toxic to aquatic organisms, in some species at concentrations as low as 1 or 2 μM. Ca^{2+} ameliorates this toxicity.

Assessments of the effects of acidification on aquatic biota can be based on the results of laboratory toxicity studies, *in situ* exposure experiments, and the results of field surveys. Model projections of future changes in surface water chemistry can be evaluated in terms of their likely biological impacts via the use of toxicity models or models based on field distributional data. An assessment must first be made of the expected fish distribution in the absence of acidification. For example, brook trout habitat in the Southern Blue Ridge was defined by Herlihy et al. (1996) as those streams having elevation greater than 1000 m, stream gradient 0.4 to 17%, and Strahler stream order (1 : 24,000 scale) less than 4. Brook trout is considered an important fish species of concern because this species is native to many upland streams in the eastern U.S. that are acid–sensitive. Thus, by using a combination of an acid–base chemistry model and a fish response model, we can estimate the potential long-term effects of changes in acidic deposition on fish communities.

2.3   MONITORING

One of the best ways to study the hydrogeochemistry of forested watersheds has been through carefully designed monitoring programs. Unfortunately, monitoring has long been viewed by many scientists and funding agencies alike as rather routine, not exciting or cutting-edge, perhaps boring. It has not helped the situation that some monitoring programs have operated for years, blindly collecting data, without any critical examination, adherence to quality assurance/quality control (QA/QC) procedures, or consideration of how the resulting data could or should be used. Only recently has the value of high-quality, long-term monitoring become somewhat more widely recognized. Monitoring of the inputs (i.e., atmospheric deposition, precipitation) and outputs (i.e., evapotranspiration, streamflow, groundwater flow) to and from the watershed system provides a means of formulating hypotheses about watershed behavior, quantifying process rates, and testing the behavior of predictive models (Cosby et al., 1996; Church, 1999). The recent results

TABLE 2.1

General Biological Effects of Surface Water Acidification

pH Decrease	General Biological Effects
6.5 to 6.0	Small decrease in species richness of phytoplankton, zooplankton, and benthic invertebrate communities resulting from the loss of a few highly acid-sensitive species, but no measurable change in total community abundance or production
	Some adverse effects (decreased reproductive success) may occur for highly acid-sensitive species (e.g., fathead minnow, striped bass)
6.0 to 5.5	Loss of sensitive species of minnows and dace, such as blacknose dace and fathead minnow; in some waters decreased reproductive success of lake trout and walleye, which are important sport fish species in some areas
	Visual accumulations of filamentous green algae in the littoral zone of many lakes, and in some streams
	Distinct decrease in the species richness and change in species composition of the phytoplankton, zooplankton, and benthic invertebrate communities, although little if any change in total community biomass or production
	Loss of a number of common invertebrate species from the zooplankton and benthic communities, including zooplankton species such as *Diaptomus silicis, Mysis relicta, Epsichura lacustris*; many species of snails, clams, mayflies, and amphipods, and some crayfish
5.5 to 5.0	Loss of several important sport fish species, including lake trout, walleye, rainbow trout, and smallmouth bass; as well as additional nongame species such as creek chub
	Further increase in the extent and abundance of filamentous green algae in lake littoral areas and streams
	Continued shift in the species composition and decline in species richness of the phytoplankton, periphyton, zooplankton, and benthic invertebrate communities; decrease in the total abundance and biomass of benthic invertebrates and zooplankton may occur in some waters
	Loss of several additional invertebrate species common in oligotrophic waters, including *Daphnia galeata mendotae, Diaphanosoma leuchtenbergianum, Asplanchna priodonta*; all snails, most species of clams, and many species of mayflies, stoneflies, and other benthic invertebrates
	Inhibition of nitrification
5.0 to 4.5	Loss of most fish species, including most important sport fish species such as brook trout and Atlantic salmon; few fish species able to survive and reproduce below pH 4.5 (e.g., central mudminnow, yellow perch, and in some waters largemouth bass)
	Measurable decline in the whole-system rates of decomposition of some forms of organic matter, potentially resulting in decreased rates of nutrient cycling
	Substantial decrease in the number of species of zooplankton and benthic invertebrates and further decline in the species richness of the phytoplankton and periphyton communities; measurable decrease in the total community biomass of zooplankton and benthic invertebrates in most waters
	Loss of zooplankton species such as *Tropocyclops prasinus mexicanus, Leptodora kindtii*, and *Conochilis unicornis*; and benthic invertebrate species, including all clams and many insects and crustaceans
	Reproductive failure of some acid-sensitive species of amphibians such as spotted salamanders, Jefferson salamanders, and the leopard frog

Source: Baker et al., 1990a.

of long-term monitoring programs initiated during the 1980s (and a few earlier) have provided some of the most useful data on quantitative watershed
response to acidic deposition and other stresses (c.f., Newell, 1993).

There are some high-quality, long-term monitoring programs that have
been in operation for more than a decade, including the Environmental Protection Agency's (EPA) Long-Term Monitoring Program (LTM) of lake and
stream-water chemistry in portions of the eastern U.S. (c.f., Newell, 1993)
and the National Atmospheric Deposition Program/National Trends Network (NADP/NTN) that monitors the chemistry of precipitation throughout the country. The U.S. Geological Survey has maintained a network of
streamflow gaging stations in headwater catchments throughout the country to quantify discharge.

Every year, pressures seem to increase to discontinue portions of these, and
other, long-term monitoring programs. As budgets tighten, routine programs
such as these often go the way of the budget axe in favor of newer and more
"innovative" research programs.

It is unfortunate that many federal program managers fail to recognize the
cumulative value of these long-term databases. With each monitoring program that is discontinued, we forever weaken our ability to make future
assessments of the impacts of acidic deposition, climate change, and other
anthropogenic or natural environmental stressors.

2.4 Historical Water Quality Assessment Techniques

2.4.1 Historical Measurements

Significant decreases in measured ANC or pH would provide the most direct
evidence of acidification, but reliable historical data are seldom available
(Schofield, 1982). Rigorous regional evaluations, however, have been conducted of historical change in pH and/or alkalinity of lake water in the
Adirondacks (e.g., Schofield, 1982; Kramer et al., 1986; Asbury et al., 1989)
and northern Wisconsin (Eilers et al., 1989a) and these were summarized by
Sullivan (1990).

Several problems are associated with the interpretation of historical surface water chemistry data. The most significant difficulty is the lack of documentation of historical sampling and analytical procedures (Kramer et al.,
1986). Prior to 1950, alkalinity and pH measurements were generally made
colorimetrically with indicator dyes. Methyl orange (M.O.) was often the
indicator for measuring alkalinity. It changes color across a pH range and
historical records are unclear regarding the exact end point used for particular studies. The most common approach for assessment of these data has
been to specify the most likely endpoints used by the original analysts and

to "correct" historical measurements prior to comparison with present-day electronic measurements (e.g., Kramer et al., 1986; Asbury et al., 1989; Eilers et al., 1989a). Additional difficulties in interpretation include a general lack of QA/QC data in most cases, potential climatological differences, land use changes, and statistical uncertainties associated with natural variability of lakes and streams. Furthermore, the "historical" measurements were typically taken in the 1930s or more recently and, thus, do not predate the occurrence of acidic deposition. Detection of significant, long-term pH changes in acidifying systems that are still in the bicarbonate buffered state cannot be made reliably because normal biologically-induced changes in CO_2 levels would be likely to obscure any pH changes resulting from decreased alkalinity (Schofield, 1982; Norton and Henriksen, 1983). Thus, interpretation of long-term pH changes for waters having pH above 6.0 must be viewed with caution. Only in cases where lakes and streams have lost all bicarbonate buffering can pH change be considered a reliable indication of acidification (Schofield, 1982).

It is difficult to draw quantitative conclusions about historical change on the basis of measurements made prior to development and common use of the glass pH electrode around 1950 (Schofield, 1982; Kramer and Tessier, 1982; Haines et al., 1983). Furthermore, pH measurements are sensitive to solution CO_2 concentration and holding time (Stumm and Morgan, 1981; Herczog et al., 1985; Small and Sutton, 1986a), and earlier use of soft-glass containers may have contributed 20 to 100 µeq/L of alkalinity to stored samples (Kramer and Tessier, 1982). It is possible that the cleaning and aging of bottles and rapid analysis reduced this effect, but the magnitude of the effect can be similar to that of the changes being measured (Kramer and Tessier, 1982). Perhaps the greatest impediment to quantification of historical acidification has been uncertainty regarding the end point used for earlier colorimetric alkalinity titrations (Kramer et al., 1986). The magnitude of inferred acidification is highly sensitive to the choice of endpoint, and Kramer et al. (1986) concluded that a wide range of M.O. endpoints could have been used in the historical measurements.

More recent chemical monitoring data, generally collected over a period of one to two decades, are not subject to the same methodological limitations as the historical data. Recent trends data are particularly useful in comparison with inferred long-term changes using other techniques, such as paleolimnology. In some cases, the recent monitoring data span a large gradient in S deposition. In such instances, valuable quantitative information can be obtained.

2.4.2 Paleolimnological Reconstructions

In the absence of long-term chemical monitoring data, inference based upon diatom and chrysophyte fossil assemblages preserved in lake sediments is the best technique available to evaluate historical chemical changes (Charles and Norton, 1986). Diatoms (Bacillariophyta) and scaled chrysophytes (Chrysophyceae, Synurophyceae) are single-cell algae composed of siliceous valves and overlapping siliceous scales, respectively. The fossil remains of these organisms are good indicators of past lake-water chemistry because

1. They are common.
2. Many taxa have rather narrow ecological (water chemistry) tolerances.
3. The remains are well preserved in sediment, usually in very large numbers.
4. They can be identified to the species level or below (Smol et al., 1984; Charles, 1985; Charles and Norton, 1986; Smol et al., 1986; Husar et al., 1991).

Paleolimnological reconstructions of past lake-water chemistry are based on transfer functions derived from relationships between current chemistry and diatom/chrysophyte remains in surface sediments. Predictive equations are developed from these relationships using regional lake data sets to infer past water chemistry. Several techniques have been developed and applied to infer pH and ANC. Calibration equations have also been developed for inferring the concentration of DOC, total Al, and monomeric Al.

Once developed, predictive equations can be applied to diatom assemblage data from lake sediment cores to infer past conditions. Trends within cores can be analyzed statistically to determine if they are significant (Birks et al., 1990a). Inferred chemical data can be dated using ^{210}Pb activity and compared with stratigraphies of other lake sediment characteristics such as pollen, charcoal, coal and oil carbonaceous particles, polycyclic aromatic hydrocarbons, Pb, Zn, Cu, V, Ca, Mg, Ti, Al, Si, S, and others that provide a record of atmospheric inputs of materials associated with the combustion of fossil fuels and watershed disturbance (Heit et al., 1981; Tan and Heit, 1981; Charles and Norton, 1986). With these data, in addition to knowledge of watershed events and some historical information on regional atmospheric emissions of S and N, it is often possible to assess with reasonable certainty whether lakes have been affected by acidic deposition, and to what extent (Husar et al., 1991; Charles et al., 1989).

A number of techniques have been used to reconstruct lake-water chemistry, particularly pH, from sedimentary diatom remains. Paleolimnology as a quantitative science has evolved extremely rapidly over the past two decades. The various techniques were reviewed Charles et al. (1989). Lake sediments can be dominated by diatom valves or chrysophyte scales, and the relative abundance and diversity of these groups will determine which will provide the most accurate information on past lake chemistry (Charles and Smol, 1988). In general, assemblages with the greatest diversity of algal remains will provide the most ecological information and the best predictive equations.

By late 1986, category-based, multiple regression techniques were being replaced by theoretically superior gradient analysis techniques. The theory has been summarized and elaborated mainly by ter Braak (c.f., ter Braak and Looman, 1986; ter Braak and Gremmen, 1987; ter Braak and Prentice, 1988). Gradient analysis theory is based on a species-packing model along environmental gradients, assuming a simple normal distribution of each species' abundance in samples along the gradient. Although much of ter Braak's research concerns multivariate analysis of several environmental gradients simultaneously (primarily by ordination techniques), the potential for application of these methods to the reconstruction of single environmental variables was recognized as a desirable goal in acidification paleolimnology (Stevenson et al., 1989; ter Braak and van Dam, 1989; Birks et al., 1990a; Kingston and Birks, 1990; Dixit et al., 1989).

Two large regional paleolimnological studies of lake acidification have been conducted in the U.S., PIRLA-I (Paleoecological Investigation of Recent Lake Acidification; Charles and Whitehead, 1986a,b) and PIRLA-II (Charles and Smol, 1990). PIRLA-I included stratigraphic analysis of about 35 lakes in four regions of the U.S. (Adirondacks and northern parts of New England, the Upper Midwest, and Florida). In addition to obtaining biological data (diatoms, chrysophytes) for inferring water chemistry, sediment measurements also included analyses of metals, S, N, C, polycyclic aromatic hydrocarbons, coal and oil carbonaceous particles, pollen, and ^{210}Pb for dating strata. The PIRLA-I project developed standardized protocols for all aspects of paleolimnological research, including QA/QC guidelines (Charles and Whitehead, 1986a). Most of the study lakes were small to moderate in size, low in alkalinity, and had forested watersheds with little or no cultural development. Alkalinity and pH inference equations were based on a calibration set of 36 lakes and involved multiple linear regression of percentages of diatoms in pH categories. The standard error of the estimates for pH and ANC were 0.26 pH units and 21 μeq/L, respectively (Charles et al., 1989). Procedures for development of the inference equations were provided by Charles (1985) and Charles and Smol (1988).

The objectives of the PIRLA-II study were to evaluate three distinct acidification issues. The first component was designed to determine the proportion of low-ANC Adirondack lakes that have become more acidic since about 1850, to quantify the ANC change that occurred, and to determine the percentage of lakes that were naturally acidic. Sediment cores from 37 Adirondack lakes that were included in the EPA's regional modeling effort, the Direct Delayed Response Project (DDRP; Church et al., 1989) were analyzed. These lakes were selected for inclusion using a statistical framework, and results of historical change estimates therefore can be extrapolated to the population of Adirondack lakes. The "tops" (0 to 1 cm depth) and "bottoms" (pre-1850, usually greater than 30 cm) of the sediment cores were analyzed for diatoms and chrysophytes (Cumming et al., 1992). A second component of PIRLA-II addressed the question of recent change in Adirondack lake chemistry during the past two decades (corresponding with decreases in acidic deposition since about 1970). Close interval sectioning (0.25 cm) sediment core analyses were performed on a subset of nonrandomly selected

Adirondack lakes for comparison with monitoring data collected during the recent past (Cumming et al., 1994). The third component of PIRLA-II was an evaluation of the historical response of seepage lakes to acidic deposition. Seepage lakes are expected to respond differently to acidic input than drainage systems because they often have less potential for enhanced base cation or Al enrichment, and their longer hydrologic retention times increase the importance of in-lake retention of S and N compounds. The seepage lake component of PIRLA-II included 10 lakes in northern Florida (Panhandle and Trail Ridge areas) and 10 lakes in the Adirondacks, analyzed for tops and bottoms of sediment cores.

For the PIRLA-II Project, weighted averaging calibration (Birks et al., 1990a) was used for development of calibration equations. Substantial differences were observed in the diatom flora between Adirondack drainage and seepage lakes, and separate calibrations were developed for each lake type. There were 71 lakes included in the drainage lake calibrations (except for monomeric Al, where $n = 62$) and 20 lakes in the seepage lake calibrations. Historical change in pH and ANC from pre-industrial time to the present was quantified for 36 statistically selected Adirondack lakes included in the historical data set, based on diatom transfer functions. Population estimates for change in pH and ANC were made for the Adirondack subregion based on the paleolimnological results (Sullivan, 1990; Sullivan et al., 1990a).

Paleolimnological inferences have been used to assess regional patterns in lakewater acidification and recovery (Sullivan et al., 1990a; Cumming et al., 1994; Smol et al., 1998). For example, Smol et al. (1998) summarized the results of diatom and chrysophyte inferences of lakewater pH from 36 lakes in the Sudbury region of Ontario, Canada, and 20 lakes in Adirondack Park, NY. In both regions, many lakes were shown to have acidified considerably since the last century, although the distribution of pre-industrial pH of the Sudbury lakes was much higher than in the Adirondacks. The Sudbury lakes have also shown greater pH recovery, probably because they had acidified to a greater extent during previous decades and because of larger declines in S deposition during recent years in response to closing the nearby smelter. All of the Sudbury lakes with present pH less than 6.6 were inferred to have acidified since the 1850s, some by over 2 pH units. The average diatom-inferred acidification was 0.6 pH units. About 40% of the lakes have shown recent pH recovery, with the average increase since the most acidic pH interval being 0.23 pH units (Smol et al., 1998). Although diatom inferences of pH recovery were not available for Adirondack lakes, chrysophyte inferences of pH suggested that recovery of Adirondack lakes has been much smaller (Cumming et al., 1994).

2.4.3 Empirical Relationships and Ion Ratios

Changes in surface water chemistry that may have occurred in response to acidic deposition can also be inferred from relationships among ionic constituents. Analyses of historical change have been presented in the form of simple empirical models (e.g., Henriksen, 1979, 1980; Wright, 1983; Eshleman and Kaufmann, 1987) or ion ratios (e.g., Schindler, 1988; Kaufmann et al., 1988). A variety of assumptions are implicit in the evaluation of empirical relationships, however, and such data must be interpreted with caution (Kramer and Tessier, 1982; Husar et al., 1991). The objective here is to summarize both the uses and limitations of empirical models and ratios. Together they constitute a valuable assessment tool, but they can also be easily misinterpreted.

Empirical evaluations are simply a logical extension of a charge balance definition of ANC. A variety of ANC definitions have been used in the literature (e.g., Stumm and Morgan, 1981; Reuss et al., 1986; Reuss and Johnson, 1985; Gherini et al., 1985; Sullivan et al., 1988; Wright, 1988). They differ principally in their treatment of organic acid anions, and metals such as Al, Mn, and Fe. Eqs. (2.1) and (2.2) correspond closely with values obtained by Gran titration determinations of ANC (Sullivan et al., 1989). The "other proton acceptors" in Eq. (2.1) include organic anions, the equivalence of Al complexed with hydroxide, and organic-Al complexes.

Eq. (2.2) can be expressed as:

$$ANC = [C_B] - [C_A] + 2[Al_m] \tag{2.4}$$

where C_B is the equivalent sum of base cations and ammonium (Ca^{2+}, Mg^{2+}, K^+, Na^+, NH_4^+), C_A is the equivalent sum of strong acid anions (SO_4^{2-}, NO_3^-, Cl^-, F^-), and Al_m is total monomeric Al, in $\mu mol/L$. Cationic Al behaves primarily as a base cation with respect to Gran titration ANC. Change in Al must be computed separately, but evaluated in conjunction with change in ANC in order to assess biologically relevant changes in surface water chemistry. Monomeric aluminum is assigned a valence of +2 in the preceding equation, corresponding to the approximate mean Al valence at the equivalence point of the Gran titration. Where organic anions are present in significant concentrations, Gran titration values will underestimate the preceding definitions of ANC (Sullivan et al., 1989).

Current spatial patterns in water chemical parameters across gradients in deposition provide useful information for evaluating historical change. Differences in surface water chemistry along a gradient of low to high deposition may represent temporal changes in lakes or streams during periods when atmospheric deposition of acids increased from low to high. This approach, called space-for-time substitution, is based on the assumption that changes in space reflect changes in time and, thus, that the parameters under investigation were relatively homogenous in the absence of deposition. A further assumption is that only the change in acidic deposition has influenced the pattern of ANC change. Such assumptions are difficult to substantiate, and spatial patterns alone are not sufficient for demonstration of temporal change. Nevertheless, spatial data provide useful information for hypothesis generation and comparison with results of other techniques (e.g., paleolimnology, monitoring data, manipulation experiments).

Spatial analyses are most easily interpreted if performed on data from waters that had relatively homogeneous chemistry in the absence of atmospheric deposition. Surface waters and watersheds vary, however, in their current chemistry and response to acidic inputs (c.f., Munson and Gherini, 1991). Where historical change is likely to have been small in magnitude, spatial analyses can be optimized by focusing on designated subsets of surface waters to reduce heterogeneity and delete systems that are unlikely to have changed or for which uncertainties are particularly large. In such cases, the results of spatial analyses are only applicable to the subset under investigation. A judicious use of screening and subsetting criteria can optimize the extraction of information for designated subsets of data. Failure to analyze appropriate subsets can obscure changes in surface water chemistry by averaging the results from a large number of systems that have changed little, or not at all, with the low-ANC systems that likely have changed appreciably in acid–base status.

Empirical models are used to quantify change in water chemistry under a particular suite of assumptions, whereas ion ratios offer a more general (qualitative) assessment of chemical change. Perhaps the most commonly used are two ratios that reflect the interrelationships between the concentrations of SO_4^{2-}, base cations, and ANC_G:

- $ANC_G/[C_B]$.
- $[SO_4^{2-}]/[C_B]$

Interpretation of both ratios is often based on the assumptions that pristine, low-DOC surface waters typically exhibit a near 1 : 1 ratio of base cations (corrected for marine contributions) to ANC (Henriksen, 1979) and that the principal determinants of ANC are base cations and SO_4^{2-} (Sullivan, 1990). The C_B term in these ratios is generally limited to $(Ca^* + Mg^*)$ (the asterisk indicates that the concentration has been corrected to remove probable marine contributions). If Na^+ and/or K^+ are associated with appreciable alkalinity sources in a particular region, especially for low ionic strength waters, then these cations should also be included (e.g., Kramer and Tessier, 1982). For example, ANC_G is approximately equal to the concentration of $(Ca^{2+} + Mg^{2+})$ in low-ANC_G drainage lakes in the Pacific Northwest subregion of the Western Lake Survey (Landers et al., 1987), whereas $(Ca^{2+} + Mg^{2-})$ concentrations are lower than ANC_G in the California subregion (Husar et al., 1991). These data and the very low K^+ concentrations of California lakes (Landers et al., 1987) indicate that Na^+ is associated with alkalinity production in California lakes, for example, via carbonic acid weathering of albite ($NaAlSi_3O_8$) (Stumm and Morgan, 1981; Melack and Stoddard, 1991).

The significance of the $ANC_G/[C_B]$ ratio has been misinterpreted in the acidic deposition literature. A ratio much less than one does not necessarily indicate that acidification [loss of ANC, as defined by Galloway (1984)] has occurred. The ratio approximates 1.0 only where surface water organic acid anion concentrations (RCOO⁻) are low and both atmospheric and watershed sources of SO_4^{2-} are minimal. Organic acid anions tend to lower ANC_G relative to base cation concentrations, as appears to be the case in northeastern Minnesota (Husar et al., 1991), although SO_4^{2-} weathering may also contribute to the observed effect for the higher ANC systems in that area (e.g., Cook and Jager, 1991). Watershed sources of SO_4^{2-} are derived from weathering reactions that yield base cations that are charge balanced by SO_4^{2-} rather than HCO_3^-. This also lowers the $ANC/[C_B]$ ratio. Even in low-DOC waters lacking watershed sources of SO_4^{2-}, a ratio much less than 1 implies only that surface water chemistry has changed. The change could be owing to increased C_B, decreased ANC, or a combination of both. Although one could argue that there is a finite limit to increased base cation release, the ratio alone does not demonstrate acidification (Husar et al., 1991).

The ratio $[SO_4^{2-}]/[C_B]$ quantifies the SO_4^{2-} concentration relative to a surface water's susceptibility to acidification. The most important factor that determines whether or not adverse effects will occur from acidic deposition is the inherent susceptibility of the watershed, as reflected in surface water base cation concentrations (Wright, 1988; Munson and Gherini, 1991). High SO_4^{2-} concentration is generally only associated with biologically significant changes in water chemistry where C_B is low. Where $[SO_4^{2-}]/[C_B] > 1$, water is acidic (ignoring Al which is not in the ratio) because of high SO_4^{2-} concentration, irrespective of organic acid anion concentrations.

In summary, a variety of empirical approaches have been used to assess current status and likely historical changes in surface water chemistry. These are generally presented in the form of ratios between ionic constituents and simple empirical models, both of which are based either implicitly or explicitly on a charge balance definition of ANC. Ion ratios constitute a useful qualitative tool for historical assessment. Empirical models are discussed next.

2.5 Models

2.5.1 Empirical Models

Steady-state models are based on ion budget calculations (input/output), empirical relationships, and first principles, especially charge balance. They do not require substantial data input and, thus, are regionally applicable. However, they lump many important processes into a few terms and thus risk overgeneralization. The historical development of this type of approach and an assessment of the strengths and weaknesses of many steady state models were presented by Church (1984) and Thornton et al. (1990). Early

developments in this field were by Almer et al. (1978) in Sweden and Henriksen (1979, 1980) in Norway. Later variations included those of Thompson
(1982), Wright (1983), Rogalla et al. (1986), Nichols and McRoberts (1986),
Small and Sutton (1986b), and others. Steady state models in the past have
often assumed that acidic deposition is the only significant source of acidity
and that base cation release accounts for only a small percentage of the total
ionic response to added sulfate.

An empirical model for estimating historical change in surface water chemistry can be derived by assuming that certain of the parameters included in
Eq. (2.4) may have changed appreciably in response to changing concentrations of SO_4^{2-} and NO_3^-, whereas other parameters are likely to have been relatively unaffected (Sullivan, 1990). For example, Henriksen (1979, 1980)
initially assumed that only the ANC change was appreciable in response to
increased SO_4^{2-} concentration. Henriksen's definition of ANC treated Al^{n+} as
an acidic cation, similar to H^+ in Eq. (2.1), and ΔAl^{n+}, therefore, was not
included explicitly as a model parameter. Subsequently, Henriksen (1982)
and Wright (1983) presented evidence for Norwegian and North American
lakes, respectively, suggesting that increased base cation release accounted
for up to 40% of added SO_4^* (asterisk designates that the concentration has
been sea salt corrected), whereas the additional 60 to 100% of SO_4^* input
replaced ANC. The proportional change in base cations relative to change in
SO_4^* is referred to as the F factor:

$$F = \Delta C_B^* / \Delta SO_4^* \qquad (2.5)$$

It is generally assumed that most or all of the base cation change (ΔC_B) is
attributed to changes in Ca^{2+} and Mg^{2+}. Husar et al. (1991) incorporated the
F-factor concept into the ANC definition presented in Eq. (2.4) and presented
an empirical model to estimate historical change in ANC as:

$$\Delta ANC = \Delta C_B^* + \Delta Al_i - \Delta SO_4^* \qquad (2.6)$$
$$= (F \times \Delta SO_4^*) + \Delta Al_i - \Delta SO_4^*$$

This estimator of change in ANC Eq. (2.6) will generally yield results
similar to that proposed by Henriksen (1979) only if F is assumed equal to
zero. It requires less restrictive assumptions, however (Eshleman and
Kaufmann, 1987), and includes the Al change in a manner consistent with
ANC_G (Husar et al., 1991). This approach assumes that Cl-, organic acid
anions, and N species have not changed, that marine sources of SO_4^{2-} and
base cations can be subtracted where appropriate using measured or estimated marine Cl- and the ionic composition of seawater, that pre-industrial concentrations of labile monomeric Al (Al_i) in surface waters were
negligible or can be estimated (Sullivan, 1991), and that pre-acidification
background SO_4^* concentration can be estimated for the waters of interest
(Husar et al., 1991).

The major uncertainties associated with estimating loss of ANC using Eq. (2.6) are

1. Estimation of an appropriate F-factor to account for mineral acid neutralization via base cation release.
2. Estimation of an appropriate regional background SO_4^* concentration from which to estimate change in SO_4^*.
3. The assumption of temporally constant organic acid anions, NO_3^-, and NH_4^+ concentrations.

Although increased deposition of N species has occurred (Gschwandtner et al., 1985), these compounds are generally rapidly assimilated by vegetation, and are present in low concentrations in most acid-sensitive lakes during autumn sampling (c.f., Landers et al., 1987; Eilers et al., 1988a). Where NO_3^- concentration is elevated, as for example in Catskill streams (Stoddard and Murdoch, 1991) and some Adirondack lakes (Driscoll et al., 1991), NO_3^- should be included in an empirical evaluation along with SO_4^{2-}, and the F factor is more appropriately defined as:

$$F = \Delta C_B^* / (\Delta [SO_4^* + NO_3^-]) \tag{2.7}$$

It is likely that base cation neutralization varies as a function of initial base cation concentration, and F will be low in watersheds where carbonic acid weathering and cation exchange yield low surface water base cation concentrations. In contrast, watersheds in which carbonic acid weathering yields relatively high surface water C_B concentrations are more likely to exhibit greater neutralization of SO_4^{2-} and NO_3^- acidity by increasing base cation release. Thus, F should approach or equal 1.0 at higher initial C_B concentrations. Unfortunately, at the time of the Integrated Assessment (NAPAP, 1991), there was little basis for describing the distribution of F-factors for regional populations of lakes, and there was no justification for the choice of a single F-factor to describe population-level change in base cation release in North American surface waters irrespective of ANC (Husar et al., 1991). Since 1990, more quantitative data have become available. In some cases, useful information was obtained in the earlier analyses by assuming that $F = 0$ in order to assess the maximum possible change in ANC attributable to SO_4^{2-} input (e.g., Eshleman and Kaufmann, 1987). In other cases, a maximum "reasonable" F was calculated for a given lake or stream by assuming a lower reasonable level for initial pre-acidification C_B. A high F-factor is then estimated as the difference between current C_B and the lower estimate for pre-industrial C_B divided by the estimated change in $(SO_4^{2-} + NO_3^-)$ concentration.

The concentration of SO_4^{2-} in precipitation has been estimated as approximately 5 to 7 µeq/L in remote areas of the world (Galloway et al., 1984, 1987). Brakke et al. (1989) used these data and an assumed 50% evapotranspiration rate to estimate an upper bound of 10 to 15 µeq/L for background SO_4^{2-} in eastern U.S. lakes having low concentrations of base cations. This assumed that watershed sources of SO_4^{2-} were inconsequential for low base cation systems, which is supported by data on the ratios of lakewater to precipitation SO_4^{2-} concentration in low-ANC northeastern lakes (Sullivan et al., 1988; Sullivan, 1991), but is probably not always the case. Watershed sources of SO_4^{2-} are generally of greater significance in lakes having higher ANC and base cation concentrations (Wright, 1983; Sullivan et al., 1988).

The issue of potential change in organic acid anion concentrations in response to acidic deposition has largely been ignored in much of the acidic deposition literature. A loss of DOC and RCOO⁻ has been hypothesized (Almer et al., 1974; Krug and Frink, 1983) and the topic was reviewed by Marmorek et al. (1988). The latter study concluded that there were inconsistencies in the available data, but most evidence suggested that organic acids have been lost from surface waters in response to increased acidic deposition. The best historical evidence available to Marmorek et al. (1988) that bears on this issue was a paleolimnological study of diatom-inferred change in DOC for two lakes in Norway by Davis et al. (1985), suggesting a historic decrease of 3 to 6 mg/L DOC. It is now known that the concentration of organic acid anions does change in response to changes in acidic deposition, although the amount of change is often small.

There are thus several problems and uncertainties associated with the use of empirical models for quantifying historical changes in surface water acid–base status. In the absence of regional quantification using an independent tool, such as paleolimnology or long-term monitoring, empirical models are of limited value except as a general indication of historical change in water chemistry.

2.5.2   Dynamic Models

A number of dynamic or mechanistic mathematical models are commonly used to further our understanding of the chemical and biological processes that affect ecosystem response to mineral acid input. Such models can be used for hypothesis testing and predictions regarding future change. Among those most commonly used during the last two decades are several that focus on soil-mediated processes, such as the Birkenes model (Christophersen et al., 1982; Rustad et al., 1986), ILWAS (Chen et al., 1983; Goldstein et al., 1984; Gherini et al., 1985), the Trickle Down model (Lin and Schnoor, 1986), and the MAGIC model (Cosby et al., 1985a,b). They contain similar assumptions regarding several key soil chemical processes, in particular, the anion mobility concept of Seip (1980), cation exchange, the carbonic acid system, Al dissolution, and weathering (Reuss et al., 1986). The Internal Alkalinity Generation (IAG) model (Baker et al., 1985, 1986; Baker and Brezonik, 1988) was developed for seepage lakes and is based on alkalinity/electroneutrality principles and input/output budgets analogous to those used in nutrient loading models (e.g., Vollenweider, 1975).

These acidification models and others have proven useful for hypothesis testing and have played an integral role in the effort to understand the net effects of the complicated processes that govern ecosystem response to acidic deposition (Stone and Seip, 1990). It has long been recognized that the hydrologic and aluminum submodels needed improvement and more emphasis was needed on model validation, however (Reuss et al., 1986; Seip et al., 1989; Stone and Seip, 1989). A major model testing and improvement effort has been underway for MAGIC since 1990 (Sullivan et al., 1994, 1996a, 1999; Sullivan and Cosby, 1995, 1998; Cosby et al., 1995, 1996). The results of the model testing and resulting changes to MAGIC are discussed in Chapter 9. Also mentioned in Chapter 9 are several more recent models that incorporate N effects.

3

Chronic Acidification

Chronic acidification of surface waters refers to loss of ANC or reduction in pH on a chronic, or annual-average, basis. Chronic acidification is often evaluated by studying changes in surface water chemistry during periods when that chemistry is expected to be relatively stable. These are generally summer or fall for lakes and spring baseflow (in the absence of storms) for streams. Attempts to measure chronic acidification focus to some extent on a moving target. Lake-water chemistry tends to be relatively stable during summer and fall, compared to other times of the year, as does spring base-flow chemistry in streams. There is still, however, often significant variability in that chemistry. Water chemistry exhibits changes on both intra and interannual time scales in response to a host of environmental factors. Key in this regard are short-term and long-term climatic fluctuations that govern the amount and timing of precipitation inputs, snowmelt, vegetative growth, depth to groundwater tables, and evapoconcentration of solutes. Many years of data, therefore, are required to establish the existence of trends in surface water chemistry, much less assign causality to changes that are found to occur.

There have been many advancements in the scientific understanding of chronic surface water acidification since 1990. Several studies that had been initiated during the original NAPAP research effort were completed post-1990 and research results from those programs continue to be published. A major research effort was conducted in Europe regarding the dynamics of N-driven acidification and related processes in both terrestrial and aquatic ecosystems. New predictive models have been developed and some previously existing models have been extensively tested and improved. Finally, the availability of increasing volumes of data from long-term monitoring programs and experimental manipulation studies have provided considerable insights regarding quantitative dose–response relationships, as well as data that provide the foundation for the establishment of standards for the protection of acid-sensitive aquatic resources.

3.1 Characteristics of Sensitive Systems

Broad areas in the U.S. that contain large populations of low-ANC lakes and streams include portions of the Northeast (particularly Maine and the Adirondack Mountains), the mid-Appalachian Mountains, northern Florida, the Upper Midwest, and the western U.S. (Figure 3.1). The Adirondack and Mid-Appalachian Mountains include many acidified surface waters that have been impacted by acidic deposition. Portions of northern Florida and to a lesser extent the Upper Midwest also contain appreciable numbers of acidic lakes and streams, although the role of acidic deposition in these areas is less clear. The western U.S. contains many of the surface waters most susceptible to potential acidification effects, but the levels of acidic deposition in the West are generally low and acidic surface waters are rare.

It was recognized relatively early in acidification research that most of the major concentrations of low ANC surface waters were probably located in areas underlain by bedrock resistant to weathering. Subsequent compilations of available water chemistry data (e.g., Omernik and Powers, 1982; Eilers and Selle, 1991) refined and expanded this image of sensitive areas in North

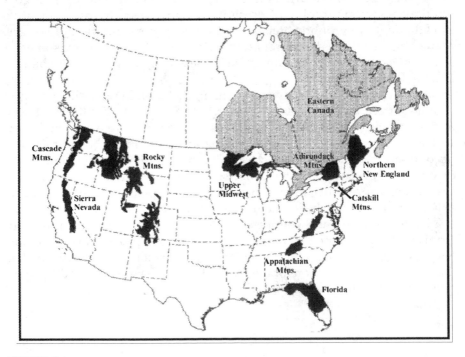

FIGURE 3.1
Major areas of North America containing low-ANC surface waters as defined by Charles (1991).

America. The extensive research programs conducted in Europe, Canada, and through NAPAP provided additional insight into factors contributing to the sensitivity of surface waters to acidic deposition by revealing the importance of soil composition and hydrologic flowpath, in addition to geology, in delineating sensitive regions.

The geologic composition of a region plays a dominant role in influencing the chemistry and, therefore, sensitivity of surface waters to the effects of acidic deposition. Bedrock geology formed the basis for a national map of surface water sensitivity (Norton et al., 1982) and has been used in numerous acidification studies of more limited extent (e.g., Bricker and Rice, 1989; Dise, 1984; Gibson et al., 1983). Analysis of bedrock composition continues to be an important element for assessing sensitivity of surface waters in mountainous regions (e.g., Stauffer, 1990; Stauffer and Whittchen, 1991; Vertucci and Eilers, 1993).

The presence of large populations of acidic and low-ANC lakes and streams in regions such as Florida that are underlain by calcareous bedrock illustrate that if the surface waters are isolated from highly weatherable bedrock minerals, acid–base status is not controlled by bedrock geology (Sullivan and Eilers, 1994). Many Karst lakes in northern Florida are situated in highly weathered marine sands that are capable of providing comparatively little neutralization of acidic inputs. For lakes located above calcareous bedrock in areas with minimal hydrologic connection with the Floridan aquifer, the surface waters can be acidic despite groundwaters saturated in carbonate minerals. Conversely, where calcareous soils have been deposited over resistant bedrock such as granite, lakes and streams draining such soils are predominantly alkaline. Thus, both soil and bedrock composition may exert strong influence on surface water acid-base chemistry and, therefore, are important factors to be considered in defining acid-sensitive regions.

The third principal factor now recognized as critical in contributing to the sensitivity of aquatic resources is watershed hydrology. The movement of water through the soils, into a lake or stream, and the interchange between drainage water and the soils and sediments regulate the type and degree of watershed response to acidic inputs. Lakes in the same physiographic setting can have radically different sensitivities to acidic deposition depending on the relative contributions of near-surface drainage water and deeper, more highly buffered groundwater (Eilers et al., 1983; Chen et al., 1984; Driscoll et al., 1991). The movement of water through natural conduits in peat can circumvent hydrologic routing through wetlands (Gjessing, 1992). Even acidic deposition that does not pass through the watershed, but instead falls as precipitation directly on the lake surface, may eventually be neutralized by in-lake reduction processes that are controlled in part by hydraulic residence time (Baker and Brezonik, 1988). Natural hydrologic events also radically alter sensitivity to acidification by bypassing normal neutralization processes during snowmelt or changing flowpaths during extended droughts (Webster et al., 1990). The importance of hydrologic factors in influencing the acid–base chemistry of surface waters across the U.S. was reinforced by Newell (1993), who identified hydrology as a key component associated with changes in the acid-base chemistry of lakes included in EPA's Long Term Monitoring Program.

3.2   Causes of Acidification

3.2.1   Sulfur

Several watershed processes control the extent of ANC generation and its contribution from soils to drainage waters as acidified water moves through undisturbed terrestrial systems. These are the major processes that regulate the extent to which drainage waters will be acidified in response to ambient levels of acidic deposition. Of particular importance is the concentration of acid anions in solution. Naturally occurring organic acid anions, produced in upper soil horizons, normally precipitate out of solution as drainage water percolates through lower mineral soil horizons. Soil acidification processes reach an equilibrium with acid neutralization processes (e.g., weathering) at some depth in the mineral soil (Turner et al., 1990). Drainage waters below this depth generally have high ANC. The addition of strong acid anions from atmospheric deposition allows the natural soil acidification and cation leaching processes to occur at greater depths in the soil profile, thereby allowing water rich in mobile anions such as SO_4^{2-} and NO_3^- to emerge from mineral soil horizons into drainage waters. If these anions are charge-balanced by H^+ and/or Al^{n+} cations, the water will have low pH and could be toxic to aquatic biota. Thus, the mobility of anions within the terrestrial system is a major factor controlling the extent of surface water acidification.

The scientific community has continued to make significant progress since 1990 in refining understanding of acidification processes and quantifying dose–response relationships. In particular, knowledge has been gained regarding the role of natural organic acidity, the depletion of base cation reserves from soils, interactions between acidic deposition and land use, and N dynamics in forested and alpine ecosystems. Each of these topics, in which significant recent advancements have been made, is discussed in the sections that follow. An expanded discussion of N dynamics is also provided in Chapter 7. It is now clear that the flux of SO_4^{2-} through watersheds is only one part of a complex set of watershed interactions that govern the response of both aquatic and terrestrial ecosystems to acidic deposition.

3.2.2   Organic Acidity

Organic acids commonly exert a large influence on surface water acid–base chemistry, particularly in dilute waters having moderate to high dissolved organic carbon (DOC) concentrations. Some lakes and streams are naturally acidic as a consequence of organic acids in solution. The presence of organic acids also provides buffering to minimize pH change in response to changes in the amount of mineral (e.g., SO_4^{2-}, NO_3^-) acid anions contributed to solution by atmospheric deposition.

The fact that there are many lakes and streams throughout the U.S. that are chronically acidic (ANC less than or equal to zero) primarily owing to the presence of organic acids is well known. NAPAP (1991) concluded that about one-fourth of all acidic lakes and streams surveyed in the National Surface Water Survey (NSWS, Linthurst et al., 1986; Kaufmann et al., 1988) were acidic largely as a consequence of organic acids. A more intensive survey of 1400 lakes in the Adirondacks by the Adirondack Lake Survey Corporation (ALSC; Kretser et al., 1989) that included lakes much smaller than those surveyed by NSWS, found a higher percentage of organically acidic lakes. Baker et al. (1990b) concluded that 38% of the lakes surveyed by ALSC had pH less than 5 owing to the presence of organic acids and that organic acids depress the pH of Adirondack lakes by 0.5 to 2.5 pH units in the ANC range of 0 to 50 μeq/L. However, the importance of organic acids in comparison with other sources of acidity has remained a subject of debate. In addition, the role of organic acids in the process of changing the acid–base character of surface waters (acidification or alkalization) is still poorly known.

Organic acids in fresh water originate from the degradation of biomass in the upland catchment, wetlands, near-stream riparian zones, water column, and stream and lake sediments (Hemond, 1994). The watersheds of surface waters that have high concentrations of organic matter (DOC greater than about 400 μM) often contain wetlands and/or extensive organic-rich riparian areas (Hemond, 1990).

Specification of the acid–base character of water high in DOC is somewhat uncertain. Attempts have been made to describe the acid–base behavior of organic acids using a single H^+ dissociation constant (pK_a), despite the fact that organic acids in natural waters are made up of a complex mixture of acidic functional groups. It has also been assumed in the past that organic acids are essentially weak acids, whereas a portion (perhaps one-third) of the acidity is actually quite strong, with some ionization occurring at pH values well below 4.0 (Hemond, 1994; Driscoll et al., 1994). A number of modeling approaches have been used to estimate the acidity of organic acids in fresh waters, often as simple organic acid analogs having different pK_a values (Oliver et al., 1983; Perdue et al., 1984; Driscoll et al., 1994).

In lakes sampled by the ALSC, estimated values of organic acid anion concentration per mol DOC ($RCOO^-$/DOC), often called the organic acid charge density, were consistent with patterns anticipated from the presence of both strong and weak organic acid functional groups (Driscoll et al., 1994; Figure 3.2). Even at pH values below 4.5, the charge density of ALSC lakes was in the range of 0.03 to 0.05, corresponding to about 25 to 30% of values found at circumneutral pH (Driscoll et al., 1994). Thus, some of the functional groups associated with naturally occurring organic acids are

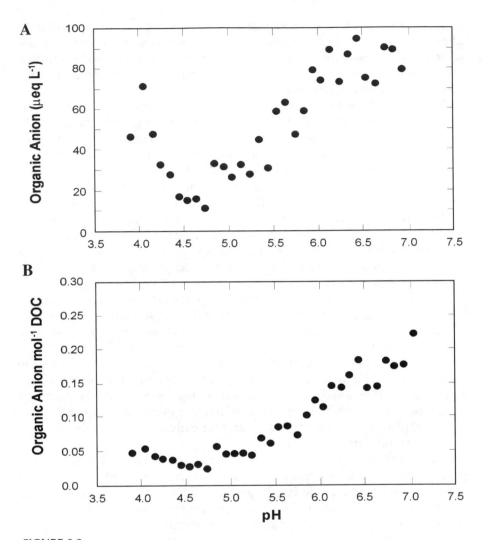

FIGURE 3.2

Mean organic anion concentration (A) estimated from anion deficit, and (B) charge density expressed as A^{n-}/DOC at 0.1 pH unit intervals, as a function of pH for the reduced ALSC data set included in the analyses of Driscoll et al. (C.T. Driscoll, M.D. Lehtinen, and T.J. Sullivan, 1994, Modeling the acid-base chemistry of organic solutes in Adirondack, NY, lakes, *Water Resour. Res.*, Vol. 30, p. 301, Figure 1; copyright by the American Geophysical Union. With permission.)

strongly acidic, and do not dissociate unless pH is below 4.0. Values of charge density in the ALSC lakes increased with increasing pH between pH values of 5.0 to 7.0 owing to the presence of weakly acidic functional groups. Thus, organic acids in surface waters include a mixture of functional groups having both strong and weak acid character. This concept was not well understood prior to 1990.

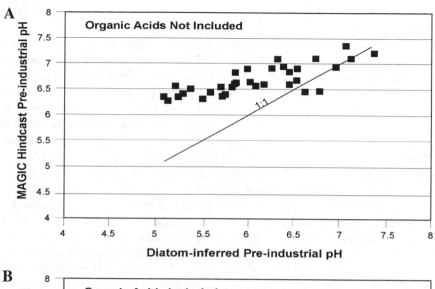

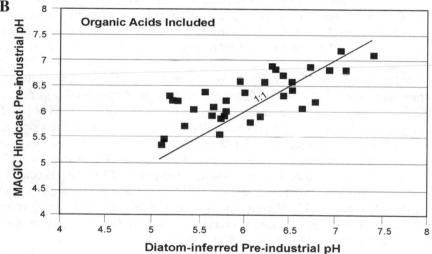

FIGURE 3.3
MAGIC model hindcast estimates of pre-industrial pH versus diatom-inferred pH for 33 statistically selected Adirondack lakes. (A) Without including organic acid representation in the MAGIC simulations, and (B) including a triprotic organic acid analog model in the MAGIC simulations. (Source: *Water Air Soil Pollut.*, Vol. 91, 1996, p. 301, Influence of organic acids on model projections of lake acidification, Sullivan, T.J., B.J. Cosby, C.T. Driscoll, D.F. Charles, and H.F. Hemond, Figure 1, copyright 1996. Reprinted with kind permission from Kluwer Academic Publishers.)

The ALSC data were fitted by Driscoll et al. (1994) to a triprotic organic acid analog representation that provided a good fit to the data ($r^2 = 0.92$), with pK_a values of 2.62, 5.66, and 5.94 to represent a range of strong to weak acid character. Inclusion of organic acidity from this analog in model calculations resulted in good agreement between measured and predicted values of lake-water pH and ANC in this large database (Driscoll et al., 1994).

The importance of naturally occurring organic acids as agents of surface water acidification has recently been substantially reinforced by several modeling studies (e.g., Sullivan et al., 1994, 1996). These have shown that inclusion of organic acids in the MAGIC model have an appreciable effect on model predictions of surface water pH, even in waters where DOC concentrations are not particularly high. Concern was raised subsequent to NAPAP's Integrated Assessment (NAPAP, 1991) regarding potential bias from the failure to include organic acids in the MAGIC model formulations used in the IA. MAGIC hindcasts of pre-industrial lake-water pH of Adirondack lakes showed poor agreement with diatom inferences of pre-industrial pH. Revised MAGIC simulations, therefore, were constructed that included the organic acid analog model developed by Driscoll et al. (1994). The revised MAGIC hindcasts of pre-industrial lake-water pH that included an organic acid representation showed considerably closer agreement with diatom inferences (Figure 3.3). The mean difference between MAGIC and diatom estimates of pre-industrial pH was reduced from 0.6 pH units to 0.2 pH units when organic acids were included in the model, and the agreement for individual lakes improved by up to a full pH unit (Sullivan et al., 1996).

Inclusion of organic acids in the MAGIC simulations for watershed manipulation data sets at Lake Skjervatjern (Norway), Bear Brook (Maine), and Risdalsheia (Norway) also had dramatic effects on model simulations of pH. In all cases, MAGIC simulated considerably higher pH values when organic acids were omitted from the model. Even at Bear Brook, where annual average DOC concentrations are very low (less than 250 μM C), incorporation of organic acids into the model reduced simulated pH by 0.1 to 0.3 pH units for the years of study. At Lake Skjervatjern and Risdalsheia, where organic acids provide substantial pH buffering, omission of the organic acid analog representation from MAGIC resulted in consistent overprediction of pH by about 0.2 to 0.5 pH units (Sullivan et al., 1994; Figure 3.4).

Rosenqvist (1978) and Krug et al. (1985) hypothesized that a significant component of the mobile acid anions contributed from atmospheric deposition (e.g., SO_4^{2-}, NO_3^-) merely replace organic anions that were previously present in solution. Under this anion substitution hypothesis, the net result of acidic deposition is not so much an increase in cations (including potentially toxic H^+ and Al^{n+}) as much as an exchange of SO_4^{2-} and NO_3^- anions for organic anions, with little or no change in ANC and pH.

Data are scarce with which to directly evaluate the hypothesis that acidic deposition causes decreased organic acidity, but a variety of indirect evidence was summarized in the review of Marmorek et al. (1988). They concluded that there were a number of inconsistencies in the available data, but most data suggested that organic acids have been lost from lake water as a consequence of acidic deposition. Hypothesized mechanisms included:

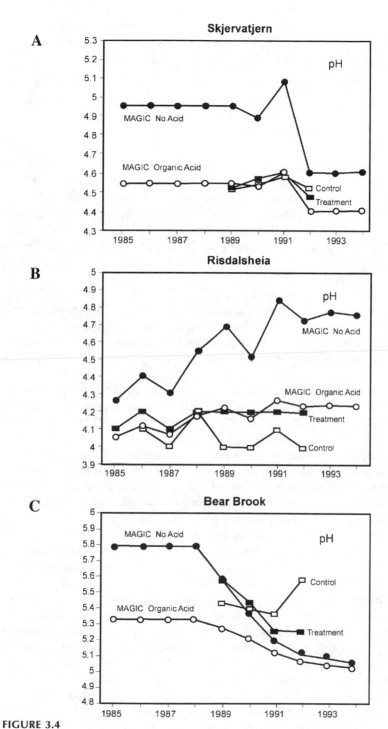

FIGURE 3.4
MAGIC simulated pH with and without inclusion of the triprotic organic acid analog, and observed pH, in the treatment and control lake/stream at (A) Skjervatjern, (B) Risdalsheia, and (C) Bear Brook.

1. Decreased mobilization of organic materials from soils and wetlands because of increased H^+ concentration.
2. Reduced microbial decomposition of organic materials in soils.
3. Changes in dissociation and/or physical structure of humics.
4. Increased loss from solution to sediments through chelation with metals (e.g., Al, Fe) mobilized by increased H^+, and subsequent precipitation of the metal–organic complex.

Of the preceding mechanisms, complexation of organic acids by metals (Almer et al., 1974; Lind and Hem, 1975; Dickson, 1978; Cronan and Aiken, 1985) and pH dependent changes in dissociation of organic acids (Oliver et al., 1983; Wright et al., 1988b) appeared most likely to be significant. Quantitative estimates of change in DOC were not possible, but based on the available data, Marmorek et al. (1988) concluded that potential DOC losses of up to 250 μM C were not unreasonable. Subsequent research has suggested, however, that decreases in DOC concentrations in surface water in response to acidic deposition have probably been less than 250 μM C (Wright et al., 1988b; Kingston and Birks, 1990; Cumming et al., 1992). Furthermore, Krug and co-workers contended that interactions between acidic deposition and organic matter can either increase or decrease DOC, depending on the nature of the organic matter interacting with the acid (e.g., Krug et al., 1985; Krug, 1991a,b).

Kingston and Birks (1990) presented diatom-based paleolimnological reconstructions of DOC for lakes studied in the Paleoecological Investigation of Recent Lakewater Acidification (PIRLA-I) project. The DOC optima and tolerances of diatom taxa in four regions (Adirondack Mountains, northern New England, northern Great Lakes states, and northern Florida) were estimated using maximum likelihood and weighted averaging regression. The cumulative fit per taxon as a fraction of the taxon's total variance revealed that few taxa were consistent in terms of their explanation of the DOC gradient from region to region. DOC explained a small, but significant, amount of taxon variance in lakes in the Adirondack Mountains, northern Florida, and the northern Great Lakes States, but the signal was much weaker in northern New England. Calculated species optima were not consistent among regions and the best indicators of DOC in the PIRLA data sets were not always in good agreement with those found in Norway and Canada (e.g., Davis et al., 1985; Anderson et al., 1986; Taylor et al., 1988). The authors, therefore, cautioned that taxa that are good indicators for one region may not be good indicators of DOC in other regions. Example reconstructions were provided for Big Moose Lake in the Adirondack Mountains, NY and Brown Lake in northern Wisconsin. The magnitudes of inferred DOC changes were small relative to the mean squared error of the predictive relation in each region (98 and 80 μM, respectively), but in each case DOC was inferred to have declined coincident with lake-water pH. For the recently acidified PIRLA-I lakes in general, inferred declines in DOC were coincident with recent acidification.

Although the magnitude of DOC change was typically small (less than 100 µM), the acid–base character of the DOC (charge density and degree of dissociation) may also have changed, so the effect on organic acid anion concentration may have been proportionately larger.

In addition to potential changes in DOC concentrations in response to acidic deposition, acidification or recovery can alter the charge density of organic solutes and, thus, influence organic contributions to acidity (e.g., Wright et al., 1988b). David et al. (in press) found that charge density of organic acids decreased by about 1 µeq/L/mg C at West Bear Brook in response to 6 years of experimental acidification, probably owing to greater protonation of organic acid anions at the lower pH. There was no evidence of a change in DOC, however, in response to the acidification. Similar results were reported by Lydersen et al. (1996) at Lake Skjervatjern in Norway. Loss of DOC in response to acidic deposition can also cause a shift in Al species composition towards lesser complexation with organic ligands. Such a shift from organic to inorganic Al increases toxicity of the Al to aquatic biota (Baker and Schofield, 1982).

Hedin et al. (1990) artificially acidified a small, moderately high-DOC (725 µM C) stream with H_2SO_4 at the Hubbard Brook Experimental Forest (HBEF) in New Hampshire. The ambient stream-water pH (4.4) was near the range of reported average pK_a values for organic acids, suggesting that the capacity of organic acids to buffer mineral acidity should be high. The loading rate of H_2SO_4 was adjusted to achieve an increased stream-water SO_4^{2-} concentration of 150 µeq/L at the downstream sampling point 108 m below the point of acid addition, and LiBr was added as a conservative tracer to adjust measured concentrations for dilution by soil water or inflow from small tributaries. Although stream-water DOC did not change significantly, the concentration of organic anions (as calculated from the charge balance) decreased by 17 µeq/L. Thus, the overall capacity of organic anions to neutralize mineral acid inputs offset about 11% of the added H_2SO_4 concentration (Hedin et al., 1990). This experiment only considered interactions between H_2SO_4 and organic matter within the stream. Any additional buffering that may have been provided within the terrestrial catchment was not represented in the experimental design. Also, any possible catchment-mediated influences of the experimental acidification on organic acid properties, DOC mobilization, and so on, were excluded from the experiment because the acid was not applied to the catchment soils.

Webster et al. (1990) reported dramatic changes in lake-water ANC in Nevins Lake, MI, in response to the effects of drought on the local hydrology. Lake-water ANC decreased by 150 µeq/L during a 5-year period of record. The pH also declined from about 7.0 to about 6.25. DOC concentrations are fairly low in Nevins Lake (approximately 250 to 300 µM C, Avis Newell, personal communication) and did not exhibit a trend coincident with the ANC and pH changes. Webster et al. (1990) did not, however, evaluate any possible change in organic acid anion concentration, using charge-balance calculations, because of a contamination problem in the laboratory analyses of some of the cations (Newell, personal communication).

Brezonik et al. (1993) acidified Little Rock Lake, WI, with sulfuric acid. The north basin of the lake was acidified from pH approximately equal to 6.1 to target values of 5.6, 5.1, and 4.7 during successive 2-year periods. The major changes in this low-alkalinity seepage lake involved increases in base cations, especially Ca^{2+}; DOC decreased slightly (approximately 50 μM C) and color decreased by half in the acidified north basin after acidification to pH equal to 4.7. Average DOC concentrations and color values of the 2 basins were significantly different at pH 4.7, but not at higher pH values. In contrast, Schindler and Turner (1982) did not find significant changes in lakewater DOC in response to the artificial acidification of Lake 223 in the Experimental Lakes Area of Ontario to pH approximately equal to 5.4.

Sullivan et al. (1994) examined the results of the three catchment manipulation experiments (Bear Brook, Maine, and Lake Skjervatjern and Risdalsheia, Norway) that were conducted by Norton et al. (1993), Gjessing (1992), and Wright et al. (1993). All three catchments showed some evidence of changes in organic acid anion concentration in response to experimental acidification or de-acidification treatment. Changes in DOC also may have occurred. Unfortunately, however, none of these manipulation experiments provided conclusive quantitative data regarding the effects of acidification on DOC mobilization from catchment to surface waters or changes in the concentration of DOC caused by acidification. There were problems in interpretation of the data regarding changes in the concentration of dissolved or total organic C (DOC/TOC) in runoff from each of the studies.

In the Watershed Manipulation Project at Bear Brook, DOC declined about 50% from 1989 to 1992 in both East (reference catchment) and West (treatment catchment) Bear Brooks. These streams were very low in DOC throughout most of the year (annual average DOC less than 300 μM C) and, therefore, are less than optimal sites for evaluation of this question. In addition, the acidification caused by the manipulation experiment was fairly modest, because a large percentage of the added S and N was retained within the catchment. It is likely that observed decreases in DOC at Bear Brook were mostly related to a pattern of generally decreasing runoff, although a small decrease in DOC in response to the chemical manipulation also seemed to have occurred (Norton et al., 1993).

Total organic carbon (TOC) concentration at Lake Skjervatjern was highly variable, thus making it difficult to quantify any changes that may have occurred in response to the experimental treatment. Data from the treatment side of the lake showed no indication of a decline in TOC relative to the control with acidification. The increase in lakewater SO_4^{2-} concentration was small, because most of the S applied to the terrestrial catchment was retained in watershed soils. Thus, the possible long-term influence of watershed acidification on TOC mobilization at Lake Skjervatjern is highly uncertain. If S retention in the watershed decreases over time, effects on TOC mobilization may become more evident.

At Risdalsheia, the annual variability in TOC was very large at both the roofed control and manipulated catchments and sufficient pretreatment data were not collected to allow a rigorous evaluation of the extent to which TOC mobilization may have been affected by the acid exclusion. Thus, none of these three experimental manipulation studies provide the kind of quantitative data on DOC/TOC responses to acidification that would be needed to justify modifying predictive models to account for hypothesized changes in the concentration of organic C in response to changes in acidic deposition.

Based on results available to date, it appears that changes in DOC concentration in response to changes in acidic deposition may occur but are generally small in magnitude. The concentration of organic acid anions is affected, however, by changes in acidic deposition, particularly in high-DOC waters. This change can be appreciable in some cases and organic acids can provide significant buffering against pH change in watersheds that receive acidic deposition. For example, results of a resurvey of 485 Norwegian lakes sampled in both 1986 and 1995 provided evidence in support of an increase in organic acid anion concentrations in association with decreased lake-water SO_4^{2-} concentration (Skjelvåle et al., 1998). On a regional basis, the organic acid anion concentration increased by an amount equal to between 9 and 15% of the decrease in SO_4^{2-} concentration in the 4 regions of the country most heavily affected by the recent decrease in S deposition. Lake-water SO_4^{2-} concentrations decreased by 9 µeq/L (western and northern Norway) to 20 to 21 µeq/L (eastern and southern Norway). Only in mid-Norway, where average SO_4^{2-} concentration decreased by only 6 µeq/L, did the organic acid anion concentration remain unchanged between 1986 and 1995 (Skjelkvåle et al., 1998).

3.2.3 Nitrogen

Nitrate (and also NH_4^+ that can be converted to NO_3^- within the watershed) has the potential to acidify drainage waters and leach potentially toxic Al from watershed soils. In most watersheds, however, N is limiting for plant growth and, therefore, most N inputs are quickly incorporated into biomass as organic N with little leaching of NO_3^- into surface waters. A large amount of research has been conducted in recent years on N processing mechanisms and consequent forest effects, mainly in Europe (Sullivan, 1993). In addition, a smaller N research effort has been directed at investigating effects of N deposition on aquatic ecosystems. For the most part, measurements of N in lakes and streams have been treated as outputs of terrestrial systems. However, concern has been expressed regarding the role of NO_3^- in acidification of surface waters, particularly during hydrologic episodes, the role of NO_3^- in the long-term acidification process, the contribution of NH_4^+ from agricultural sources to surface water acidification, and the potential for anthropogenic N deposition to stimulate eutrophication of freshwaters and estuaries.

Concern was raised in the mid-1980s about the possible adverse effects on soils, forests, and drainage waters from atmospheric deposition of inorganic N compounds. Prior to that time, atmospheric deposition effects research focused almost exclusively on S. Within the 1980 to 1990 NAPAP research program, relatively little attention was paid to N research.

Concern for chronically elevated NO_3^- concentrations in aquatic ecosystems received considerably greater attention in 1988 following the publication of the resurvey of Norwegian lakes (SFT, 1987). Over 1000 lakes, 305 of which were originally sampled in 1974/75 (Wright and Henriksen, 1978), were sampled again in 1986 (SFT, 1987; Henriksen and Brakke, 1988). Even though the average SO_4^{2-} concentration declined in the lakes, the pH remained virtually unchanged because of increased NO_3^- and decreased base cation concentrations. In the southern portions of Norway, NO_3^- concentrations in the lakes doubled between 1974–1975 and 1986, reaching county-wide average concentrations as high as 14 µeq/L in Rogaland County and up to 50 µeq/L in individual lakes (Henriksen and Brakke, 1988). An analysis of fisheries in the study lakes showed an increase in the number of fishless lakes, perhaps attributable to the concomitant increase in labile Al and decrease in $(Ca^{2+} + Mg^{2+})$ (SFT, 1988). Analysis of selected lakes and streams with longer-term records also showed increases in NO_3^- concentrations, providing additional evidence for an increasing trend in NO_3^-. Although SO_4^{2-} remained the dominant anion in most systems, the ratio of $NO_3^-/(NO_3^- + SO_4^{2-})$ reached 0.54 on an equivalent basis in some lakes and rivers in southwestern Norway (Henriksen and Brakke, 1988). These authors summarized the ratio of $NO_3^-/(NO_3^- + SO_4^{2-})$ for many acidified waters in Europe and North America, illustrating that the relative importance of NO_3^- in acidified surface waters can be substantial, particularly in central Europe.

Some of the concerns raised by the results of the 1987 Norwegian lake survey have been lessened in response to more recent data. In 1995, 485 lakes were again resurveyed throughout Norway. The concentration of NO_3^- changed little, on average, in the various regions that were surveyed. Only in western Norway did the average lakewater NO_3^- concentration increase between 1986 and 1995, and the average increase was only 1 µeq/L (Skjelkvåle et al., 1998).

Increased atmospheric deposition of N does not necessarily cause adverse environmental impacts. In most areas, added N is taken up by terrestrial biota and the most significant effect is an increase in forest productivity (Kauppi et al., 1992). However, in some areas, especially at high elevation sites, terrestrial ecosystems have become N saturated* and high levels of deposition cause elevated levels of NO_3^- in drainage waters (Aber et al., 1989, 1998; Stoddard, 1994). This enhanced leaching of NO_3^- causes depletion of

---

* The term nitrogen-saturated has been defined in a variety of ways, all reflecting a condition whereby the input of nitrogen (e.g., as nitrate, ammonium) to the ecosystem exceeds the requirements of terrestrial biota and a substantial fraction of the incoming nitrogen leaches out of the ecosystem in groundwater and surface water.

Ca^{2+} and other base cations from forest soils and can cause acidification of drainage waters in base-poor soils.

Chronically high concentrations of lake or stream-water NO_3^-, which in some cases may be indicative of ecosystem saturation, have been found in recent years at a variety of locations throughout the U.S., including the San Bernardino and San Gabriel Mountains within the Los Angeles Air Basin (Fenn et al., 1996a), the Front Range of Colorado (Baron et al., 1994; Williams et al., 1996a), the Allegheny Mountains of West Virginia (Gilliam et al., 1996a), the Catskill Mountains of New York (Murdoch and Stoddard, 1992; Stoddard, 1994), and the Great Smoky Mountains in Tennessee (Cook et al., 1994).

Nitrate concentrations during the fall sampling season were low in most western lakes sampled in the Western Lake Survey. Only 24 sampled lakes were found to have NO_3^- concentrations greater than 10 µeq/L. Of those, 5 lakes were situated at low elevation (less than 500 m) in the state of Washington and had relatively high ANC (greater than about 200 µeq/L). Because of the high neutralization capacity, the N concentrations did not have a significant impact on chronic acid–base status of these lakes. The other 19 high NO_3^- lakes were all situated at high elevation, most above 3000 m. Cold temperatures in such lakes undoubtedly play a major role in maintaining chronically elevated NO_3^- concentrations, largely by limiting biological uptake processes in both the aquatic and terrestrial environments. The high NO_3^- concentrations are most likely to have significant impacts on the acid–base chemistry of the lakes only where ANC is low. Of the lakes, 8 showed high NO_3^- (greater than 10 µeq/L) and low ANC (less than 50 µeq/L), all of which were extremely low in DOC (less than 1 mg/L) and occurred at elevations higher than 3100 m. There were four located in Colorado, two in Wyoming, and one each in California and Utah. In all cases, pH was above 6.5 and ANC greater than or equal to 15 µeq/L, suggesting that chronic biological impacts were unlikely to have occurred as of the sampling date. Such lakes are likely highly sensitive, however, to episodic pulses of NO_3^- acidity which could be very important biologically.

The Uinta Mountains of Utah and the Bighorn Mountains of central Wyoming had the greatest percentages of high NO_3^- lakes in the West, irrespective of lake-water ANC, with 19% of the lakes included within the Western Lakes Survey having NO_3 more than 10 µeq/L. This is a high percentage of lakes with measurable NO_3^- for fall samples and indicates that NO_3^- deposition in these areas may have exceeded the capability of these systems to assimilate N. It is unknown if these concentrations of NO_3^- represent impacts from anthropogenic sources or if this constitutes an unusual natural condition associated with inhibited NO_3^- assimilation in extremely cold alpine environments.

Williams et al. (1996a) contended that nitrogen saturation is occurring throughout high-elevation catchments of the Colorado Front Range at N deposition levels considered quite low by European standards. Total N deposition is 4 to 7 kg N/ha per year in this region, about double that in most other mountainous areas of the West and approaching the deposition levels found in parts of the East, but still well below the 10 kg N/ha per year threshold found for inducing NO_3^- leaching in most European forests (Dise and Wright, 1995). Many lakes in the Colorado Front Range have chronic NO_3^- concentrations greater than 10 µeq/L and concentrations during snowmelt are frequently much higher. The observed high concentrations of NO_3^- in lake and streamwaters of the Front Range are likely owing to leaching from tundra, exposed bedrock, and talus areas at high elevations. Although biological N uptake appears to be high in subalpine forests, such uptake is limited in non-forested alpine watersheds by large N inputs from snowmelt, steep watershed gradients, rapid water flushing, and possibly limitations on the growth of phytoplankton in alpine lakes by factors other than N (e.g., P, temperature; Baron et al., 1994). See further discussion of this topic in Chapters 7 and 11.

Recent research results regarding N dynamics and the effects of elevated N deposition are considered in greater detail in Chapter 7.

3.2.4 Base Cation Depletion

Calcium and other base cations are important nutrients that are taken up through plant roots in dissolved form. Base cations are typically found in abundance in rocks and soils, but a large fraction of the base cation stores are bound in mineral structures and are unavailable to plants. The pool of soluble base cations resides in the soil as cations that are adsorbed to negatively charged exchange sites. They can become desorbed in exchange for H^+ or Al^{3+} and are, thus, termed exchangeable cations. The process of weathering gradually breaks down rocks and minerals, returning their stored base cations to the soil in dissolved form and, thereby, contributing to the pool of adsorbed base cations. Base cation reserves are gradually leached from the soils in drainage water, but are constantly being resupplied through weathering.

It has long been recognized that elevated leaching of base cations by acidic deposition might deplete the soil of exchangeable bases faster than they are resupplied via weathering (Cowling and Dochinger, 1980). However, base cation depletion of soils had not been demonstrated at the time of the Integrated Assessment. Scientific appreciation of the importance of this response has increased with the realization that watersheds are generally not exhibiting ANC and pH recovery in response to recent decreases in S deposition. In many areas, this lack of recovery can be at least partially attributed to decreased base cation concentrations in surface water. The base cation response is quantitatively more important than was generally recognized in the 1980s.

The development of this understanding has evolved slowly. During most of the 1980s, the generally accepted paradigm of watershed response to acidic deposition was somewhat analogous to a large-scale titration of ANC (Henriksen, 1980). It was widely believed that atmospheric input of acidic anions (mainly SO_4^{2-}) resulted in movement of those anions through soils into drainage waters with near stoichiometric loss of surface water ANC.

This view was tempered somewhat by Henriksen (1984), who suggested that a modest component of the added SO_4^{2-} (up to a maximum of about 40%) could be charge balanced by increased mobilization of base cations from soils, and the remaining 60 to 100% of the added SO_4^{2-} resulted in loss of ANC in surface waters.

During the latter part of the 1980s, it became increasingly clear that a larger component (much greater than 40%) of the added SO_4^{2-} was, in fact, neutralized by base cation release in most cases and the ANC (and, therefore, also pH) of surface waters typically did not change as much as was earlier believed. This understanding developed in large part from paleoecological studies (e.g., Davis et al., 1988; Charles et al., 1990; Sullivan et al., 1990a) that concluded that historical changes in lake-water pH and ANC were small relative to estimated increases in lake-water SO_4^{2-} concentrations since pre-industrial times.

After passage of the Clean Air Act in 1970 and subsequent amendments in 1990, emissions and deposition of S were reduced and the concentrations of SO_4^{2-} in lake and stream-water in the eastern U.S. and Canada decreased (Dillon et al., 1987; Driscoll et al., 1989a; Sisterson et al., 1990). Long-term monitoring data confirmed that much of the decrease in surface water SO_4^{2-} concentration was accompanied by rather small pH and ANC recoveries (Driscoll and van Dreason, 1993; Kahl et al., 1993b; Driscoll et al., 1995; Likens et al., 1996). The most significant response, on a quantitative basis, was decreased concentrations of Ca^{2+} and other base cations. Similarly, long-term monitoring data from four small watersheds in Norway illustrated substantial declines in both S deposition and stream-water SO_4^{2-} concentration since the late 1970s. Reductions in SO_4^{2-} concentration in runoff at these sites have been approximately balanced by reductions in Ca^{2+} and Mg^{2+} concentrations. As a result, stream-water pH and Al concentrations have not shown significant recoveries (Kirchner and Lydersen, 1995). The authors concluded that the observed long-term declines in base cation concentrations in runoff were quantitatively consistent with depletion of exchangeable bases in the soil by accelerated leaching under decades of high acid loading. Kirchner and Lydersen (1995) also contended that, even though water quality had not recovered in response to reduced S deposition throughout southern portions of Norway, reductions in deposition have been valuable because they have prevented significant further acidification that would otherwise have occurred under continued high acid loading.

A paradigm shift has occurred. The earlier belief that changes in SO_4^{2-} were accompanied mainly by changes in ANC and pH has been replaced by the realization that changes in SO_4^{2-} were accompanied mainly by changes in base cations. This means that surface waters have not been acidified as much by historical deposition as was widely believed only 10 years earlier. It also suggests that surface water ANC and pH will not recover so quickly upon reduced emissions and deposition of S and N.

Thus, as SO_4^{2-} concentrations in lakes and streams have declined so, too, have the concentrations of Ca^{2+} and other base cations. There are several apparent reasons for this. First, the atmospheric deposition of base cations has decreased in recent decades (Hedin et al., 1994), likely owing to a combination of air pollution controls, changing agricultural practices, and the paving of roads (the latter two affect generation of dust that is rich in base cations). It has been estimated that more than half of the supply of Ca^{2+} to cation pools of forest soils in the northeastern U.S. may be derived from atmospheric inputs. Similarly, Driscoll et al. (1989a) estimated that between 77 and 85% of the decline in the concentration of base cations in stream water at Hubbard Brook Experimental Forest (HBEF) could be attributed to decreased base cation deposition. However, atmospheric deposition of base cations has increased in Maine since 1982 concurrent with large declines in the concentration of base cations in drainage lakes. Moreover, the concentrations of base cations in groundwater recharge seepage lakes have not declined, which suggests that watershed processes have been altered and that changes in base cation deposition are not responsible for changes in the concentration of base cations in lake waters in Maine (Kahl, personal communication). Second, decreased movement of SO_4^{2-} through watershed soils causes reduced leaching of base cations from soil surfaces. Third, soils in some sensitive areas have experienced prolonged base cation leaching to such an extent that soils may have been depleted of their base cation reserves. Such depletion greatly prolongs the acidification recovery time of watersheds and may adversely impact forest productivity (Kirchner and Lyderson, 1995; Likens et al., 1996).

As aquatic effects researchers have revised their understanding of the quantitative importance of the various acidification processes, terrestrial effects researchers have also turned greater attention to the importance of the response of base cations to acid deposition and the interactions between base cations (especially Ca^{2+} and Mg^{2+}) and Al. Likens et al. (1996) concluded that acidic deposition enhanced the release of base cations from forest soils at HBEF from the mid-1950s until the early 1970s, but that, as the labile pool of base cations in soil became depleted, the concentrations in stream water decreased from 1970 through 1994 by about one-third. The marked decrease in base cation inputs and concomitant increase in net soil release of base cations at HBEF have likely depleted soil pools to the point where ecosystem recovery from decreased S deposition will be seriously delayed. Moreover, Likens et al. (1996) suggested that recently observed declines in forest biomass accumulation at HBEF might be attributable to Ca^{2+} limitation or Al-toxicity, which can be expressed by the Ca^{2+} to Al^{n+} ratio in soil solution (Cronan and Grigal, 1995).

Lawrence et al. (in press) investigated base cation dynamics in soils in the Neversink River Basin in the Catskill Mountains, NY. They found that S deposition increased along an elevational gradient, whereas the concentrations of soil exchangeable bases decreased with elevation. A large quantity of soil was collected from a low-elevation site, bulked, and then redistributed to about 30 sites along the elevational gradient. At each site, soil was placed in mesh bags, buried, and then retrieved and analyzed after 1 year. Results of chemical analyses confirmed that the concentration of exchangeable bases and the base saturation* decreased with increasing elevation (Lawrence et al., in press). Field data and laboratory analyses of soil samples were consistent with the interpretation that observed decreases in ANC of stream water in the Neversink River watershed since 1984 have been the result of decreased base saturation of soils caused by acidic deposition (Lawrence et al., in press).

Lawrence et al. (1995) proposed that the dissolution of Al in the mineral soil by mineral acid anions supplied by acidic deposition (SO_4^{2-}, NO_3^-) can decrease the availability of Ca^{2+} in the overlying forest floor. This conclusion was based on the results of a survey in 1992 and 1993 of soils in red spruce forests that had been acidified to varying degrees throughout the northeastern U.S. The proposed mechanism for Ca^{2+} depletion is as follows. Acidic deposition lowers the pH in the mineral soil, thereby increasing the concentration of dissolved Al in soil solution. Some of the Al is then taken up by tree roots and transported throughout the trees, eventually to be recycled to the forest floor in leaves and branches. Additional dissolved Al is transported to the forest floor by rising water table during wet periods and by capillary movement during dry periods. Because Al^{3+} has a higher affinity for negatively charged soil surfaces than Ca^{2+}, introduction of Al into the forest floor, where root uptake of nutrients is greatest, causes Ca^{2+} to be displaced from the cation exchange complex and, therefore more easily leached into drainage water (Lawrence et al., 1995; Lawrence and Huntington, 1999).

3.2.5 Land Use

The influence of landscape processes, such as forest succession and watershed disturbance, on surface water acid–base chemistry have not been well-integrated into acidic deposition assessments. Land use practices and resulting vegetation patterns have changed more or less continuously in the northeastern U.S. for about the past 250 years. These changes in human activity, and consequent changes in forest structure and dynamics, can influence the response of forested ecosystems to external stressors, such as atmospheric deposition of S or N, exposure to ozone, natural disturbance factors such as wind and fire, and climatic changes.

Landscape processes affect the acid–base chemistry of drainage waters in a variety of ways. Some processes contribute to the acidification of soil and surface waters or reduce the base saturation of the soils thereby increasing their sensitivity to acidic deposition. Other processes cause decreased acidity (Sullivan et al., 1996b; Table 3.1).

Disturbances such as logging, blowdown, and fire affect surface water pH and ANC. Watershed disturbance disrupts the normal flow of water, in some cases causes increased contact between runoff water and soil surfaces, and often leads to increased base cation concentration and ANC in surface waters. Recovery from disturbance will, in most cases, lead to a decrease in

---

* Base saturation is the concentration of exchangeable base cations as a percentage of the total cation exchange capacity, which also includes H^+ and Al^{n+}.

TABLE 3.1

Overview of Selected Major Processes by Which Landscape Change Can Alter
Drainage Water Acid-Base Chemistry

Landscape Change	Impact on Acid–Base Chemistry
Logging, blowdown	Dilution
	Lower deposition, less acidity
	Pulse of nitrate acidity initially
	Less base cation neutralization, more acidity
	Less water contact with mineral soils, less neutralization of acidic deposition inputs
Road building and construction	More base cation neutralization, less acidity initially
	Depletion of base cation reserves in soils, more acidity long term
Drainage of wetlands	Re-oxidation of stored sulfur, pulses of acidity with increased discharge
Drought	Reduced groundwater inputs to seepage lakes with consequent increased acidity
	Increased relative baseflow to drainage waters with consequent decreased acidity
Lake shore development	Decreased acidity
Insect damage	Pulse of nitrate acidity initially

pH and ANC as the system returns to predisturbance conditions. A short-term investigation of an ecosystem in the process of recovering from a watershed disturbance might erroneously conclude that acidification was occurring in response to changes in atmospheric deposition or some other cause external to the watershed.

The influence of historical forest management on the ability of a given forest ecosystem to process N is not well understood. Nevertheless, forest management practices, especially those that have occurred over many generations, can have important effects on soils (i.e., erosion), nutrient supplies (i.e., harvesting), organic material (i.e., litter raking), and, thereby, many aspects of N cycling and effects. For example, by introducing Norway spruce in high-elevation areas on nutrient-poor soils, forest management in the Vosges Mountains of France may have exacerbated the impacts of acidic deposition on forests (Landmann, 1991).The introduced Norway spruce likely contributed to increased dry deposition to the forest and also increased cation uptake relative to the original forest stands of mixed birch and silver fir. The observed needle yellowing in Norway spruce in the Vosges Mountains has been attributed to Mg^{2+} deficiency, which can be influenced by land management and by acidic deposition. European forests have typically been harvested for many generations, have been changed in species composition or community type (e.g., conversion from heathland to forest), and managed or manipulated in a variety of ways. The interactions between these activities and atmospheric deposition have not been well quantified.

Numerous investigators have dismissed change in land use as a primary causal factor in regional surface water acidification, based on the observation that lakes have become acidic even in high elevation, pristine watersheds where land use changes presumably have not occurred. Whereas it is true that landscape change often occurs as a manifestation of land use change, equally dramatic landscape changes can occur in response to natural factors without change in the way that humans use the land. Also, the documented occurrence of acidification in the absence of either land use or landscape change does not negate the importance of other questions concerning the interactions between such changes and acidic deposition (Sullivan et al., 1996b).

Land management activities, particularly removal of forest or change in forest structure, have important effects on hydrology and the total deposition of S, N, and marine salts. An important land use change during the last 60 years in the British uplands has been the widespread afforestation of acid moorland with conifers. Streams draining the afforested areas are more acidic and contain higher concentrations of dissolved Al than adjacent moorland catchments (Ormerod et al., 1989). It is likely that both the increased dry deposition of S to tree surfaces in the afforested catchments and the enhanced base cation uptake by the growing trees contribute to this difference. These changes have been implicated in the decline of fisheries. Subsequent clear cutting of these afforested catchments can result in short-term pulses of NO_3^- and inorganic Al in streamwaters, thereby exacerbating the biological effects of acidification (Reynolds et al., 1992).

Forests are very efficient at scavenging S from the atmosphere. Differences in forest canopy, particularly between deciduous and coniferous stands, can cause large differences in dry deposition, and, therefore total deposition, of S and N compounds. Thus, in polluted regions, forests exacerbate acidification by enhancing total deposition of acid-forming precursors. In some cases dry and occult deposition can contribute significantly more S to a forest ecosystem than precipitation (Rustad et al., 1994). In addition to the enhanced deposition caused by older and larger trees, there are pronounced differences in nutrient uptake among trees of different age classes. Younger stands take up larger quantities of N and other nutrients.

The removal or cutting of the forest has immediate effects on drainage water quality in several respects. Deposition of S and N to the site are reduced. Leaching of NO_3^- increases and, in some cases, causes a pulse of surface water acidification. Base cations are lost from the system. The subsequent regrowth of the forest following deforestation may further affect drainage water quality through vegetation uptake processes. This is because trees accumulate base cations to a greater degree than anions. In order to balance the resulting charge discrepancy, roots release an equivalent amount of protons. This is an acidifying process. Base cation accumulation by growing trees is strongly age dependent. Young, fast-growing forests are more acidifying than older forests (Nilsson et al., 1982; Nilsson, 1993) and retain greater amounts of N inputs. For example, Reynolds et al. (1994) found concentrations of NO_3^- in 136 streams in upland Wales were significantly correlated ($p < 0.001$) with the average age of conifers.

It has been proposed that forest blowdown affects surface water acid–base chemistry via changes in hydrologic flow (Dobson et al., 1990). Pipes formed in the soil by tree roots can alter hydrologic flow so that less water enters the soil matrix, where neutralization processes buffer the acidity of incoming rainwater and snowmelt. Pipes tend to be located in near-surface soil horizons where most tree rooting occurs, and contact between drainage waters and mineral soil is reduced when runoff is routed through them. If enhanced pipeflow affects a large portion of any watershed, stream and lake chemistry may be expected to reflect the chemical characteristics of surface and near-surface soil waters more so than the characteristics of deeper groundwater and more so than would be the case in the absence of such pipeflow.

During the 1980s, the prevailing scientific consensus held that the majority of lakes in eastern North America that had pH less than about 5.5 to 6.0 had been acidified by acidic deposition. Temporal/spatial correlations were claimed to support this contention. Reports that acidic surface waters were rare or absent in "equivalent areas" not receiving acidic deposition were used as illustrations of acidification by acidic deposition in many regions (e.g., Neary and Dillon, 1988; Sullivan et al., 1988; Baker et al., 1990a).

Rosenqvist and Krug proposed that land use changes could explain recent lake acidification in southern Norway and the northeastern U.S. (Rosenqvist, 1978; Krug and Frink, 1983; Krug, 1989, 1991b). According to this hypothesis, natural soil processes that respond to vegetation change have the potential to generate far more acidity than is received from atmospheric deposition. For example, an increase in acidic humus formation in response to decreased upland agriculture was purported to be responsible for regional acidification in southern Norway, rather than acidic deposition. Subsequent acidic deposition effects research in some cases seemed to be designed to refute this hypothesis.

Evaluation of the quantitative importance of land use changes in influencing lake-water acid–base chemistry has been seriously hampered by a tendency among acid deposition researchers to pose scientific questions that were intended to discriminate between acidic deposition and land use as the *major cause of acidification*. Not surprisingly, such studies generally concluded that acidic deposition was the principal cause of regional acidification in certain areas of North America and Europe. Perhaps a more appropriate research question might focus on quantifying the relative importance of land use activities in exacerbating or ameliorating acidic deposition effects. The importance of acidic deposition as an agent of acidification does not preclude the fact that land use and landscape changes may also be important and, in some cases, more important than acidic deposition (Sullivan et al., 1996b).

It is now clear that acidic deposition causes acidification of some sensitive waters. It is no longer appropriate to phrase scientific research questions as "acid deposition or land use." Unfortunately, such a shift in the approach of scientific investigations has been slow to occur, and the advancement of science has suffered as a consequence.

As discussed previously, land use changes and disturbances within the drainage basins of lakes and streams can influence water chemistry, but the regional acidification of surface waters in parts of Europe and North America has not been attributed to changes in land use practices. In many cases, such disturbances increase ANC and pH, and cause water quality problems other than acidification. Where land use changes have been substantial, it may be difficult to quantify the effects of acidic deposition on a regional scale. A critical limitation of much of the acidic deposition effects research conducted to date has been, however, the nature of the questions being asked. In the majority of cases, land use has been addressed only as a potential alternative explanation for acidification, rather than being evaluated in an open and objective fashion (e.g., Havas et al., 1984; Birks et al., 1990b).

There has not been a rigorous regional evaluation of land use changes in areas of the U.S. susceptible to acidic deposition effects. In the absence of such an investigation, it has not been possible to quantify the extent or magnitude of land use related effects on water quality within the regions of concern. It is clear, however, that such changes can have important effects on acid–base status.

Renberg et al. (1993) evaluated sediment composition, pollen, radiocarbon and lead dating, and diatom reconstructions to ascertain the effects of changing land use in 14 widely separated lakes in southern Sweden over the past 10,000 years. Lake-water pH declined from about 7.0 to 5.5 in the first few thousand years after deglaciation in response to natural processes. During the Iron Age, the area was deforested as an agrarian economy developed in the region: this caused an increase in lake-water pH of 0.5 to 1.4 pH units in 12 of the 14 study lakes. Subsequently, pH declined during the nineteenth and twentieth centuries, following abandonment of agriculture, to levels less than 5.0 in some lakes. The acidification can be attributed partially to reforestation (recovery from disturbance) and partially to atmospheric deposition. Prior to the study of Renberg et al. (1993), conventional wisdom held that all of the observed or inferred acidification in such systems would be attributable to acidic deposition.

Modeling studies and calculations performed for selected watersheds in Europe have suggested that acidic deposition and landscape processes are of approximately equal importance as regulators of surface water acid–base chemistry within the watersheds investigated (Jenkins et al., 1990; Cosby et al., 1990; Nilsson, 1993). In the U.S., however, the importance of landscape processes in influencing surface water acid–base chemistry and the response of surface waters to acidic deposition have not been well-studied. Model predictions for NAPAP (1991) of the response of acid-sensitive watersheds in the U.S. implicitly assumed that changes in landscape processes either do not occur or are not important in determining surface water chemical response to changes in atmospheric deposition of S. Such an omission may have biased model projections of acidification and/or recovery of some surface waters in response to changing levels of S deposition (Sullivan et al., 1996b). Such an omission could become more problematic as efforts shift more heavily into model-based assessments of N effects. This is because NO_3^- leaching from forested watersheds is largely controlled by age-dependent forest N uptake processes as well as atmospheric deposition of N.

3.2.6   Climate

Climate can have a large influence on acid sensitivity and the effects of elevated S or N deposition in several ways. Drought can alter hydrologic flowpaths and change the relative contribution of near-surface runoff vs. deeper baseflow. Because these source areas typically generate different levels of ANC, such changes in hydrologic input can profoundly influence surface water acid–base chemistry (Webster et al., 1993; Newell, 1993).

The volume of annual precipitation received by a watershed, especially during winter, has been shown to dramatically affect the total annual wet deposition of S and N to that watershed. Because a relatively large proportion of the snowpack ionic load is released during the early phases of snowmelt, high-elevation western watersheds are potentially exposed to greater episodic acidification during years with greater precipitation.

Climate warming can influence the response of surface waters to past and future acidic deposition. Under cool, moist conditions, a sizable component of the atmospheric S inputs can be stored as reduced S in soils, especially in wetland areas. This storage protects surface waters from acidification (Rochefort et al., 1990). However, under warmer and drier climatic conditions, this stored S can be reoxidized and, consequently, released to drainage waters during periods of rainfall or snowmelt (Bayley et al., 1992; LaZerte, 1993; Schindler, 1998).

Temperature can also have a variety of effects on S and N dynamics. The timing and rapidity of snowmelt are important factors governing the delivery of ionic loads from the snowpack to surface waters. Temperature also has a large influence on biological uptake of N within both terrestrial and aquatic ecosystems.

Drought conditions in the Sierra Nevada were judged by Melack et al. (1998) to be responsible for increasing the proportion of runoff derived from shallow groundwater in the Ruby Lake basin, as evidenced by an increase in SO_4^{2-} concentration from about 6 to 12 μeq/L from 1987 through 1994. Melack et al. (1998) also speculated that drought may be responsible for recent increases in N retention in the Emerald Lake catchment. The monitoring data illustrated a 25 to 50% reduction in annual NO_3^- maxima and minima in Emerald Lake, with a concomitant shift in the lake phytoplankton community from P limitation toward N limitation (Melack et al., 1998).

3.2.7   Fire

The effects of fire on NO_3^- mobilization in chaparral watersheds in the San Gabriel Mountains subject to a high level of chronic atmospheric N deposition were investigated by Riggan et al. (1994). Each watershed was burned with fires of different intensity. Then, after rainfall occurred, NO_3^-, NH_4^+, and SO_4^{2-} were measured in watershed streams. The amount and concentration of N release were found to be related to fire intensity. N release was up to 40 times greater in burned watersheds than in unburned watersheds. Similarly,

Chorover et al. (1994) evaluated the effects of fire on soil and stream-water chemistry in Sequoia National Park. Burning increased concentrations of NO_3^- and SO_4^{2-} in soils and stream water. Sulfate increased 100 fold, and NO_3^- remained higher in soils and stream water for 3 years. Fenn and Poth (1998) hypothesized that successful fire suppression efforts may have contributed to the development of N saturation in fire-adapted ecosystems in southern California by allowing N to accumulate in soil and in the forest floor, and by maintaining dense overmature stands with reduced N demand.

3.2.8 Hydrology

The impacts of atmospheric deposition on high-elevation aquatic systems are strongly controlled by the flowpaths of water through the catchments. Hydrology is an important controlling factor for deposition impacts in virtually all environments (Turner et al., 1990), but hydrology is of overriding importance in alpine and subalpine ecosystems, such as are found throughout the West. The depth and make-up of soils, talus, and colluvium, and the slope of the watershed collectively determine the residence time of subsurface water within the watershed, the extent to which snowmelt and rainfall runoff interact with soils and geologic materials, and consequently the extent of NO_3^- uptake by biota versus NO_3^- leaching and acid neutralization within the watershed.

Chemical hydrograph separation techniques (e.g., Hooper and Shoemaker, 1985; Hinton et al., 1994) have been used to trace the movement of water through alpine and subalpine basins (Caine, 1989; Mast et al., 1995; Sueker, 1995). New water (snowmelt) often contributes more than half of the streamflow after seasonal peak flows have been achieved, but old water (stored from the previous year) typically dominates the hydrograph early in the snowmelt process. Sueker (1995) used chemical hydrograph separation to estimate stream-water contributions from snowmelt and subsurface sources during the period from early snowmelt through autumn 1994 in 3 headwater basins in Rocky Mountain National Park, CO. All 3 basins were located on the east side of the continental divide above 2500 m elevation. Such separations are problematic, however, because they require the assumption that the source waters maintain constant chemistry over time. Unfortunately, the chemical composition of the major source waters (soils, talus fields, snowpack) change at the same time that their mixing ratio in streams change, confounding use of end-member mixing models to describe the controls on ionic contributions to stream waters (Campbell et al., 1995).

Mast et al. (1995) evaluated the mechanisms that control streamflow at Loch Vale using the concentrations of ^{18}O and dissolved Si as chemical tracers. They concluded that streamflow is generated approximately as follows. Streamflow at the beginning of the melt period has a large component of "old water" that was displaced into the stream by the piston effect as meltwater infiltrated the soil and talus areas. After the pre-event soil waters have been flushed into streams, streamflow is mainly generated by snowmelt discharge from subsurface reservoirs, with increasing amounts of surface flow during the main melting event in June. During the initial stages of snowmelt, therefore, stream chemistry reflects the months of weathering and decomposition products that accumulated in soils and talus areas under the snowpack and that were pushed into streams by the piston effect. As the melt continues, the contribution of pre-event water declines and the composition of stream water is increasingly controlled by relatively rapid geochemical reactions between soils and talus and the infiltrating snowmelt (Mast et al., 1995).

3.3   Effects of Acidification

Acidification from acidic deposition has a number of important chemical and biological effects. The most noteworthy relate to changes in the acid–base status of surface and soil water, sometimes resulting in short-term or long-term toxicity of the water to aquatic or terrestrial biota. Acidic deposition generally increases the concentration of acid anions in solution. These anions are to some extent charge-balanced by cations derived from the soil cation exchange complex or released through mineral weathering. The latter can include base cations such as Ca^{2+} and Mg^{2+} and acid cations such as H^+ and Al^{n+}. When the concentration of H^+ and/or Al^{n+} in solution increases, toxic conditions may result. Increased H^+ concentration (reduced pH) affects different species of aquatic biota at different levels. Some species are affected at pH levels near 6.0, whereas others can be quite tolerant of pH values well below 5.0. Many species of fish are adversely affected at H^+ concentrations greater than about 10 µeq/L (pH 5.0).

3.3.1   Aluminum

Aluminum is abundant in soils, derived from weathering reactions, and has a pH-dependent solubility in water. Solubility increases dramatically at pH values below about 5.5, and also is enhanced by the formation of soluble organic complexes (Schnitzer and Skinner, 1963; Lind and Hem, 1975). Based upon data obtained in the northeastern U.S., Canada, Norway, Sweden, and Germany, Cronan and Schofield (1979) concluded that one of the most dramatic effects of acidic deposition upon watershed ecosystems has been increased mobilization of Al from soils to surface waters. Al mobilization is widely believed to be one of the most important ecological effects of surface water acidification (Mason and Seip, 1985). The aluminum response has been well documented in numerous studies (Schofield, 1976, 1982; Dickson, 1978, 1980; Seip, 1980; Wright and Henriksen, 1978; Cronan and Schofield, 1979; Driscoll, 1984; Johnson et al., 1981; Driscoll et al., 1980;

LaZerte, 1984, 1986; Driscoll and Newton, 1985; Campbell et al., 1984; Sullivan et al., 1986, 1989; Hooper and Shoemaker, 1985). Aqueous Al concentrations in acidified drainage waters are often an order of magnitude higher than in circumneutral waters. Potential effects of Al mobilization to surface and soil waters include alterations in nutrient cycling (Dickson, 1978; Eriksson, 1981), pH buffering effects (Driscoll and Bisogni, 1984), toxicity to aquatic biota (Schofield and Trojnar, 1980; Muniz and Leivestad, 1980; Baker and Schofield, 1982; Driscoll et al., 1980), and toxicity to terrestrial vegetation (Ulrich et al., 1980).

Inorganic Al is mobilized from soils in response to the increased mineral acidity associated with dissolved SO_4^{2-} or NO_3^- (Cronan and Schofield, 1979). Thus, inorganic Al concentrations generally increase with decreasing pH, and often reach appreciable concentrations (greater than 1 to 2 μM) in surface drainage waters having pH less than about 5.5.

Processes that control the mobilization, speciation, and toxicity of Al in forested ecosystems have been intensively studied during the past 15 years. Although much valuable information has been gained, these processes are still not well understood. Our ability to predict the Al component of the acidification response remains limited (Sullivan et al., 1998). This is unfortunate, given the biological importance of this parameter and its central role in acidification processes (Sullivan, 1994).

The products of Al hydrolysis, $Al(OH)^{2+}$ and $Al(OH)_2^+$, are often considered to be the most toxic to fish. This is based on the observation that dissolved Al seems to be particularly toxic in the pH range between 5 and 5.5, a range in which Al hydrolysis products often constitute a significant fraction of the dissolved inorganic Al (Heliwell et al., 1983; Leivestad et al., 1987; Lydersen et al., 1994). Al^{3+} is also considered highly toxic to both aquatic organisms and plant roots. Increased fish mortality has been reported shortly after liming acid waters (McCahon et al., 1989; Hutchinson et al., 1989) and this may be owing either to high toxicity of cationic Al-hydroxide species, toxicity of intermediate hydrolysis products formed during Al polymerization, or alternatively to the formation of $Al(OH)_3(s)$ that may precipitate out of solution on fish gill tissue or egg membranes. Wigington et al. (1993) found that pH was positively correlated with fish mortality during *in situ* bioassays. Thus, for given concentrations of Al_i and Ca^{2+}, mortality was higher at higher pH. This finding provides further evidence that either the Al hydroxide species or the occurrence of oversaturated conditions contribute disproportionately to the toxicity of Al to fish (Sullivan, 1994).

Recent studies have shown that Al is often especially toxic in mixing zones, immediately subsequent to pH neutralization of acidic, Al-rich water (McCahon et al., 1989; Lydersen et al., 1990, 1994; Rosseland et al., 1992; Polèo et al., 1994). This higher Al toxicity has been attributed to Al-polymerization by Lydersen et al. (1994), but could also be owing at least in part to greater toxicity of Al-hydroxide species, which are formed in greater amounts as pH increases.

Aluminum toxicity to fish is species and life-stage specific. Wigington et al. (1993) found 50% brook trout mortality for *in situ* exposure experiments at median Al_i concentrations of about 200 µg/L. For prolonged exposures, greater than 8 days, having Al_i continuously above various threshold levels, 50% mortality was found for Al_i threshold values between 100 and 200 µg/L. Blacknose dace were somewhat more sensitive to Al exposure; 50% mortality was observed at a median Al_i concentration of 120 µg/L. Sculpin were less sensitive and showed 50% mortality at Al_i concentrations ranging from 200 to 300 µg/L.

Aluminum disrupts the functioning of fish gills and inhibits respiration and ion regulatory activities (Howells et al., 1990). The mechanisms of acute Al toxicity are not clear. Toxicity is most severe at pH values between 5.0 and 6.0, especially when pH is rapidly increased (Poléo et al., 1994). Aluminum hydroxides constitute a significant fraction of the inorganic Al species in this pH range. It has been suggested that the process of Al polymerization is the major cause of acute Al toxicity to fish (Poléo and Muniz, 1993; Poléo et al., 1994; Poléo, 1995). According to this theory, positively charged Al-hydroxide species bond with negatively charged sites of the gill surface that act as polymerization nuclei. Growth of Al polymers on the gill surface then increases mucus secretion and causes clogging of the interlamellar spaces, leading to acute hypoxia (Poléo, 1995).

Aluminum is not only toxic to aquatic biota. Aqueous Al is also toxic to tree roots, although much higher concentrations of Al^{3+} in soil solution are required in order to elicit a toxic response as compared with the toxicity of Al to fish in surface water. Plants affected by high levels of Al in soil solution typically exhibit reduced root growth. Stunting of the root system restricts the ability of the plant to take up water and nutrients (Parker et al., 1989). Ca^{2+} is well known as an ameliorant for Al toxicity to roots, as well as to fish. Mg^{2+}, and to a lesser extent the monovalent base cations, Na^+ and K^+, have also been associated with reduced Al toxicity. Neither the molecular basis for Al toxicity to plant roots nor the basis for the reduction in toxicity found for base cations is well understood. Efforts to estimate critical levels of atmospheric S or N deposition that will protect sensitive forest resources from damage often use the molar ratio of Ca to Al in solution as an indicator of potential toxic effects. It has been suggested that damaged forest stands often exhibit Ca : Al < 1.0 (Ulrich, 1983; Schulze, 1989; Sverdrup et al., 1992).

3.3.2 Effects on Aquatic Biota

Relatively little progress has been made since 1990 regarding our understanding of the chronic effects of acidification on biota. Somewhat greater effort has been placed on improving our understanding of episodic biological effects. Nevertheless, one major study on fisheries response was conducted in Shenandoah National Park, VA, and some noteworthy research has been conducted in the vicinity of Sudbury, Ontario to document the responses of in-lake biota to recent large decreases in local S deposition. In addition, more progress has been made in documenting the central role of Al as the principal toxic agent associated with acidification effects on biota, both chronically and episodically.

The Shenandoah National Park—Fish in Sensitive Habitats Project documented adverse effects on fish populations and communities in chronically acidified streams of Shenandoah National Park. Fish species richness, population density, condition factor, age distribution, size, and bioassay survival were all reduced in low-ANC streams when compared to intermediate-ANC and high-ANC streams (Bulger et al., 1995; Dennis and Bulger, 1995; Dennis et al., 1995; MacAvoy and Bulger, 1995). The results of this study provided a database that will allow estimation of biological impacts that might be associated with various future changes in acid–base chemistry.

The Sudbury region of Ontario, Canada has been an important location for studying the chemical and biological effects of S deposition for decades. Mining and smelting of copper-nickel ore began in the 1880s. By the 1950s and 1960s, SO_2 emissions from these operations peaked at over 5000 tons per day and extensive acidification of surface waters occurred (Beamish and Harvey, 1972). The Sudbury region has been the focus of a great deal of chemical and biological effects work ever since (Keller, 1992). Emissions of SO_2 decreased markedly after 1970 to less than one-third of the peak values of earlier decades. These emission reductions were accompanied by improved water quality in many lakes (Keller and Pitblado, 1986; Keller et al., 1986). Some fisheries recovery has also now been documented (Gunn and Keller, 1990; Keller and Yan, 1991). Griffiths and Keller (1992) documented changes in the occurrence and abundance of benthic fauna consistent with a direct effect of reduced lakewater acidity.

Whitepine Lake, located 90 km north of Sudbury, was low in pH (5.4) and ANC (1 μeq/L) in 1980 and its fish populations exhibited symptoms of acid stress. Acid-tolerant yellow perch (*Perca flavescens*) were abundant, whereas acid-sensitive species such as lake trout (*Salvelinus namaycush*) and white sucker (*Catostomus commersoni*) were rare and not reproducing. Fish populations were assessed by Gunn and Keller (1990) from 1978 through 1987 and zooplankton were sampled at least monthly during the open-water periods of 1980 through 1988. Water quality improved significantly from 1980 to 1988; pH and ANC increased to 5.9 and 11 μeq/L, respectively. Young lake trout first appeared in 1982 and became increasingly abundant throughout the study. The number of taxa of benthic invertebrates increased from 39 in 1982–1983 to 72 in 1988 and the relative abundance of many of the invertebrates found in 1982 changed dramatically (Gunn and Keller, 1990).

This research at Sudbury has been important in several ways. It was clearly documented that chemical recovery of lakes was possible upon reduced emissions and deposition of S. Furthermore, this work illustrated that biological recovery, involving many trophic levels, would soon follow. These findings have reinforced the efforts to curtail emissions and deposition elsewhere.

Specific information about the sensitivity of aquatic biota in the West is limited both because of the inadequate characterization of taxa that occur in the mountain ranges of concern and also the lack of extensive testing of the response of western taxa to acidic conditions. Furthermore, many of the high-elevation sensitive lakes historically lacked fish. The fish, including nonnative eastern species such as brook trout, were introduced and are suspected of altering the nonfish biota. There are three issues:

1. Characterization.
2. Evaluation of sensitivity.
3. Evaluation of the effects of stocking.

These issues need to be resolved before a more complete assessment of biological effects can occur (Eilers et al., 1994a).

Several studies have recently examined the sensitivity of native salmonids in the western U.S. to acid and Al stresses. Various subspecies and strains of cutthroat trout (*Oncorhynchus clarki*) occupy, or historically occupied, high-elevation drainages that are sensitive to potential episodic and chronic acidification from S and N deposition. Many of the drainages occur in national parks and wilderness areas. It is, therefore, important to understand the pH and Al levels at which these native fish would be adversely impacted in order to protect the fisheries resource against significant deterioration. Although toxicity studies earlier documented such thresholds for species native to the northeastern U.S. and Europe (e.g., Baker, 1982; Brown, 1983; Baker and Schofield, 1982), only recently have native western salmonids been the focus of such studies (e.g., Woodward et al., 1989, 1991; Farag et al., 1993).

Native western trout are sensitive to short-term increases in acidity. For example, Woodward et al. (1989) exposed native western cutthroat trout to pH depressions (pH 4.5 to 6.5) in the laboratory. Freshly-fertilized egg, eyed embryo, alevin, and swim-up larval stages of development were exposed to low pH for a period of seven days. Fish life stages were monitored for mortality, growth, and development to 40 days posthatch. The test fish were taken from the Snake River in Wyoming. Reductions in pH from 6.5 to 6.0 in low-calcium water (70 μeq/L) did not affect survival, but did reduce growth of swim-up larvae. Eggs, alevins, and swim-up larvae showed significantly higher mortality at pH 4.5 as compared to pH 6.5. Mortality was also somewhat higher at pH 5.0, but only statistically higher for eggs.

Woodward et al. (1991) conducted laboratory bioassays of greenback cutthroat trout (*O. c. stomias*) exposed to 7-day pH depressions to simulate episodic acidification. Low-calcium (65 μeq/L) water at pH 6.5 was reduced to pH values of 6.0, 5.5, 5.0, and 4.5. Exposed were four life stages: freshly fertilized egg, eyed embryo, alevin, and swim-up larva. Alevin survival was reduced at pH 5.0, whereas survival of eggs, embryos, and swim-up larvae was reduced at pH 4.5. After the exposure of swim-up larvae, feeding inhibitions were noted at pH 4.5. The authors concluded that the threshold for effects of H⁺ ion concentration on greenback cutthroat trout in the absence of Al (which increases toxicity) was pH 5.0 (Woodward et al., 1991).

Survival of various life stages of Yellowstone cutthroat trout (*O. c. bouveri*), exposed to 7-day pH depressions in order to simulate episodic acidification, was studied by Farag et al. (1993). They also evaluated the added toxicity associated with elevated concentrations of Al. However, there were procedural problems associated with their Al exposures (i.e., Al added in excess of the amount soluble at a given pH) and therefore the data for evaluating fish response to pH depressions is only considered here. Eggs were most sensitive to low pH of the four life stages studied. Eggs exposed to pH 5.0 experienced a statistically significant reduction in survival when compared with eggs exposed for 7 days to pH 6.5 water. Survival of alevin and swim-up larvae were significantly reduced from near 100% at pH 6.5 to near 0% at pH 4.5. Intermediate pH values (6.0, 5.5) in all cases showed reduced survival compared with the control (6.5) but not by a statistically significant amount. Eyed embryos were not sensitive to any of the exposures.

Much of the work conducted to date on the biological effects of acidification has been focused on fish. However, this focus has largely been the result of the value people place on fish and fishing, rather than any ecological consideration. Algal, inverberate, and other vertebrate communities are also sensitive to adverse impacts of acidification. Furthermore, aquatic ecosystems are heterogeneous and exhibit pronounced temporal variability in response to a host of biotic and abiotic factors. Against this background of variability, it is exceedingly difficult to detect changes in biological communities in response to specific environmental stresses without detailed, long-term biological data (Schindler, 1990; Lancaster et al., 1996). High-quality, long-term data sets are rare, and there are few well-documented instances of temporal changes in biological communities in response to changes in water chemistry. It is clear, however, that surface water acidification affects virtually all trophic levels (e.g., Flower and Battarbee, 1983; Økland and Økland, 1986; St. Louis et al., 1990; Rundle and Hildrew, 1990; Simonin et al., 1993; Ormerod and Tyler, 1991; Lancaster et al., 1996).

Several studies have been conducted of the acid-sensitivity of aquatic invertebrates in the Sierra Nevada. *In situ* enclosure acidification studies were conducted for 35 days during the summer of 1987 at Emerald Lake by Barmuta et al. (1990). In contrast to previous studies (e.g., Melack et al., 1987), the lake sediments were included within the enclosures, thereby allowing the response of zoobenthos to be evaluated as well as zooplankton. Treatments included a control (pH 6.3) and pH levels of 5.8, 5.4, 5.3, 5.0, and 4.7. The zooplankton assemblage was sensitive to acidification, but there was no evidence that the zoobenthos were affected by the artificial acidification. *Daphnia rosea* and *Diaptomus signicauda* decreased in abundance below pH 5.5 to 5.8 and virtually disappeared below pH 5.0. *Bosnia longirostris* and *Keratella taurocephala* became more abundant with decreasing pH, although *B. longirostris* was rare in the pH 4.7 treatment. Barmuta et al. (1990) concluded that even slight acidification of high-elevation lakes in the Sierra Nevada can alter the structure of the zooplankton community.

Kratz et al. (1994) examined the responses of aquatic macroinvertebrates to pulsed acidification experiments in 12 streamside channels along the Marble Fork of Kaweah River in Sequoia National Park. Replicated treatments (4 reps per treatment) included a control (pH 6.5 to 6.7) and pH levels of 5.1 to 5.2 and 4.4 to 4.6. Invertebrate drift was monitored continuously and benthic densities were determined before and after acid addition. For sensitive taxa, drift was enhanced and benthic densities reduced by single 8 h acid pulses; second acid pulses, 2 weeks after the first, had no additional effect on benthic density. *Baetis* showed reduced density post-treatment to less than or equal to 25% of control densities in both pH treatments (5.2, 4.6) and 2 different experimental exposures. Densities of *Paraleptophlebia* appeared to be reduced by the acidification, but most treatment effects were not statistically significant. Kratz et al. (1994) hypothesized that the effects of acid inputs on the densities of individual benthic species depended on microhabitat preferences. For example, *Baetis* nymphs are epibenthic and active, spending most of their time on the upper surfaces of rocks directly exposed to acidified water.

Engle and Melack (1995) reported the results of multi-year sampling of zooplankton in seven high-elevation lakes in the Sierra Nevada. The lakes contained 5 copepods, 6 cladocerans, and at least 20 rotifers. No evidence was found for long-term trends in species richness in any of the lakes throughout the six-year period of record. Overall, zooplankton species richness can be expected to decline during lake acidification (Locke, 1992). Engle and Melack (1995) found many species in the lakes that had been classified by other studies as acid-intolerant. In general, *Daphnia* spp. and cyclapoid copepods, especially of the genus *Cyclops* are much reduced or absent in acidic lakes (Brett, 1989). *Daphnia rosea* was present in all seven of the study lakes, although experimental evidence suggests that it will be one of the first taxa to be eliminated if Sierran lakes become acidified (Engle and Melack, 1995).

Studies in both the U.S. and Canada have provided new understanding of the feasibility and complexity of biological recovery in response to chronic de-acidification. Biological recovery of previously acidified lakes can be a slower process than chemical recovery because of several factors. Examples of such factors are:

1. Other environmental stresses, such as metal contamination (e.g., Sudbury, Ontario) (Gunn, 1995; Havas et al., 1995; Jackson and Harvey, 1995; McNicol et al., 1995; Yan et al., 1996a,b).

2. Barriers imposed by water drainage patterns between lakes that hinder recolonization by some fish species (Jackson and Harvey, 1995).

3. The influence of predation by nonacid-sensitive fish species on the recovery of zooplankton and macroinvertebrate communities (McNicol et al., 1995).

4. Predation on tributary-spawned young trout when they move downstream into lakes inhabited by predatory fish and birds (Schofield and Keleher, 1996).

Results from the Lake Acidification Mitigation Program involving liming of the catchments of the tributaries to an acidified Adirondack lake with limited in-lake spawning habitat indicated that re-establishment of tributary spawning populations of brook trout may be possible, with future reductions in acidic deposition. However, restoration of tributary spawning habitat may not be sufficient to produce self-sustaining populations, because of high rates of predation on young trout after they move downstream into lakes (Schofield and Keleher, 1996).

3.3.3 Effects on Amphibians

Populations of many species of amphibians have declined or become eradicated throughout the world in recent decades (Baringa, 1990; Wake, 1991). These declines have been particularly alarming because the causes have not been evident and because declines have occurred in remote pristine areas. In the Sierra Nevada, at least two of five species of aquatic-breeding amphibians, *Rana muscosa* (mountain yellow-legged frog) and *Bufo canorus* (Yosemite toad) have been declining (Phillips, 1990).

A number of hypotheses have been proposed for recent amphibian declines, including acidic deposition. In the western U.S., however, acidic deposition has been discounted as the primary cause of the decline of *R. muscosa* and *B. canorus* in the Sierra Nevada and of *R. pipiens* and *B. boreas* in the Rocky Mountains (Corn et al., 1989; Bradford et al., 1992; Corn and Vertucci, 1992).

Bradford et al. (1993) suggested that recent declines in *R. muscosa* populations in the Sierra Nevada have been owing to population fragmentation as a consequence of fish predation. It is generally recognized that *R. muscosa* was eliminated by introduced fish early in this century in many lakes and streams in Sequoia and Kings Canyon National Parks. The amphibians have been eliminated from nearly all waters inhabited by fish, presumably by predation on tadpoles.

Prior to 1870, virtually all of the high-elevation (greater than 2500 m) lakes in the Sierra Nevada were barren of fish. Populations of *R. muscosa* were presumably connected to one another via waterways because the species inhabits both streams and lakes. *R. muscosa* has disappeared from many waters not containing fish during the past three decades. Fragmentation of populations may have caused or contributed to these recent extinctions because populations are now much more isolated from each other than they were formerly. Bradford et al. (1993) surveyed 312 lake sites in 95 drainage basins in the parks. For the 109 sites containing *R. muscosa*, they delineated networks of sites connected to one another via fishless streams, and compared those present fishless networks with presumed former fishless networks. Most present fishless networks consisted of only 1 site (mean 1.4), whereas former networks averaged 5.2 sites. This difference represents approximately a 10-fold decline in connectivity (mean number of potential dispersal links per network; Bradford et al., 1993).

Bradford et al. (1994a) conducted a survey of 235 potential breeding sites for *R. muscosa*, *B. canorus*, and *Pseudacris regilla* (Pacific treefrog) in 30 randomly selected survey areas in the Sierra Nevada. There were no significant differences in water chemistry parameters between sites with and sites without each of these three species of amphibians. Moreover, the water chemistry parameters did not differ among sites inhabited by amphibians in a manner paralleling their degree of acid tolerance. These findings were interpreted by Bradford et al. (1994a) as a contra-indication of acidic deposition as a cause of recent amphibian population declines in the Sierra Nevada at high elevation.

Similar findings were reported by Soiseth (1992), who surveyed 14 ponds in the Emerald Lake watershed to evaluate whether pH or ANC was related to the occurrence of *P. regilla*. The pH and ANC of ponds containing larvae of the species ranged from 5.3 to 7.2 and 0 to 132 µeq/L, respectively. No significant difference in pH or ANC was observed between ponds lacking vs. ponds containing *P. regilla* in each year of the study (Soiseth, 1992).

A comparison of 21 historical (1955–1979) and recent (1989–1990) records for sites scattered throughout Sequoia and Kings Canyon National Parks showed that *R. muscosa* remained at only 11 of these sites in 1989–1990. A similar comparison of 24 historical and recent records elsewhere in the Sierra Nevada showed that *R. muscosa* remained at only 3 sites (Bradford et al., 1994b).

4

Extent and Magnitude of Surface Water Acidification

For the regions of the U.S. identified as having sensitive aquatic resources, some relevant information has been compiled and evaluated subsequent to the NAPAP Integrated Assessment (IA) regarding the relationship between deposition loading (N and S) and the estimated (or expected) extent, magnitude, and timing of aquatic effects (c.f., Sullivan and Eilers, 1994; van Sickle and Church, 1995; EPA, 1995a; NAPAP, 1998). These studies have generally employed for this task a weight of evidence evaluation of the relationships between deposition and effects, as followed by NAPAP in the IA (NAPAP, 1991).

There were six types of evidence used in the IA to assess the extent and magnitude of acidification in sensitive regions and the sensitivity of aquatic resources to changes in deposition magnitude and timing:

1. Watershed models that project or hindcast chemical changes in response to changes in sulfur deposition (particularly the MAGIC model).
2. Biological response models linked to the outputs from watershed chemistry models.
3. Inferences from current surface water chemistry in relation to current levels of deposition.
4. Trend analyses based on comparing recent and past measurements of chemistry and fishery status during the past one or two decades in regions that have experienced large recent changes in acidic deposition.
5. Paleolimnological reconstructions of past water chemistry using fossil remains of algae deposited in lake sediments.
6. Results from watershed or lake acidification/deacidification experiments.

Evidence of each type contributes to our understanding of the quantitative importance of the various acidification and neutralization processes for surface waters in the areas of interest.

The total concentration of the mineral acid anions in surface waters that are derived from atmospheric deposition of air pollutants (e.g., SO_4^{2-} and NO_3^-) has changed over time throughout the northeastern U.S. In response to such changes, the concentrations of other ions must also have changed in order to satisfy the electroneutrality constraint. The total amount of positively charged cations must equal the total amount of negatively charged anions in any solution. Therefore, if the sum of SO_4^{2-} and NO_3^- increases, the other anions (e.g., bicarbonate) must decrease and/or some cations (e.g., base cations, hydrogen ion, or aluminum) must also increase in order to maintain the charge balance.

The only way in which acidification results quantified using different approaches can be compared on a quantitative basis is by normalizing surface water response as a fraction of the change in SO_4^{2-} concentration (or SO_4^{2-} + NO_3^- concentration where NO_3^- is also important). This is often done using the F-factor (Henriksen, 1982), which is defined as the fraction of the change in mineral acid anions that is neutralized by base cation release [Eq. (2.7)]. Where acidification occurs in response to acidic deposition, changes in ANC and/or Al_i concentration comprise an appreciable percentage of the overall surface water response and, therefore, the F-factor is substantially less than 1.0 (Sullivan, 1990). The F-factor provides the quantitative linkage between inputs of acid anions (e.g., SO_4^{2-}, NO_3^-) and effects on surface water chemistry.

The sensitivity to acidification of surface waters in a region is a function of regional deposition characteristics, surface water chemistry, and watershed factors. The following section attempts to integrate these three elements to provide a qualitative assessment of watershed sensitivity to acidification and a quantitative assessment of the magnitude of acidification currently experienced within the study regions. These results are further integrated in Chapter 5 to provide an assessment of the likely dose–response relationships for the regions of interest and a discussion of the feasibility of adopting one or more acid deposition standards.

4.1   Northeast

4.1.1   Monitoring Studies

The concentration of SO_4^{2-} in precipitation has declined for the past two decades in the northeastern U.S., consistent with decreased atmospheric emissions of SO_2. At Huntington Forest in the Adirondack Mountains in New York, the concentrations of strong acid anions in precipitation have decreased to a greater extent than the concentrations of base cations since 1978, resulting in a marked decrease in the acidity of precipitation. Sulfate concentrations in precipitation have decreased about 2 µeq/L per year. The annual volume-weighted pH of precipitation at Huntington Forest increased from 4.10 during the period 1978 to 1981 to 4.42 during the period 1990 to 1993 (Driscoll et al., 1995).

Monitoring data are available since the early 1980s for many lakes and streams in acid-sensitive areas of the U.S., including the Northeast. In particular, EPA's Long-Term Monitoring (LTM) Program has provided a wealth of important information in this regard. Available LTM data allow scientists to evaluate trends and variability in key components of lake or stream-water chemistry prior to, during, and subsequent to Title IV implementation. LTM data have shown that, in many areas of the U.S., the concentration of SO_4^{2-} in surface waters has decreased dramatically during the last one to two decades (Figure 4.1). This decrease has been caused by decreases in the emissions and atmospheric deposition of S on a regional basis throughout many parts of the U.S. during that time period. To some extent, these changes may be related to partial implementation of Title IV; to some extent, they were already occurring without Title IV. Decreased concentrations of SO_4^{2-} in surface waters have been most pronounced in portions of the northeastern U.S., where approximately 15% decreases commonly have been observed.

Analyses of wet deposition monitoring data illustrate that S deposition has declined in the northeastern U.S. in response to emissions reductions in the Midwest and Northeast (Lynch et al., 1996; NAPAP, 1998). A seasonal trend model was developed by Lynch et al. (1996) to explain the historical declines in S deposition from 1983 through 1994. The model was used to estimate that an additional 10 to 25% reduction in the concentration of SO_4^{2-} in precipitation was realized in 1995, presumably owing at least in part to implementation of emissions reductions required by Title IV of the Clean Air Act Amendments of 1990 (NAPAP, 1998).

Clow and Mast (1999) reported the results of trends analysis of precipitation data from eight sites and stream-water data from five headwater catchments throughout the Northeast. The precipitation data covered the period 1984 to 1996 and the stream-water data 1968 to 1996. Stream-water SO_4^{2-} concentrations declined ($p < 0.1$) at 3 of the sites throughout the period of record and at all sites from 1984 to 1996. Sulfate concentration in precipitation declined at 7 of 8 sites since 1984 and the magnitudes of decline (-0.7 to -2.0 µeq/L per year) were similar to those of stream-water SO_4^{2-} concentration. In most cases, stream-water ($Ca^{2+} + Mg^{2+}$) concentrations declined by similar amounts (Clow and Mast, 1999).

A relatively uniform rate of decline has been observed in lake-water SO_4^{2-} concentrations in Adirondack lakes since 1978 (1.81 ± 0.25 µeq/L per year), based on analyses of 16 lakes included in the Adirondack Long Term Monitoring Program (ALTM, Driscoll et al., 1995). These observed declines in lake-water SO_4^{2-} concentrations undoubtedly have been owing to the decreased S emissions and deposition. There has been no systematic increase in lake-water pH or ANC, however, in response to the decreased SO_4^{2-} concentrations. In contrast, the decline in lake-water SO_4^{2-} has been charge-balanced by a near stoichiometric decrease in the concentrations of

Aquatic Effects of Acidic Deposition

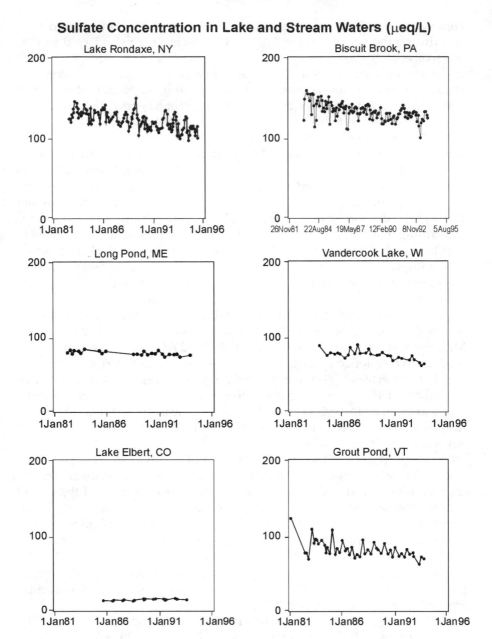

FIGURE 4.1

Measured concentration of SO_4^{2-} in selected representative lakes and streams in 6 regions of the U.S. during the past approximately 15 years. Data were taken from EPA's Long Term Monitoring (LTM) program.

base cations in low-ANC lakes (Figure 4.2; Driscoll et al., 1995). *F*-factors were calculated by Driscoll et al. (1995) for the 9 ALTM lakes that showed significant declines in both C_B and $(SO_4^{2-} + NO_3^-)$ during the period of study.

The resulting F-factors ranged from 0.55 to greater than 1.0, with a mean of 0.93. These high F-factor values for acidification recovery were similar to results of historical acidification obtained by Sullivan et al. (1990a), based on paleolimnological analyses of historical change for 33 Adirondack lakes.

Stoddard et al. (1998) presented trend analysis results for 36 lakes having ANC less than or equal to 100 µeq/L in the Northeast from 1982 to 1994. Trend statistics at each site were combined through a meta-analytical technique to determine whether the combined results from multiple sites had more significance than the individual Seasonal Kendall Test statistics. All lakes showed significant decline in SO_4^{2-} concentration (ΔSO_4^{2-} = -1.7 µeq/L per year; $p \leq 0.001$). Lakes in New England showed evidence of ANC recovery (ΔANC = 0.8 µeq/L per year; $p \leq 0.001$), whereas Adirondack lakes exhibited either no trend or further acidification. As a group, the ANC change for Adirondack lakes was -0.5 µeq/L per year ($p \leq 0.001$). Stoddard et al. (1998) attributed this intraregional difference to declines in base cation concentrations that were quantitatively similar to SO_4^{2-} declines in Adirondack lakes, but smaller in New England lakes.

Although recent widespread changes in the concentration of SO_4^{2-} in surface waters over the past one to two decades have been driven primarily by changes in S emissions and deposition, concurrent changes in the concentration of other chemical parameters have been generally less clear and consistent, and also have been influenced more strongly by factors other than atmospheric deposition. For example, the observed changes in the concentration of NO_3^- in some surface waters have likely been owing to a variety of factors, including N deposition and climate.

During the 1980s, a pattern of increasing lake-water NO_3^- concentration had been observed in surface waters in the Adirondack and Catskill Mountains in New York (Driscoll and van Dreason, 1993; Murdoch and Stoddard, 1993). There was concern that increasing N saturation of northeastern forests was leading to increased NO_3^- leaching from forest soils throughout the region and, consequently, negating the benefits of decreased SO_4^{2-} concentrations in lake and stream waters. This trend was reversed in about 1990, however, despite relatively constant levels of N deposition during the past 15 years. This is because the amount of NO_3^- that leaches through soils to drainage waters is the result of a complex set of biological and hydrological processes that include N uptake by plants and soil microbial communities, microbial transformations between different forms of inorganic and organic N, rates of organic matter decomposition, amount of rain and snow received, and the amount (and form) of N that enters the ecosystem as atmospheric deposition. Most of these important processes are strongly influenced by climatic factors such as temperature, moisture, and snowpack development. The end result is that NO_3^- concentrations in surface waters, although clearly influenced by atmospheric N deposition, respond to many factors and can be difficult to predict. There has been a decline in lake-water NO_3^- concentrations since 1991. Overall, throughout the period of record for ALTM lakes, there has been no significant trend in lake-water NO_3^- concentration. Nitrate

Base Cations Concentration in Lake and Stream Waters (μeq/L)

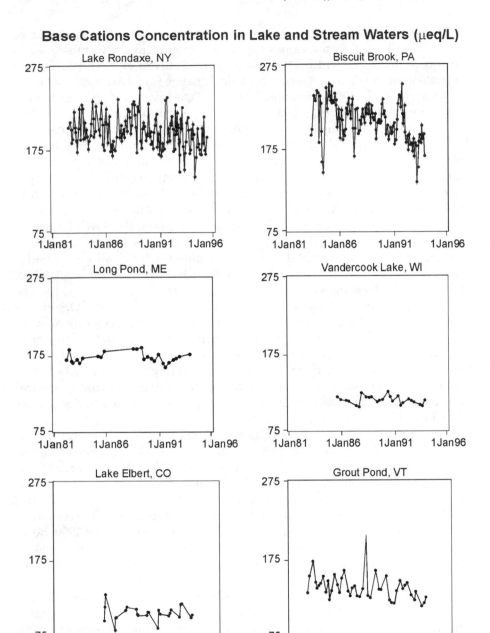

FIGURE 4.2
Measured concentration of base cations in selected representative lakes and streams in 6 regions of the U.S. during the past approximately 15 years. Data were taken from EPA's Long Term Monitoring (LTM) program.

leaching is clearly governed by a more complex set of processes than N deposition alone. As a consequence, monitoring programs of several decades will likely be needed to elucidate trends in NO_3^- leaching in forested watersheds (Driscoll et al., 1995).

Stoddard and Kellog (1993) found that many lakes in Vermont exhibited significant decreasing trends in SO_4^{2-} and base cation concentrations from 1980 through 1989 (n = 24). Few of the monitored lakes showed significant changes in pH or ANC, although examination of all trend results (significant and insignificant) suggested small increases in both. The most consistent response of surface water chemistry in the northeastern U.S. to the recent observed decrease in SO_4^{2-} concentration has been a decrease of approximately the same magnitude in the concentration of Ca^{2+} and other base cations (Figure 4.2). With few exceptions, pH, Al, and ANC have not responded in a systematic fashion (Figures 4.3 and 4.4).

One must be cautious in interpreting the observed surface water chemistry as a direct response to estimated changes in S and/or N deposition, however. Some effects of changing deposition can exhibit significant lag periods before the ecosystem comes into equilibrium with the changed or cumulative amount of S and N inputs. For example, watershed soils may continue to release S at a higher rate for an extended period of time subsequent to a decrease in atmospheric S loading. Thus, concentrations of SO_4^{2-} in surface waters may continue to decrease in the future as a consequence of deposition changes that have already occurred. Also, if soil base cation reserves become sufficiently depleted by long-term S deposition inputs, base cation concentrations in some surface waters could continue to decrease irrespective of any further changes in SO_4^{2-} concentrations. This would cause additional acidification. Nevertheless, the observed patterns of change, and lack thereof, in the chemistry of the lakes and streams included in the long-term monitoring data sets provide valuable information regarding the response of surface waters to an approximate 15 to 25% decrease in S deposition in many areas of the U.S. over the past 1 to 2 decades.

Thus, the status of sensitive (to acidic deposition) aquatic receptors in the U.S. has not changed much since the 1980s. Chemical conditions that are most important biologically, especially pH and Al concentrations, have not changed appreciably in most cases during that time period. This is in spite of fairly large changes in S deposition and SO_4^{2-} concentrations in many lakes and streams in some areas. Calcium concentrations have generally decreased in concert with the decreases in SO_4^{2-} concentration. Overall, the water quality has probably declined slightly since the early 1980s. The recovery that was anticipated by many has not been realized.

It is too early to judge the extent to which reductions in acid deposition in response to implementation of Title IV of the Clean Air Act Amendments of 1990 have or have not affected aquatic chemistry or biology in the northeastern U.S. Chemical effects owing to changes in atmospheric deposition exhibit lag times of one to many years. Lags in measurable effects on aquatic biota can be longer. Continued monitoring of water quality for several years will be required to assess potential improvements that may occur as a consequence of emissions reductions already realized. The concentrations of SO_4^{2-}

pH in Lake and Stream Waters

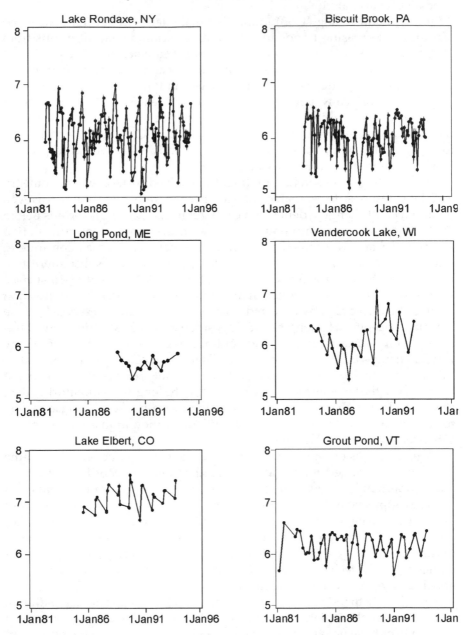

FIGURE 4.3

Measured concentration of pH in selected representative lakes and streams in 6 regions of the U.S. during the past approximately 15 years. Data were taken from EPA's Long Term Monitoring (LTM) program.

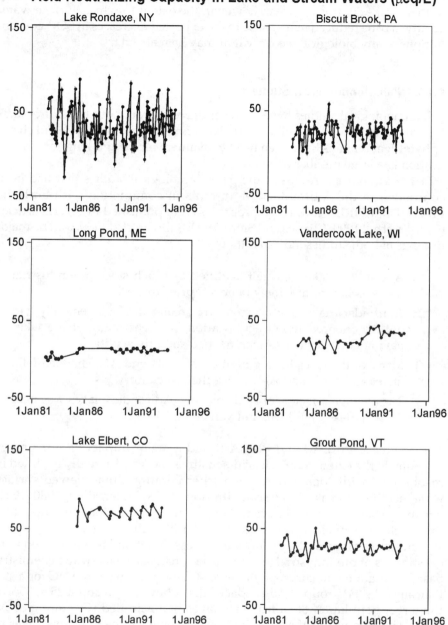

FIGURE 4.4

Measured concentration of ANC in selected representative lakes and streams in 6 regions of the U.S. during the past approximately 15 years. Data were taken from EPA's Long Term Monitoring (LTM) program.

in surface waters will probably continue to decline in many areas, especially in the Northeast. It is not clear, however, the extent to which surface water acidity may be reduced in response to the expected decreases in SO_4^{2-} concentrations or any biological recovery that may be realized.

4.1.2   Paleolimnological Studies

Paleolimnological studies have been conducted throughout the Adirondack Mountains in New York and northern New England. Both diatom and chrysophyte algal remains have been used to evaluate recent and long-term acidification in a large number of lakes.

In 1990, important results of the paleolimnological studies that had been conducted in the Adirondack Mountains in conjunction with both the PIRLA-I and PIRLA-II research programs were published in several articles (Charles et al., 1990; Charles and Smol, 1990; Sullivan et al., 1990a). The major findings of both studies indicated that

1. Adirondack lakes had not acidified as much since pre-industrial times as had been widely believed prior to 1990.
2. Adirondack lakes with current pH greater than 6.0 generally had not experienced recent acidification, whereas many of the lakes having current pH less than 6.0 had recently acidified.
3. Many of the lakes having high current pH and ANC had actually increased in pH and ANC since the last century.
4. The average *F*-factor for acid-sensitive Adirondack lakes was near 0.8 (Charles et al., 1990; Sullivan et al., 1990a).

The results of PIRLA-I and PIRLA-II had a major impact on our understanding of the extent to which acid-sensitive lakes had actually acidified in response to acidic deposition. The earlier paradigm that viewed surface water acidification as a large scale titration of ANC (Henriksen 1980, 1984) began to disappear from the scientific community. This does not imply that the conclusions of Henriksen were flawed; rather they represented an early step in a rather long and complicated process that is still being worked out.

Estimates of pre-industrial to present-day changes in lake-water chemistry, based on diatom and chrysophyte reconstructions of pH and ANC for a statistically selected group of Adirondack lakes, showed that about 25 to 35% of the target population of Adirondack lakes had acidified (Cumming et al., 1992). The magnitude of acidification was greatest in the low-ANC lakes of the southwestern Adirondacks. Lakes in this area generally have low buffering capacity and receive the highest annual rainfall and deposition of S and N in the Adirondack Park. Cumming et al. (1992) estimated that 80% of the population of lakes with current pH less than or equal to 5.2 have undergone large declines in pH and ANC since the last century. An estimated 30 to 45% of the lakes with current pH between 5.2 and 6.0 were similarly affected.

Paleolimnological methods were also developed for estimating historical lake-water concentrations of inorganic monometric Al (Al_i) in Adirondack lakes (Kingston et al., 1992). Canonical correspondence analysis (CCA, ter Braak, 1986) was used to quantify relationships between modern diatoms in lake sediments and recent lake-water chemistry. Fossil historical samples from dated down core slices of sediment cores were added to the CCA axes and used to obtain inferred values of historical lake-water concentrations of Al_i. The effects of other chemical variables (e.g., pH, DOC, Secchi depth) were partitioned out in a series of partial CCA's and the significance of the Al_i effect was tested with unrestricted Monte Carlo permutation tests. In all cases, the Al_i signal was significant ($p \leq 0.01$). In other words, there was a significant contribution from Al_i in explaining the observed variations in the diatom data, and this contribution was independent of the effects of pH, DOC, and Secchi depth transparency. The historical inferences developed by Kingston et al. (1992) for Big Moose Lake suggested a major increase in the concentration of Al_i between 1953 and 1982; this agreed with the observed fishery decline in this lake since the 1940s. Diatom-inferred pre-industrial Al_i concentrations were compared with estimates generated by Sullivan et al. (1990a) using an empirical relationship between Al_i and pH (Table 4.1). The agreement between these estimates was generally good. Both suggested approximately two- to four-fold historical increases in Al_i concentrations in these four lakes.

Cumming et al. (1994) examined the question of acidification timing in the Adirondacks, on the basis of chrysophyte inferences of pH in recently deposited lake sediments in 20 low-ANC Adirondack lakes. About 80% of the study lakes were inferred to have acidified since pre-industrial times. Lakes that acidified about 1900 were generally smaller, higher elevation lakes with lower pre-industrial pH values than the group of study lakes as

TABLE 4.1

Observed, Present-Day Inferred, and Pre-1850 Inferred Monomeric Al Concentrations Based on the Direct Diatom Relationship Developed by Kingston et al. (1992) Compared with Values Inferred by an Empirical Relationship Between Monomeric Al and pH (Sullivan et al., 1990a) Using the pH Reconstructions from Charles et al. (1990) (Units are in µM.)

Lake	Observed Calibration Monomeric Al	Recent (1982) Diatom Inferred	Recent (1982) from Empirical Relationship	Pre-1850 Diatom Inferred	Pre-1850 from Empirical Relationship
Big Moose Lake	5.3	7.4	4.2	1.1	1.2
Deep Lake	10.7	9.4	9.5	2.7	2.4
Upper Wallface Pond	5.3	7.1	5.4	3.9	2.8
Windfall Pond	1.0[a]	0.12	0.9	0.3	0.3

[a] RILWAS data from the outlet stream, approximately 1.5 km downstream from Windfall Pond.

a whole. These were apparently among the most acid-sensitive lakes and were, therefore, the first to acidify with increasing acidic deposition, probably in response to S deposition levels around 4 kg S/ha per year (c.f., Husar et al., 1991). They are located in the high peaks area and in the southwestern portion of Adirondack Park. Cumming et al. (1994) also identified several other categories of acidification response, including lakes that were very low in pH (less than 5.5) historically but acidified further beginning in about 1900. These lakes are also located in the high peaks area. The third identified type of response was lakes with pre-industrial pH in the range of about 5.7 to 6.3 that started to acidify around 1900 but showed their greatest pH change around 1930 to 1950. The fourth and final category was lakes that have not acidified; these had pre-industrial pH around 6.0 and are located at relatively low elevation where levels of acidic deposition are somewhat lower.

Davis et al. (1994) selected 12 lakes in northern New England for paleolimnological study that were expected to have been sensitive to acidification from acidic deposition. Histories of logging, forest fire, and vegetation composition in the watersheds were pieced together from oral and written historical information, aerial photographs, and tree ring analyses. Sediment cores were analyzed for pollen, diatoms, and chemistry to reconstruct past conditions for several hundred years in each lake. All 12 lakes were naturally low in pH and ANC, with diatom-inferred ANC of -12 to 31 µeq/L. The pH and ANC of the lakes were relatively stable throughout the one to three centuries of record prior to watershed disturbance by Euro-Americans. From the early nineteenth into the twentieth century, however, all of the lakes exhibited periods of increased diatom-inferred pH of about 0.05 to 0.6 pH units and increased diatom-inferred ANC of about 5 to 40 µeq/L. Most of these changes correlated temporally with watershed logging. Following recovery to prelogging acid–base conditions, all of the lakes were inferred to have continued to decline in pH and ANC, presumably in response to acidic deposition. The post-recovery decreases in pH ranged from 0.05 to 0.44 pH units and less than 10 to 26 µeq/L of ANC. The 12-lake mean decreases in pH and ANC were 0.24 pH units and 14 µeq/L, respectively (Davis et al., 1994). Assuming a background SO_4^{2-} concentration of 13% of present-day values (c.f., Husar et al., 1991) combined with the mean lake-water SO_4^{2-} concentration for the 12 lakes (53 µeq/L), an estimated 30% of the recent increase in lake-water SO_4^{2-} concentration resulted in a stoichiometric decline in lake-water ANC.

Uutala (1990) described a paleolimnological technique for reconstructing fisheries status on the basis of invertebrate remains in lake sediments. Different species of *Chaoborus* (Diptera: Chaoboridae) can be used to determine whether or not fish were present because of differential fish predation on diurnal vs. nocturnal *Chaoborus*. Kingston et al. (1992) evaluated the *Chaoborus* data of Uutala (1990) for four Adirondack lakes. The timing of major diatom-inferred increases in Al concentration matched the known history of fishery decline and the *Chaoborus*-based assessment of fisheries changes.

4.1.3 Experimental Manipulation

The Bear Brook Watershed project in Maine was established in 1986 as part of the Environmental Protection Agency's Watershed Manipulation Project (WMP). The goals of the project were to:

1. Assess the chemical response of a small upland forested watershed to increased loadings of SO_4^{2-}.

2. Determine interactions among biogeochemical mechanisms controlling watershed response to acidic deposition.

3. Test the assumptions of the Direct/Delayed Response Project (DDRP) computer models of watershed acidification.

The two Bear Brook watersheds (East and West) are located on the upper southeast-facing slope of Lead Mountain, Hancock County, ME, approximately 45 km east of the University of Maine, Orono. The brooks are tributaries to the inlet of Bear Pond. The elevation at the top of the watersheds is 450 m; the total relief is approximately 210 m. The adjacent watersheds are East Bear, 10.95 ha, and West Bear, 10.26 ha in area. The East Bear and West Bear watersheds are similar in most respects including slope, aspect, elevation, area, geology, hydrology, soils, vegetation, and water chemistry.

A total of 6 bimonthly applications per year of $(NH_4)_2SO_4$ fertilizer was applied in dry form by helicopter to the West Bear Brook watershed since November 1989. Of the applications, two were applied each year to the snowpack (if present), two were applied during the summer growing season, and one each was applied in the spring and fall. Each application consisted of 220 kg of $(NH_4)_2SO_4$. The total 1320 kg $(NH_4)_2SO_4$ per year approximately tripled the annual flux of SO_4^{2-} and quadrupled the N flux to the watershed. The target loading for each application was 20.6 kg/ha $(NH_4)_2SO_4$.

The effect on stream-water chemistry in West Bear Brook from experimental watershed acidification with $(NH_4)_2SO_4$ has been pronounced, and has involved multiple ionic responses (Norton et al., 1992, in press; Kahl et al., in press). After 1 year of treatment, the watershed retention of the added SO_4^{2-} was about 88%. Nevertheless, stream-water SO_4^{2-} concentration in West Bear Brook during Year 1 of the manipulation increased significantly in response to the treatment, as compared with the reference stream. During subsequent years, the watershed retention of SO_4^{2-} has declined to about 35% (Norton et al., in press). The increase in exported SO_4^{2-} was primarily compensated by increased base cation and Al concentrations in stream water and lower pH and ANC (Norton et al., 1992, 1994, in press).

A number of ionic constituents changed in concentration in response to the measured change in volume-weighted $[SO_4^{2-} + NO_3^-]$ at West Bear Brook. The change in base cation concentration was largest, and after correcting for base cations charge-balanced by Cl^- (marine contribution), accounted for 54% of the change in $[SO_4^{2-} + NO_3^-]$ during the first 2 years of watershed manipulation and about 80% after 3 years of manipulation (Norton et al., 1994). The base cation response subsequently decreased to about 50% of the change in (SO_4^{2-} + NO_3^-) concentration by 1995 (Kahl, personal communication). Substantial changes also occurred, as proportions of the change in [SO_4^{2-} + NO_3^-], for Al^{n+} and ANC. During the first year of treatment, 94% of the added N was retained by the Bear Brook watershed. Percent retention subsequently decreased to about 82% in subsequent years (Kahl et al., 1993a, in press). Although the forest ecosystem continued to accumulate added N, a substantial amount of added N was reflected in increased NO_3^- leaching throughout the experimental treatment.

Data from the paired-catchment manipulation at Bear Brook watershed were used by Cosby et al. (1996) to evaluate MAGIC model projections of bio-geochemical response. Model output was compared with three years of experimental data. The model was calibrated to pretreatment data from the manipulated catchment and also to four years of data from the reference catchment. The trends in variables simulated by the model paralleled the observed trends in West Bear Brook: increased concentrations of SO_4^{2-}, NO_3^-, base cations, Al and H^+, and decreased alkalinity and DOC. Problems were noted in the model simulation, however, by Cosby et al. (1996) related to interanual variability, S adsorption by watershed soils, and calibration of Al solubility.

4.1.4   Model Simulations

MAGIC model simulations of the response of lakes and streams in the northeastern U.S. to changing levels of S deposition were conducted for the NAPAP Integrated Assessment in 1990 and reported by NAPAP (1991), Sullivan et al. (1992), and Turner et al. (1992). Results of these model simulations suggested that the projected median change in lake-water or stream-water ANC during 50-year simulations were quite similar from region to region. The major difference among subregions was that the projected ANC change, as a function of change in S deposition, for surface waters in the Southern Blue Ridge and mid-Atlantic Highlands were shifted downward relative to the other regions. This was owing to the fact that the MAGIC model projected substantial acidification (approximately 20 µeq/L) of aquatic systems in the Southern Blue Ridge and mid-Atlantic Highlands under scenarios of constant (from 1985) deposition. This reflected a delayed response in the model to the deposition histories of these systems caused by S adsorption on watershed soils. If deposition was held constant at 1985 levels, MAGIC projected little future loss of ANC in most northeastern watersheds, ranging from a projected median decline of 1 µeq/L in New England to 4 µeq/L in the Adirondacks over 50 years. These modeled changes were owing to slight depletion of the supply of base cations from soils (Turner et al., 1992). The percentage of acidic Adirondack lakes, which were modeled to be more sensitive to change than the nonacidic lakes, was projected to increase by 8% even though SO_4^{2-} concentrations were projected to continue to decline as the soils attain a new steady-state equilibrium between S input and output under prolonged constant deposition.

On average, each kg/ha per year change in S deposition was projected by MAGIC to cause a 3 to 4 µeq/L median change in surface water ANC. Such projected changes in ANC, while considerably smaller than was generally thought to occur in the 1980s, nevertheless suggested widespread sensitivity of surface water ANC to changes in S deposition throughout the regions modeled.

Since 1990, a number of changes has been made to the MAGIC model and its method of application. These changes have been made in response to extensive testing of the model using paleolimnological data (Sullivan et al., 1992, 1996a) and the results of acidification and deacidification experiments (Norton et al., 1992; Cosby et al., 1995, 1996) and empirical studies (Sullivan and Cosby, 1998). These model testing exercises and changes to the model are discussed in Chapter 9. The cumulative impact of these model changes has only been evaluated for the Adirondack region, where the net effect has been that the model projects somewhat lesser sensitivity of Adirondack lakes to change in S deposition as compared to the version of MAGIC applied in 1990 (Sullivan and Cosby, 1995).

Church and van Sickle (1999) used the MAGIC model to simulate the response of the 36 statistically selected watersheds in the Adirondack Mountains to changing levels of S and N deposition. Model results for the year 2040 were reported, representing 50 years after passage of the 1990 Clean Air Act Amendments. Each simulated watershed was weighted to reflect the number of watersheds in the target population that it represented. Various assumptions were made for different model scenarios to represent N dynamics under constant and changing N deposition. Net N uptake was estimated for each watershed as the proportion of total NO_3^- and NH_4^+ inputs that are removed by uptake, based on 1984 estimates or measurements of deposition, annual runoff, and lake-water chemistry. Nitrogen uptake was modeled at constant fractional uptake rates throughout the simulation period and at declining net uptake on three different time scales. It was assumed for these model scenarios that N uptake would be reduced to 5% or less of N input within 50 years, 100 years, and 250 years. The results of this modeling exercise illustrated that the assumed time to N saturation had a dramatic effect on watershed response to future acidic deposition.

4.2 Applachian Mountains

The Appalachian Mountain region constitutes an important region of concern with respect to the effects of acidic deposition. Many streams at higher elevation, particularly in the mid-Appalachian portion of the region, have chronically low-ANC values and the region receives one of the highest rates of acidic deposition in the U.S. (Herlihy et al., 1993). The acid–base status of stream waters in forested upland watersheds in the mid-Appalachian Mountains has been extensively investigated in recent years (e.g., Church et al., 1992; Herlihy et al., 1993; Webb et al., 1994; van Sickle and Church, 1995).

Sulfur adsorption by soils is an important aspect of watershed acid neutralization in the southeastern U.S. Where S adsorption is high, even relatively high levels of S deposition have little or no impact on surface water chemistry, at least in the short term. Over long periods of time, however, this S adsorption capacity can become depleted under continued high levels of S deposition, causing a delayed acidification response. Stream-water SO_4^{2-} concentrations and stream discharge estimates suggest that S outputs approximate inputs in some of the watersheds of the Appalachian Plateau. Sulfur adsorption in soils is highest in the Southern Blue Ridge, where about half of the incoming S is retained, and is somewhat lower in the Valley and Ridge watersheds (Herlihy et al., 1993). Thus, there is a general pattern of increasing S adsorption as you move to the south in the mid and southern Appalachian regions.

Perhaps the most important study of acid–base chemistry of streams in the Appalachian region in recent years has been the Virginia Trout Stream Sensitivity Study (VTSSS, Webb et al., 1994). Water quality assessment and modeling efforts of the Southern Appalachian Mountain Initiative (SAMI) are also highly relevant. The results to date of both of these programs are discussed next.

Based on measurements of visibility impairment, acid-deposition, and ground level ozone, the National Park Service has determined that air quality problems in the Great Smoky Mountains and Shenandoah National Parks are among the most serious in the national parks system. These two parks have been more intensively studied with respect to acidic deposition effects than other parts of the southern Appalachian Mountain region, and also contain some of the watersheds that have been most impacted. Data from intensively studied watersheds in these two parks, therefore, receive somewhat greater coverage here than other parts of the region.

SAMI was established in 1992 to provide a regional strategy for assessing and improving air quality through public and private cooperation. SAMI focuses on air quality issues in the southern Appalachian Mountains and their effects on resources, including visibility, water, soils, plants, and animals. SAMI is somewhat unique because it is a voluntary regional initiative unlike those mandated by the Clean Air Act. Its membership includes the environmental regulatory agencies of eight states, federal agencies, industry, academia, environmental organizations, and other stakeholders across the region.

The SAMI region includes three physiographic provinces that are oriented as southwest to northeastern bands: Blue Ridge Mountains, Valley and Ridge, and Appalachian Plateau. There are no historical data available on stream-water chemistry in the region. However, the Eastern Lakes Survey (Linthurst et al., 1986) sampled lakes in the southern Blue Ridge and the National Stream Survey (Kaufmann et al., 1988) sampled streams throughout the region. Only 5% of the southern Blue Ridge lakes had ANC less than 50 µeq/L and none were acidic. In the Valley and Ridge Province, low ANC streams are generally absent from the valleys which frequently contain limestone bedrock. Ridge streams are often acid sensitive, however, and about one-fourth are low in ANC (less than or equal to 50 μeq/L) in their upper reaches. The highest proportion of acidic (5%) and low ANC (31%) streams are found in the Appalachian Plateau Province (Herlihy et al., 1996), even after excluding those affected by acid mine discharge (Herlihy et al., 1990). Acidic and low ANC streams are more prevalent in the northern part of the region, in Virginia and West Virginia, than in the south. This gradient is owing, at least in part, to the higher rates of S and N deposition and the lower S adsorption of soils in the northern part of the region. Throughout the region, acidic and low-ANC stream water is confined to small (less 20 km^2) upland, forested watersheds in areas of base-poor, weathering-resistant bedrock (Herlihy et al., 1993).

4.2.1 Monitoring Studies

The VTSSS conducted a synoptic survey of stream-water chemistry for 344 (approximately 80%) of the native brook trout (*Salvelinus fontinalis*) streams in western Virginia. Subsequently, a geographically distributed subset of the surveyed streams were selected for long-term monitoring and research (Webb et al., 1994). About one-half of the streams included in the VTSSS had ANC less than 50 μeq/L, suggesting widespread sensitivity to acidic deposition impacts. In contrast, the ANC distribution obtained by the National Stream Survey (NSS; Kaufmann et al., 1988) for western Virginia suggested that only about 15% of the streams in the NSS target population had ANC less than 50 μeq/L. Webb et al. (1994) attributed these chemical differences to the smaller watershed size, more mountainous topography, and generally more inert bedrock of the VTSSS watersheds. Thus, the VTSSS focused on a subset of watersheds that were somewhat more acid sensitive than the population of watersheds represented by the NSS.

Water chemistry data are available for a great many upland streams in Class I wilderness areas and national parks within the SAMI region. Those data were summarized by Herlihy et al. (1996, Table 4.2). Acidic streams appear to be especially prevalent in Dolly Sods and Otter Creek Wilderness areas on the West Virginia Plateau. The lower quartile of measured stream-water ANC values was also below 25 μeq/L in Shenandoah (Virginia) and Great Smoky Mountains (Tennessee) National Parks and James River Face Wilderness (Virginia). The wilderness areas with higher ANC (Table 4.2) are all located in the southern half of the SAMI region, in the Southern Blue Ridge and Alabama Plateau.

The Dolly Sods and Otter Creek Wilderness Areas are found about 25 km apart in an area of base-poor bedrock in the Appalachian Plateau of West Virginia. Most streams draining these wilderness areas are acidic or low in ANC and have concentrations of H$^+$ and Al$_i$ that are high enough to be toxic to many species of aquatic biota.

TABLE 4.2

Median Values (with First and Third Quartiles in Parentheses) for Major Ion
Chemistry in Streams in Class I Wilderness Areas and in the Entire Southern
Appalachians; Year(s) of Data Collection and Number of Observations (N) are Given
Below the Wilderness Area Name

Wilderness Area	ANC (µeq/L)	pH	Sulfate (µeq/L)	Nitrate (µeq/L)	Chloride (µeq/L)	DOC (mg/L)
Dolly Sods	-18	4.7	105	4	11	2.2
1994 (*n* = 34)	(-53– -3)	(4.3–5.1)	(91–115)	(2–6)	(9–11)	(1.7–3.1)
Otter Creek	-28	4.6	129	6	9	2.0
1994 (*n* = 63)	(-82–11)	(4.1–6.0)	(111–153)	(1–14)	(8–10)	(0.9–3.1)
Shenandoah	82	6.7	85	7	28	—
National Park	(21–120)	(6.0–6.9)	(66–103)	(3–23)	(25–32)	
1981–1982 (*n* = 47)						
James River Face	25	6.3	68	0	19	
1991–1994 (*n* = 8)	(22–44)	(6.1–6.5)	(54–74)	(0–0)	(18–20)	
Great Smoky Mt.	44	6.4	31	15	14	—
National Park	(24–64)	(6.2–6.6)	(18–46)	(6–29)	(12–16)	
1994–1995 (*n* = 337)						
Joyce	70	—	—	7	—	—
Kilmer/Slickrock	(53–80)			(6–11)		
1992–1995 (*n* = 9)						
Shining Rock	70	6.8	—	7	—	—
1992–1993 (*n* = 9)	(65–80)	(6.7–7.0)		(6–7)		
Cohutta	41	6.5	35	14	24	1.8
1992–1994 (*n* = 16)	(26–56)	(6.2–6.6)	(25–53)	(9–210)	(21–28)	(1.4–2.5)
Sipsey	245	7.3	94	2	33	2.2
1991–1993 (*n* = 30)	(120–699)	(6.8–7.6)	(83–106)	(1–3)	(32–34)	(1.6–2.7)
SAMI Regional	172	7.1	135	16	36	1.0
Streams[a]	(65–491)	(6.5–7.5)	(62–229)	(4–34)	(18–68)	(0.7–1.7)
1986 NSS						
(*n* = 19,940)						
Acidic SAMI	-24	43.7	142	0.3	16	1.4
Streams[a]	(-35– -24)	(4.5–4.7)	(117–229)	(0.2–3.5)	(12–25)	(1.0–1.7)
1986 NSS (*n* = 730)						
South Blue Ridge	152	6.8	29	1	25	1.0
Lakes[a]	(87–246)	(6.7–7.0)	(23–36)	(0–6)	(18–42)	(1.2–1.5)
1984 ELS (*n* = 71)						

[a] Regional estimate for SAMI region is calculated using National Stream Survey (NSS) data
for the upstream segment end population (extrapolated from 154 sample streams). The
Southern Blue Ridge lake estimate is extrapolated from 45 lakes sampled in the Eastern
Lake Survey (Baker et al., 1990a).
— Not measured, no data found.
Source: Herlihy et al., 1996.

There is a strong relationship between stream-water ANC and geology in
Shenandoah National Park (Cosby et al., 1991). The geologic formations in
the southwestern part of the park are most resistant to weathering and have
the streams with lowest ANC. These are the Hampton (phyllite, shale, sandstone, and quartzite) and Antietam (sandstone and quartzite) formations.
About one-fourth of the streams in Shenandoah National Park and almost all

of the streams in James River Face wilderness have ANC less than or equal to 50 μeq/L (Cosby et al., 1991; Webb et al., 1994).

In Great Smoky Mountains National Park, the acidic streams are found at higher elevations in watersheds that are likely influenced by sulfide mineral weathering. Whereas, a high proportion of the SO_4^{2-} received in deposition is retained in the soils of most of the studied watersheds, SO_4^{2-} concentrations tend to be relatively high (greater than 65 μeq/L) in streams that are acidic (Elwood et al., 1991; Cook et al., 1994; Webb et al., 1996). Low-ANC streams (less than or equal to 50 μeq/L) are common throughout the park, however, and are sensitive to future acidification to the extent that the watershed retention of atmospherically deposited S or N declines in the future under continued high levels of acidic deposition.

Webb et al. (1994) devised a watershed classification scheme for western Virginia based on ecoregion maps, geologic maps, and stream-water chemistry data. Watershed response classes were designated, in decreasing order of acid sensitivity, as siliclastic, minor carbonate, granitic, basaltic, and carbonate classes. Median stream-water ANC in the siliclastic class was only 3 to 4 μeq/L in the Blue Ridge Mountains and Allegheny Ridges subregions. The minor carbonate and granitic classes were somewhat less acid sensitive, with median ANC values of 20 and 61 μeq/L, respectively.

Results of chemical analyses of water samples collected between October 1987 and April 1993 in VTSSS headwater streams ($n = 78$) showed that ANC values tend to be lower by about 10 μeq/L (in acidic and near-acidic streams) to 40 μeq/L (in intermediate ANC streams) during winter and spring than they are during summer and fall.

Studies at a few stream sites in the mid-Appalachian Mountains have documented toxic stream-water chemistry conditions during episodes, fish kills, and loss of fish populations as a result of increased acidity. An estimated 18% of potential brook trout streams in the mid-Appalachian Mountains are too acidic for brook trout survival (Herlihy et al., 1996).

An effort to assess the effects of acid–base chemistry on fish communities in upland streams of Virginia was initiated in 1992 (Bulger et al., 1995). The study streams experience both chronic and episodic acidification. A number of differences are apparent between the low- and high-ANC streams included in this study. These include differences in such factors as age, size, and condition factor of individual fish, bioassay survival, fish species richness, and population size. Young brook trout exposed to chronic and episodic acidity experienced increased mortality (MacAvoy and Bulger, 1995); the condition of blacknose dace was poor in the low-ANC streams compared to the high-ANC streams (Dennis and Bulger, 1995).

NO_3^- concentrations in upland streams of Great Smoky Mountain National Park are very high in some locations (approximately 100 μeq/L) and are correlated with elevation and forest stand age (Cook et al., 1994). The old growth sites at higher elevation showed the highest NO_3^- concentrations, likely owing to the higher rates of N deposition and flashier hydrology at high elevation, as well as decreased vegetative N demand in the more mature forest stands. High N deposition at these sites has likely contributed to both chronic and episodic acidification (Flum and Nodvin, 1995; Nodvin et al., 1995).

Adverse effects on aquatic biota have also been found in Great Smoky Mountains National Park. A steady decline in brook trout range has been reported since the 1930s (Herlihy et al., 1996). In addition, invertebrate density and species richness were higher in high-pH streams (Rosemond et al., 1992).

Water chemistry data collected as part of the VTSSS between 1987 and 1993, and presented by Webb et al. (1994), provide an excellent example of complex interactions between terrestrial biota and drainage water chemistry. Since its introduction to North America during the last century, the gypsy moth has expanded its range to include most of the northeastern U.S. Since about 1984, the area of forest defoliation by the gypsy moth has expanded southward about 30 km per year along the mountain ridges of western Virginia. Infestation and accompanying forest defoliation occur at a given site over a period of several years.

Webb et al. (1994) compared quarterly pre- and post-defoliation stream-water chemistry for 23 VTSSS watersheds. NO_3^- concentrations increased dramatically in most of the streams, typically to 10 to 20 µeq/L or higher. The most probable source of the increased stream-water NO_3^- concentration was the N content of the forest foliage consumed by the gypsy moth larvae (Webb et al., 1994). Additional observed changes in stream-water chemistry included decreased SO_4^{2-} concentrations and ANC, which were also hypothesized to be attributable to the gypsy moth defoliation. Increased nitrification in response to the increased soil N pool may have caused soil acidification, which in turn would be expected to increase S adsorption in soils (c.f., Johnson and Cole, 1980). In addition, declines in S deposition during the comparison period may have played a role in the observed SO_4^{2-} response.

Stream-water chemistry in two headwater catchments in Shenandoah National Park (White Oak Run and Deep Run) showed trends of increasing SO_4^{2-} concentrations in the 1980s (Ryan et al., 1989). In the 1990s, however, the SO_4^{2-} concentrations have been altered as a consequence of gypsy moth defoliation. These changes induced by insect damage have masked any continued change in SO_4^{2-} concentration that may have been occurring in response to atmospheric inputs of S and progressive saturation of the S-adsorption potential of watershed soils (Webb et al., 1995).

Eshleman et al. (1998) examined NO_3^- fluxes from five small (less than 15 km²) forested watersheds in the Chesapeake Bay Basin of the Appalachian Highlands physiographic province from 1988 to 1995. Of the watersheds, four are located in Shenandoah National Park, within the Blue Ridge Province, and the fifth in Savage River State Forest in western Maryland, within the Appalachian Plateau Province. The five watersheds vary in geology and acid sensitivity, with baseflow ANC typically in the range of 0 to 10 µeq/L in Paine Run to the range of 150 to 350 µeq/L in Piney River. Forest vegetation is also variable. The composition of oak species (*Quercus* spp.) that are a preferred food source of gypsy moth larvae, ranged from 100% in Paine Run to about 60% in 3 of the other watersheds. Nitrate concentrations increased markedly in at least 3 of the watersheds during the late 1980s to early 1990s, with peak annual average NO_3^- concentrations of about 30 to 55 µeq/L. The increased leakage of NO_3^- occurred contemporaneously with a period of intense defoliation by the gypsy moth larva. Leakage was shown to occur primarily during storm flow conditions.

4.2.2 Model Simulations

One aspect of the SAMI effort to date has been a preliminary application of the MAGIC and NuCM models to assess the sensitivity of three watersheds to future increases in S and/or N deposition. The watersheds selected for study include Noland Divide in Great Smoky Mountains National Park and White Oak Run and North Fork Dry Run in Shenandoah National Park. All three watersheds were judged to be sensitive to acidification from S deposition, whereas sensitivity to N deposition was most pronounced at Noland Divide (Cosby and Sullivan, 1999). The latter is a high-elevation spruce–fir forest, and this is probably the cause of the model-estimated greater sensitivity to N effects. Spruce-fir forests are relatively rare in the southern Appalachian Mountains, and are found above about 1370 m elevation in scattered locations. Great Smoky Mountains National Park contains about three-fourths of the spruce–fir forests in the region.

Empirical model analyses by Webb et al. (1994) of VTSSS streams in western Virginia, suggested that an approximately 70 to 80% reduction in the anthropogenic component of S deposition would be required to maintain the current acid–base status of these acid-sensitive streams. These estimates are generally in agreement with the results of MAGIC model simulations. However, additional modeling will be required before any conclusions can be reached regarding regional responses to future changes in S and N deposition loading.

4.3 Florida

Florida lakes are located in marine sands overlying carbonate bedrock and the Floridan aquifer, an extensive series of limestone and dolomite that underlies virtually all of Florida. In the Panhandle and northcentral lake districts, the Floridan aquifer is separated from the overlying sands by a confining layer known as the Hawthorne formation. The major lake districts are located in karst terrain, and lakes probably formed through dissolution of the underlying limestone followed by collapse or piping of surficial deposits into solution cavities (cf. Schmidt and Clark, 1980; Arrington and Lindquist, 1987). Flow of water from the lakes is generally downward, recharging the Floridan aquifer. Historical changes in lake stage have differed from lake to lake in response to long-term trends in precipitation, and lakes with direct hydraulic connections with the Floridan aquifer have shown considerably broader ranges in stage compared to lakes where the connection is impaired (cf. Clark et al., 1964a; Hughes, 1967). Base cation enrichment appears to be small in most study lakes in Florida and ANC generation is owing primarily to in-lake anion reduction (SO_4^{2-} and NO_3^-; Baker et al., 1988; Pollman and Canfield, 1991). Retention of SO_4^{2-} by watershed soils also may be important. Where groundwater interactions with the deeper aquifers are present, surface waters can be highly alkaline. However, those lakes with hydrologic contributions from shallow aquifers in highly weathered sands can be quite acidic and presumably sensitive to acidic deposition. As is the case elsewhere, the key to understanding the potential response of Florida lakes to acid inputs is related largely to knowledge of the hydrologic flowpaths (Sullivan and Eilers, 1994).

Topographic relief in Florida is minimal and attempts to relate groundwater contributing areas to specific lakes have been problematic (Pollman and Canfield, 1991). Detailed studies of low-ANC seepage lakes in northern Florida show that, unlike low-ANC seepage lakes in the upper Midwest, groundwater contributions can represent the major hydrologic input. For example, Lake Five-O in the Panhandle receives the majority of its annual inflow from groundwater sources. An additional anomaly with regard to the flowpath is that water does not exit the lake through the opposing shoreline, but rather passes vertically downward through the lake bottom. Despite the considerable groundwater contributions to Lake Five-O, the pH (5.4), ANC (-4 µeq/L), and nonmarine base cation concentrations are low (Pollman et al., 1991). This reflects the highly weathered nature and low base saturation of the sands through which the groundwater flows before entering the lake.

Although evaporation plays a role in most regions in concentrating acidic inputs from atmospheric deposition, the effect of evaporation is much greater in Florida than other low-ANC regions of the U.S. Annual pan evaporation measured at several stations ranged from 149 to 175 cm, increasing in a southerly direction. As a consequence, the net precipitation in the Panhandle is 50 to 100% greater than that in the Central Trail Ridge (Pollman and Canfield, 1991).

In-lake processes are also important components influencing the chemistry of Florida lakes. Baker and Brezonik (1988) illustrated the importance of in-lake anion retention in generating ANC for Florida lakes. Retention of inorganic N is nearly 100% and ANC generation from SO_4^{2-} retention may approach 100 µeq/L in some Florida lakes (Pollman and Canfield, 1991). Base cation deposition and NH_4^+ assimilation are additional important influences on the acid–base status of clearwater lakes in Florida.

Current deposition in Florida is moderately acidic with volume-weighted mean (VWM) pH ranging from 4.55 to 4.68 for the 4 northern FADS (Florida Acid Deposition Study) sites. Nonmarine SO_4^{2-} VWM concentrations ranged from 19.8 to 22.9 µeq/L and NO_3^- VWM concentrations ranged from 9.5 to 11.1 µeq/L. Ammonium VWM concentrations ranged from 4.2 to 6.3 µeq/L. Based on regional estimates of dry : wet deposition ratios for Florida, dry deposition of S and N are 70 and 96%, respectively, of wet values (Baker, 1991). Total S and N deposition in parts of the north-central peninsula are, therefore, approximately 10 and 9 kg/ha per year, respectively (Sullivan and Eilers, 1994).

For the southeastern U.S., both S and N emissions showed only modest increases from 1900 to 1960. However, from 1960 to 1980, emissions increased approximately four-fold and then began decreasing in the 1980s (Gschwandt-ner et al., 1985). More detailed analyses of recent trends in regional emissions indicated that emissions peaked about 1978 and declined slightly during the following decade (Placet et al., 1990). Model analyses suggest that within-state sources contributed about two-thirds of the total deposition of S in 1983 (FCG, 1986). Emissions of SO_2 in Florida fluctuated around 900,000 tonnes per year from 1976 through 1984, but have been projected to increase (FDER, 1984). If the population continues to increase in Florida, it appears reasonable to assume that NO_x emissions also will continue to increase.

Northern Florida contains the highest percentage of acidic lakes of any lake population in the U.S. (Linthurst et al., 1986). Of the Panhandle lakes, 75% were acidic, as were 26% of the lakes in the northern peninsula in 1984. This large population of acidic lakes, combined with increasing emissions of S and N for the state, stimulated investigations of the acid–base chemistry of these lakes (Pollman and Canfield, 1991). Most of these acidic lakes are clearwater (DOC less than 400 μmol) seepage lakes in which the dominant acid anions are Cl^- and SO_4^{2-}. The most dilute group of lakes is found in the Panhandle which Pollman and Canfield (1991) attributed to higher precipitation, lower evaporation, and lower watershed disturbance. The regional difference in evapoconcentration for Florida can have two opposing effects (Pollman and Canfield, 1991). Concentrating an acidic solution increases its acidity. However, increasing evaporation may have an opposing effect on lake chemistry by affecting lake hydrology. As evaporation increases, groundwater inflow might also increase in importance and provide a proportionally greater supply of base cations. Increasing evaporation also increases the lake hydraulic residence time (τ_w), thus increasing the opportunity for dissimilatory SO_4^{2-} reduction (Baker and Brezonik, 1988). Nitrate and ammonium concentrations in lakes that do not have agricultural contributions of N (as estimated by K^+ less than 15 μeq/L) are generally not measurable (Sullivan and Eilers, 1994). Retention of inorganic N is highly efficient in Florida lakes and contributing areas, similar to lakes in the upper Midwest.

Although concentrations of DOC are high in many Florida lakes, organic anions are generally less important than SO_4^{2-} in the low-ANC and acidic lakes (Pollman and Canfield, 1991). Aluminum concentrations are very low in Florida lakes despite their high acidity. Although Al^{n+} is mobilized in surficial soils (e.g., less than 15 cm depth) by the acid loading from atmospheric deposition, most of the Al^{n+} is removed from solution by precipitation and ion exchange reactions within 75 cm depths (Graetz et al., 1985), and relatively little Al^{n+} is transported in solution to lake waters.

Evidence for recent acidification of some Florida lakes has been supported by historical analyses of lake chemistry (Crisman et al., 1980; Baker, 1984;

Battoe and Lowe, 1992), inferred historical deposition (Husar et al., 1991; Hendry and Brezonik, 1984), and paleolimnological reconstructions of lake pH (Sweets et al., 1990; Sweets, 1992). However, the case for acidification by acid deposition is equivocal with respect to all lines of evidence (cf., Pollman and Canfield, 1991; Sullivan and Eilers, 1994) and the interpretation is complicated by profound regional and local changes in land use and hydrology. For example, an alternative explanation (other than acidic deposition) for the apparent acidification of Lakes Barco and Suggs (Sweets et al., 1990) is that the apparent recent decline in pH may have been caused by a regional decline in the potentiometric surface of the groundwater. Large groundwater withdrawals of the Floridan aquifer for residential and agricultural purposes may have contributed to reduced groundwater inflow of base cations into seepage lakes, thereby causing lake-water acidification (Sullivan and Eilers, 1994).

Other land use changes have probably increased lake pH by providing increased inputs of fertilizer, thus increasing the productivity of many lakes. Paleolimnological evidence of this process was provided by Brenner and Binford (1988) and Deevey et al. (1986). The importance of assessing land use changes in Florida is further indicated by the high percentage (57%) of the lakes having evidence of disturbance based on ion chemistry deviations from expected geochemistry (Pollman and Canfield, 1991). Battoe and Lowe (1992) attributed a recent decline in the pH of Lake Annie in central Florida to acidic deposition. However, preliminary analyses of aerial photographs show that the watershed of Lake Annie has been subjected to numerous land use changes including construction of extensive ditches that might explain all or part of the observed changes in acid–base chemistry (Eilers, unpublished data).

4.3.1 Monitoring Studies

Historical data on the water chemistry of lakes in the Trail Ridge area of northcentral Florida have been evaluated by Crisman et al. (1980), Hendry and Brezonik (1984), and Pollman and Canfield (1991). Analyses by Crisman et al. (1980) and Hendry and Brezonik (1984), were based on comparison of recent data with data collected by Clark et al. (1964a) and Shannon (1970). The more recent work by Pollman and Canfield (1991) also included water chemistry data from ELS-I (Linthurst et al., 1986) and PIRLA-I (Sweets et al., 1990). Of the seven lakes analyzed by Pollman and Canfield (1991), four showed significant increases in H^+ concentration with time. The other lakes showed either significant declines (two lakes) or no trend (one lake).

The most extensive monitoring data base available was for McCloud Lake, an undeveloped seepage lake that had also been sampled from 1980 to 1982 by Baker (1984). The pH of McCloud Lake decreased about 0.3 pH units from 4.9 in 1968–1969 to 4.6 in 1986, but the apparent trend was driven by Shannon's data collected in 1968. Later surveys (1978–1986) suggested short-term variability, but no consistent trend (Pollman and Canfield, 1991).

4.3.2 Paleolimnological Studies

Diatom-inferred pH reconstructions are available for 16 lakes in the Florida Panhandle and northcentral Florida, nearly all of which lie in upland or ridge regions where soils are deeply weathered quartz sands and quite acidic (Carlisle et al., 1978). Paleolimnological reconstructions of the chemistry of six seepage lakes in Florida were calculated as part of the PIRLA-I project and reported by Sweets et al. (1990). In addition, ten Florida seepage lakes were cored as part of PIRLA-II, and results of these analyses were reported by Sweets (1992). Paleolimnological study lakes in Florida have been located in the Panhandle, the Trail Ridge Lake District, and Ocala National Forest, generally in terraces of loose sand (Entisols) that were deposited on top of the clay confining layer (Hendry and Brezonik, 1984). The sands are highly weathered and soils have very low cation exchange capacity (Carlisle et al., 1981). Many of the low-ANC Florida seepage lakes serve as recharge areas for the Floridian aquifer (Baker and Brezonik, 1988).

Paleolimnological data collected for Florida lakes in PIRLA-I were not extensive, and sedimentary evidence for changes in lake-water chemistry in response to atmospheric deposition was not conclusive. Partial mixing of sediment layers of cores was also common. Of the six lakes analyzed in PIRLA-I, two (Lakes Barco and Suggs) were inferred to have acidified since 1950, three lakes were inferred to have remained stable or have fluctuated with no steady change in pH (Sweets et al., 1990). The acidification of Lake Barco by 0.3 to 0.8 pH units began about 1950. Lake Suggs was inferred to have decreased 0.5 pH units between 1880 and 1920, and a second pH decrease of 0.4 units occurred between 1950 and 1970. The timing of the onset of inferred acidification after 1950 correlated with increases in SO_2 emissions and S deposition that has been estimated to have increased steadily since about 1945 (Husar et al., 1991). Also, sedimentary accumulation of Pb, Zn, and PAH increased greatly between 1940 and 1950, indicating increased deposition of atmospheric pollutants.

Sweets et al. (1990) provided quantitative estimates of diatom-inferred change in ANC since pre-1900 for Lakes Barco and Suggs. The diatom-inferred ANC of Lake Barco decreased by 36 μeq/L (average of 3 cores) since about 1950, coincident with increases in acidic deposition in the region. The total loss of ANC inferred by the diatoms since pre-industrial times at Lake Suggs, was 19 μeq/L. Perhaps half of this change might be attributed to acidic deposition since 1940 (Sweets et al., 1990). If it is assumed that essentially all of the current lake-water concentration of SO_4^{2-} in Lakes Barco and Suggs is of atmospheric, anthropogenic origin, then approximately 27% of the increase in SO_4^{2-} in both lakes has caused a stoichiometric decrease in ANC.

Diatom-inferred changes in pH have been derived for 16 Florida seepage lakes studied to date in both PIRLA projects. There were 5 lakes studied in or near the Trail Ridge region, and all showed some evidence of recent acidification (greater than 0.2 pH unit decrease; Sweets, 1992). With the exception of

Lake Five-O, lakes in the Panhandle region and Ocala National Forest did not show evidence of recent acidification. Although Lake Five-O was inferred to have decreased by 2 pH units, this acidification cannot be attributed to acidic deposition. The paleolimnological data suggest that this large decline was associated with a sudden change in lake-water chemistry, probably caused by a catastrophic disturbance such as sinkhole activity, rather than by acidification from atmospheric deposition (Pollman and Sweets, 1990; Sweets, 1992).

4.3.3 Model Simulations

Pollman and Sweets (1990) conducted hindcast simulations for 15 lakes in the Florida Panhandle and northcentral Florida, using the Internal Alkalinity Generation (IAG) model (Baker et al., 1986). Study lakes were those included within the PIRLA-I and PIRLA-II paleolimnological investigations. Model hindcast projections were constructed assuming a range of base cation neutralization of SO_4^{2-} in groundwater from 0 ($F = 0$) to 100% ($F = 1.0$). IAG model results were compared with diatom inferences of historical change in pH. Although Pollman and Sweets (1990) concluded generally good agreement between the two approaches, this agreement was primarily confined to those lakes inferred by both techniques to have not experienced acidification. Those lakes inferred by either technique to have experienced acidification of more than the standard error of the diatom inference equation (0.31 pH units) showed generally poor agreement between the IAG model and the diatom hindcast approaches (Sullivan and Eilers, 1994).

As pointed out by Pollman and Sweets (1990), there are a number of assumptions required for conducting IAG hindcast simulations, and these assumptions can have a very large influence on model output. Assumptions regarding base cation deposition are more important for IAG modeling of seepage lake chemistry than for modeling drainage systems, using a model such as MAGIC. Pollman et al. (1990) found IAG hindcast simulations to be extremely sensitive to assumed pre-industrial base cation deposition. For example, changing this estimate by 10% resulted in factor-of-two differences in estimates of historical acidification for Lake Barco. Thus, although modeling studies may aid in improving our understanding of the response of seepage lakes in Florida to acidic deposition, basic geochemical and hydrological data with which to parameterize the models are generally lacking.

4.4   Upper Midwest

The upper Midwest is characterized by numerous lakes created by repeated glaciations. The region shows little topographic relief and the deep glacial overburden results in little or no exposed bedrock. Sensitive aquatic resources in the upper Midwest are largely comprised of seepage lakes (Eilers et al., 1983). Limited studies of stream systems have shown them to be insensitive to acidic deposition because groundwater discharge constitutes a major component of stream flow. Furthermore, because of the low relief and permeable till, streams in this region are not subject to abnormally high discharge in the spring such as observed in the West (Eilers and Bernert, 1989). The seasonal depression of pH observed in some Wisconsin lakes cannot be attributed to snowmelt runoff, but rather is caused by CO_2 accumulation from respiratory processes under the ice. Most drainage lakes (and some seepage lakes) in the Midwest receive much of their inflow from groundwater which is generally enriched in ANC from dissolution of carbonates and silicate minerals. Those seepage lakes that have low base cation concentrations receive nearly all of their hydrologic inputs as precipitation directly on the lake surface (Baker et al., 1991a). Consequently, such lakes generally have long hydraulic residence times (e.g., τ_w is approximately equal to 10 years). The long residence time provides opportunity for in-lake reduction and assimilation processes to neutralize much of the acidic inputs that would otherwise be concentrated from evaporation. Baker and Brezonik (1988) quantified the importance of in-lake processes for SO_4^{2-} reduction as a function of τ_w in the IAG Model. For lakes that receive the vast majority of their hydrologic input from on-lake precipitation, even small additions of groundwater can be expected to neutralize acidic deposition from atmospheric sources indefinitely (Kenoyer and Anderson, 1989). Only where groundwater contributions approach zero can one expect that acidification from atmospheric sources would be a serious concern in the upper Midwest.

The critical importance of hydrologic flowpaths in the upper Midwest has been refined and reiterated in a number of regional assessments (Eilers et al., 1983; Nichols and McRoberts, 1986; Linthurst et al., 1986; Sullivan and Eilers, 1994), synthesis analyses of this region (Baker et al., 1990a; Cook and Jager, 1991), and process studies of low-ANC lake systems (Lin and Schnoor, 1986; Kenoyer, 1986; Anderson and Bowser, 1986; Kenoyer and Anderson, 1989; Wentz and Rose, 1989; Webster et al., 1990, 1993; Webster and Eilers, 1994).

Emissions of SO_2 and NO_x in the upper Midwest appear to have increased dramatically in the twentieth century. Recent trends indicate that emissions reached a maximum in about 1978 and have decreased since that period (Placet et al., 1990). In Wisconsin and Minnesota, S emissions decreased in the mid-1980s, following enactment of pollution control legislation. However, the greatest reduction in regional emissions was attributed to a reduction of SO_2 emissions in Michigan, which declined from about 1.3 M tons per year in 1975 to about 0.6 M tons per year in 1988 (Webster et al., 1993). Recent declines in NO_x emissions appear to be substantially less than those observed for SO_2. The latter may continue to decline as older stationary emission sources are retired; however, NO_x emissions appear to have stabilized and can be expected to follow trends related to future consumption of fossil fuels within the region.

Wet S deposition ranges across the region from about 3 to 4 kg S/ha per year in Minnesota to near 5 kg S/ha per year in eastern portions of the Upper Peninsula of Michigan. Nitrogen deposition follows a similar pattern and ranges across the region from about 3 to 4 kg N/ha per year in wet deposition. These levels of deposition are moderate compared to other regions of the country.

Deposition of base cations is an important component influencing the acidity of deposition in the upper Midwest and the major ion chemistry of seepage lakes. One of the major sources of atmospheric base cations is dust derived from unpaved roads (Placet et al., 1990). As more rural roads are paved, it is likely that atmospheric concentrations of base cations will continue to decline, thus partially negating the positive effects of expected continued declines in SO_2 emissions (Sullivan and Eilers, 1994).

Acidic lakes in the Upper Midwest are primarily small, shallow seepage lakes that have low concentrations of base cations and Al and moderate SO_4^{2-} concentrations. Organic anions, estimated by both the Oliver et al. (1983) method and the anion deficit, were less than one-half the measured SO_4^{2-} concentrations (Eilers et al., 1988b).

Groundwater flow-through lakes in the upper Midwest were defined by Baker et al. (1991a) as those having SiO_2 greater than 1 mg/L. Classification of seepage lakes into groundwater flow-through and recharge categories on the basis of SiO_2 concentration has been shown to result in a very clear distinction of upper Midwestern subpopulations and marked differences in base cation concentrations and ANC between these subpopulations. Groundwater flow-through lakes in the upper Midwest were generally high pH/ANC systems, largely owing to substantial groundwater inputs of base cations (e.g., Baker et al., 1991a). Only 6% of these lakes had ANC less than or equal to 50 µeq/L and none were acidic. They are, therefore, of relatively little interest with respect to potential changes in acid–base status in response to acidic deposition. Groundwater recharge lakes in the upper Midwest were more common (71% of the seepage lakes), and were more frequently low pH/ANC. Of these lakes, 5% were acidic and 9% had pH less than or equal to 5.5. Nearly 90% of upper Midwestern lakes that had ANC less than or equal to 50 µeq/L were in this hydrological category (see e.g., Baker et al., 1991a). The modest influence of watershed processes on the chemistry of groundwater recharge lakes results in minimal weathering contributions, as reflected by the low concentrations of SiO_2 and Ca^{2+} (Schnoor et al., 1986). Such lakes are, therefore, susceptible to acidification from acidic deposition.

Lakes in the upper Midwest exhibit considerable spatial gradients in pH, ANC, base cations, and DOC, all of which decrease from west to east (Table 4.3). Sulfate concentrations in lakes, however, do not show a comparable change across the region when all lakes are included in the analysis, despite a 40% increase in wet SO_4 deposition from Wisconsin to Michigan. Cook and Jager (1991) attributed the major west-to-east pattern in lake chemistry in the upper Midwest to increasing frequency of seepage lakes in the eastern portion of the region, rather than a gradient in the deposition of S. They attributed the absence of a more pronounced gradient in lake-water SO_4^{2-}

TABLE 4.3

Population Estimates of the 5th and 50th Percentiles of Selected Parameters for Lakes in the Upper Midwest (Units for ANC, C_B, and SO_4^{2-} are in µeq/L. DOC is in mg/L[a].)

Population	n	N	pH		ANC		C_B		SO_4^{2-}		DOC	
			P_5	P_{50}	P_5	P_{50}	P_5	P_{50}	P_5	P_{50}	P_5	P_{50}
Northwestern Wisconsin (2c)	117	1519	6.11	6.68	31	126	113	232	45	76	9.2	15.3
Northcentral Wisconsin (2d)	218	2388	5.31	7.03	-1	278	80	429	62	223	5.1	13.8
Upper Peninsula of Michigan (2e)	62	540	4.53	7.41	-34	904	55	1028	83	246	6.8	20.1

[a] Number of sampled lakes, n; estimated number of lakes in the population, N.

concentration across the region to a combination of processes, including watershed sources of S (primarily in Minnesota) and variable anion retention in lakes, related to the supply of iron and organic C. The retention of SO_4^{2-} by dissimilatory reduction is particularly high for seepage lakes. For example, a seepage lake with mean depth of 3 m and τ_w of 7.5 years would be expected to lose about 50 µeq/L of SO_4^{2-} from the water column by this process (Cook and Jager, 1991). In contrast, a typical drainage lake (2.5 m deep, τ_w equal to 1 year) in the upper Midwest would be expected to lose only about 10 µeq/L of SO_4^{2-} (Sullivan and Eilers, 1994).

Concentrations of inorganic N are uniformly low throughout the upper Midwest and are efficiently retained in both terrestrial and aquatic systems. Snowmelt does not provide any significant NO_3^- influx to lakes in the upper Midwest because most snowmelt first infiltrates into the soil prior to reaching the drainage lakes. Snowmelt input of N into seepage lakes is limited mainly to the snow on the lake surface. Aluminum concentrations are far lower in the upper Midwest than observed in lakes of similar pH in the Northeast.

The large number and extent of wetlands in the upper Midwest contribute to high production of organic matter which is reflected in high DOC concentrations in many surface waters. Nevertheless, base cation production is the dominant ion-enrichment process in most lakes in the upper Midwest. Even in low-ANC groundwater-recharge seepage lakes, base cation production accounts for 72 to 86% of total ANC production (Cook and Jager, 1991). Sulfate is the dominant anion in the low-ANC (less than or equal to 50 µeq/L) groundwater-recharge lakes.

The upper Midwest has a large population of low ANC lakes, but relatively few acidic (ANC less than or equal to 0) lakes (Linthurst et al., 1986). Paleolimnological evidence suggests slight acidification of selected lakes (Kingston et al., 1990) consistent with the modest historical and current rates of S deposition. Time-trend analysis of 28 lakes in the region showed decreasing SO_4^{2-} concentrations in the lakes from 1983 to 1989, consistent with decreases in regional SO_2 emissions and S deposition during the period (Webster et al., 1993). The clearest trends in lake-water SO_4^{2-} concentration were observed in the higher ANC lakes, which were generally short-residence time drainage lakes. Most lakes failed to show significant trends in acid–base chemistry other than SO_4^{2-} concentration, perhaps because:

1. The period of the time series was too short to allow for the chemistry to reach equilibrium with changing atmospheric deposition in the long-residence time lakes.

2. Changes in ANC were minimized by other changes in the acid–base chemistry of the lakes (e.g., organic acid anions, aluminum, or base cations) and each was too small to be detected in the limited database.

Most of the acidic and low-ANC lakes in the upper Midwest are seepage type. Some drainage lakes with ANC less than or equal to 50 µeq/L were sampled in ELS-I, however, and they exhibited a pattern of increasing SO_4^{2-}/C_B (where C_B is the base cation sum: Ca^{2+} + Mg^{2+} + K^+ +Na^+) and decreasing (HCO_3^- - H^+) concentration across a longitudinal gradient from eastern Minnesota to eastern Michigan (Sullivan, 1990). Atmospheric deposition of S and hydrogen ion increase along the same gradient (Eilers et al., 1988b). The decline in (HCO_3^- - H^+) could not be explained by differences in concentration of DOC, but was driven by four acidic lakes in eastern Michigan that had very low (Ca^{2+} + Mg^{2+}) concentration. Of the 4 lakes, 3 had SO_4^{2-}/C_B between 0.7 and 0.9 and only moderate concentrations of DOC (approximately 3 to 6 mg/L). High SO_4^{2-} relative to C_B concentration is largely responsible for the current acidity of these systems. The fourth lake had SO_4^{2-}/C_B of only about 0.5 and much higher DOC (approximately 13 mg/L). Organic acidity plays a major role in the acidity of this lake and also of the lowest ANC lake sampled in ELS-I in central Wisconsin (DOC approximately 16 mg/L). These spatial distributions for low ANC drainage lakes in the upper Midwest are consistent with the hypothesis of historical acidification of some lakes with very low Ca^{2+} + Mg^{2+} in the easternmost portion of the region. For some of the lowest ANC systems, however, organic acidity is also an important factor (Sullivan, 1990).

The distribution of selected chemical concentrations across a longitudinal gradient in the upper Midwest is presented in Figure 4.5 for low-ANC (less than or equal to 50 µeq/L) groundwater recharge seepage lakes (Sullivan, 1990). This hydrologic lake type was defined on the basis of low (less than 1 mg/L) silica, which is indicative of minimal weathering (thus groundwater) inputs. Lakes having high Cl⁻ concentration (greater than 20 µeq/L) were deleted to minimize road salt influence and other anthropogenic watershed disturbances. Although biological uptake of SiO_2, for example by diatoms, influences surface water concentrations, SiO_2 concentration in seepage lakes is generally indicative of groundwater input (Sullivan, 1990; Baker et al., 1991a).

Across an increasing S depositional gradient from eastern Minnesota eastward to eastern Michigan, (HCO_3^- - H^+) decreases and the ratio SO_4^{2-} to

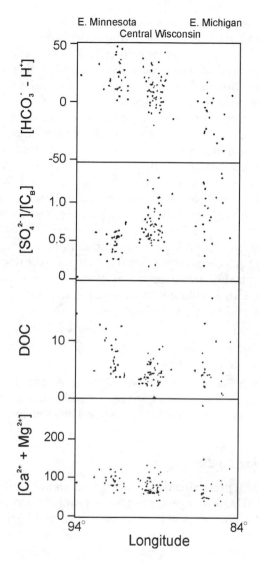

FIGURE 4.5

$[HCO_3^- - H^+]$, $[SO_4^{2-}]/[C_B]$, DOC, and $[Ca^{2+} + Mg^{2+}]$ in water of seepage lakes having SiO_2 less than 1 mg/L across a longitudinal gradient in the upper midwestern U.S. Data are presented for low ANC (less than or equal to 50 µeq/L) lakes that have low concentrations of Cl⁻ (less than or equal to 20 µeq/L). Deposition of H⁺ varied across this gradient from approximately 250 equiv./ha per year in eastern Michigan to 100 equiv./ha per year in eastern Minnesota. (Source: Sullivan, 1990.)

C_B increases in these groundwater recharge seepage lakes (Figure 4.5). In Michigan and Wisconsin, many lakes have SO_4^{2-} greater than C_B and are currently acidic because of high SO_4^{2-} relative to C_B concentration. There are also many lakes that have high concentrations of DOC, and organic acidity undoubtedly accounts for many of these lakes having ANC less than or equal to 0, particularly where base cation concentrations are low (e.g., less than 50 µeq/L). The spatial pattern in (HCO_3^- - H^+) is not attributable to DOC, which generally shows a decreasing trend with increasing deposition. The concentration of (Ca^{2+} + Mg^{2+}) also decreases with increasing deposition, and this is probably attributable to lower levels of base cation deposition and greater amounts of precipitation in the eastern portion of the region. Atmospheric deposition is an important source of base cations for groundwater recharge seepage lakes because of minimal groundwater inputs. In the eastern portion of the region, these lakes are more sensitive to pH and ANC depression in response to either elevated SO_4^{2-} or DOC. The spatial pattern for low-ANC groundwater recharge lakes in the upper Midwest are consistent with the following hypotheses (Sullivan, 1990):

1. Sensitivity to mineral and organic acidity as mechanisms for depressing ANC below zero increases from west to east because of decreasing lake-water base cation concentrations, and this may be owing, in part, to changes in base cation deposition and precipitation volume along this gradient.

2. High concentrations of DOC are responsible for the acidic conditions in some of the lakes, and DOC may have decreased in response to acidic deposition.

3. Many of the lakes in eastern Michigan, and some in Wisconsin, are currently acidic because of high SO_4^{2-} relative to base cation concentration, and have been probably acidified by atmospheric deposition.

4.4.1   Monitoring Studies

Webster et al. (1993) presented data on temporal trends in the chemistry of 28 upper Midwestern lakes sampled 3 times per year between 1983 and 1989. The lakes are located in Minnesota (4), Wisconsin (13), and Michigan (11). Statistically significant trends in SO_4^{2-} concentration were found for eight lakes and were negative in direction, consistent with a recent regional decline in S emissions and deposition. Lakes exhibiting significant declines in SO_4^{2-} concentration were primarily drainage type and had ANC greater than 100 µeq/L; only one, McNearney Lake, was seepage type and low in ANC (-39 µeq/L). Furthermore, of the eight lakes that showed decreases in SO_4^{2-} concentration, only Buckeye Lake also showed increases in ANC (+2.4 µeq ANC/L per year), and this lake was relatively high in pH (approximately 7.1) and ANC (168 µeq/L). For the most part, the anticipated trends of decreasing SO_4^{2-} and increasing pH and ANC that were expected to occur in the more sensitive acidic and low-ANC upper Midwestern seepage lakes were not found. Although 15 seepage lakes having pH less than 6 were included in the monitoring program in Wisconsin and Michigan, none of them showed the expected patterns of "recovery" from acidic deposition effects. During the study period, SO_4^{2-} deposition decreased about 9 to 14 eq/ha per year at 3 NADP monitoring stations in the region (about 4% per year decrease; Webster et al., 1993).

A second group of five lakes, primarily acidic or low-ANC seepage or closed-basin lakes, exhibited significant trends of decreasing ANC that were associated with concurrent declines in base cation concentrations rather than changes in SO_4^{2-} concentration. Nevins Lake, MI, exhibited the most marked changes; ANC declined from 178 µeq/L in 1983 to 21 µeq/L in 1988 and pH declined 0.75 pH units (Webster et al., 1990, 1993). These changes were attributed by the authors to climatic fluctuations. Lower than normal precipitation volumes reduced lake and groundwater levels, resulting in diminished inflows of ANC-rich groundwater into these (primarily seepage) lakes. The magnitude of the ANC decrease exceeded the magnitude of lake acidification from atmospheric deposition by a considerable margin. Thus, if the acidic and low-ANC seepage lakes were responding to decreases in acidic deposition, such responses were overshadowed by changes in acid–base chemistry related to climate and hydrology.

4.4.2 Paleolimnological Studies

In the upper Midwest region, diatom-inferred pH reconstructions have been completed for 15 lakes, and summarized by Cook and Jager (1991). There were 4 lakes that had pH less than or equal to 5.7 and showed a pH decline of 0.2 to 0.5 pH units during the past 50 to 100 years. Diatom-inferred pH increased in 1 lake by 0.2 pH units. No change was inferred for the other 10 lakes, including 4 lakes with pH greater than 6.0. Of the lakes, nine were included in PIRLA-I, and results from these were described by Kingston et al. (1990). No major, recent, regional acidification was indicated by the diatom-inferred pH reconstructions. Changes in most lakes were apparently small, and were no greater during the industrial period than during the pre-industrial period. In Michigan, two lakes were studied in PIRLA-I, McNearney Lake and Andrus Lake. McNearney Lake is highly acidic (ANC is equal to -38 µeq/L) and has experienced little change in pH for thousands of years. It is atypical for upper Midwestern lakes because of its extremely low pH and ANC and high concentrations of labile monomeric Al. Its pH is strongly buffered by Al (Cook et al., 1990). The pH of Andrus Lake was inferred to have declined about 0.3 pH units from 1840 to the turn of the century, followed by a rise of 0.2 units after major logging and fire in the watershed around 1920. Subsequently, the pH declined about 0.2 pH units to the present (Kingston et al., 1990). The floristic changes in Andrus Lake in recent decades were consistent with lake acidification from atmospheric deposition and decline in DOC and/or response to watershed disturbance.

In Wisconsin, 4 lakes were studied by PIRLA-I, all of which have ANC near 0 and pH 5.2 to 5.7. Brown Lake showed the most consistent decline in diatom-inferred pH (approximately 0.5 units) of any upper Midwestern lake studied. None of the pH reconstructions for other Wisconsin lakes suggested recent acidification except possibly the one for Denton Lake, but it fluctuated widely. The diatom-inferred pH of Otto Mielke Lake increased 0.2 units in recent decades, coincident with changes in fisheries management. Camp 12 Lake also exhibited evidence of slight pH increase, shortly after major logging and slash fire in the watershed about 1912 (Kingston et al., 1990).

Paleolimnological evidence for recent lake-water acidification in the upper Midwest as a result of acidic deposition is strongest for Brown and Andrus Lakes. However, Kingston et al. (1990) estimated that the change in alkalinity was only -6 to -8 µeq/L for these lakes. Such changes are small relative to natural variability (Schnoor et al., 1986; Kingston et al., 1990) and relative to changes inferred for many Adirondack lakes (Sullivan et al., 1990a).

Although paleolimnological data suggest that some upper Midwestern lakes have acidified since pre-industrial times, there is little paleolimnological evidence indicating substantial widespread acidification in this region (Kingston et al., 1990; Cook et al., 1990). It is likely that land use changes and other human perturbations of upper Midwestern lakes and their watersheds have exerted more influence on the acid–base chemistry of lakes than has acidic deposition (Eilers et al., 1989a; Kingston et al., 1990; Sullivan, 1990). This result is not unexpected because acidic deposition has been much smaller in magnitude in the upper Midwest than in most areas of the eastern U.S. (Husar et al., 1991). It is clear, however, that the portion of the region most likely to have experienced acidification from acidic deposition is the Upper Peninsula of Michigan, where acidic seepage lakes are particularly numerous (Baker et al., 1990a), acidic deposition is highest for the region, and the $[SO_4^{2-}]/[C_B]$ ratio is commonly greater than 1.0 (Figure 4.5). The percentage of acidic lakes in the eastern portion of the Upper Peninsula of Michigan (east of longitude 87°) is 18 to 19% (Schnoor et al., 1986; Eilers et al., 1988b).

Paleolimnological data are available for only two lakes in upper Michigan, McNearney and Andrus Lakes. McNearney Lake may have been naturally acidic and appears to be atypical for the region. Andrus Lake is inferred to have experienced declines in pH and DOC since pre-industrial times that could be related to acidic deposition (Kingston et al., 1990). It is likely that other lakes in this subregion have also experienced recent acidification. Data are lacking, however, with which to quantify the amount of acidification that may have occurred in the past or the dose–response relationships of these systems. In addition to the scarcity of paleolimnological data within the portion of the upper Midwest most likely to have experienced widespread historical acidification, there is also a paucity of basic biogeochemical data on the response of the major lake type in this region (seepage) to atmospheric inputs of S and N.

4.4.3   Experimental Manipulation

Quantitative information on the acidification response of a seepage lake is available from the Little Rock Lake Acidification Project. This whole-lake manipulation project was initiated in 1983, and involved the controlled acidification of a two-basin low-ANC seepage lake in northcentral Wisconsin. The lake was divided into treatment and reference basins using a flexible vinyl curtain and the treatment basin was acidified with sulfuric acid in steps of 0.5 pH units every 2 years, from an initial pH of 6.6 in 1985.

The study was described by Brezonik et al. (1986) and Watras and Frost (1989). Chemical results of the manipulation were presented by Sampson et al. (1994) and Brezonik et al. (1993). Acidification to target values of pH 5.1 and 4.7 was achieved after addition of sufficient S to raise the lake-water SO_4^{2-} concentration in stages from 53 to 116 and 147 µeq/L. In response to the increased SO_4^{2-} concentration, ANC decreased from 27 µeq/L to -4 and -14 µeq/L, respectively (Brezonik et al., 1993). In both cases, increased mobilization of base cations, presumably from the lake sediments, neutralized 53% of the added SO_4^{2-} (Sampson et al., 1994). It is important to note that this degree of base cation neutralization ($F = 0.53$) does not include any watershed neutralization that might have occurred if the acid had been applied beyond the boundaries of the lake itself, although groundwater contributions to Little Rock Lake during the study approached zero. This estimate of base cation neutralization of atmospheric acidic inputs should be viewed as a conservative estimate for a seepage lake.

4.4.4 Model Simulations

Pollman and Sweets (1990) conducted IAG model hindcasts for seven upper Midwestern lakes. The magnitude of simulated historical acidification followed the gradient in S deposition across the region, with the largest declines in ANC predicted for lakes in the Upper Peninsula of Michigan, -20 µeq/L for Andrus Lake, and -37 µeq/L for McNearney Lake, assuming $F = 1.0$ in groundwater and a groundwater concentration factor of 1.25. In northern Wisconsin, simulated historical acidification ranged from near zero (Denton Lake) to slight (Otto Mielke Lake, -8 µeq/L; Brown Lake, -4 µeq/L; Camp 12 Lake, -4 µeq/L). Agreement between IAG hindcast and diatom-inferred pH change was reasonable for five of the lakes. But for the lakes with the largest simulated declines in pH (Andrus -0.96 and Otto Mielke -0.68 pH units), the model estimates of historical acidification were substantially larger than the diatom inferences [-0.4 and +0.4 pH units, respectively; Pollman and Sweets, (1990)].

4.5 West

Current water quality conditions in the eastern U.S. are much the same as they were in 1990, and the scientific understanding of that chemistry is also similar. Substantial progress in understanding current chemical conditions has been made in the West, however. An overview is provided here of what is known about acid-sensitivity and the extent and magnitude of surface water acidification in the West. Greater detail is given in the case study of Class I areas in the West (Chapter 11).

In the far West, the areas that are sensitive to adverse effects from acidic deposition form two nearly continuous ranges, the Sierra Nevada in California and the Cascades starting in California and extending through Washington (Figure 4.6). The Rocky Mountains, in contrast, are discontinuous ranges with highly variable geological composition. Consideration of the Rocky Mountains as a single range for the purpose of evaluating sensitivity to acidic deposition is merely a convenience. Precise assessments of the sensitivity of Rocky Mountain resources need to be specific to individual ranges (Turk and Spahr, 1991; Sullivan and Eilers, 1994). Portions of the mountainous West are similar to the highly sensitive areas in Norway where watersheds contain large areas of exposed bedrock, with little soil or vegetative cover to neutralize acidic inputs. This is particularly true of alpine regions of the Sierra Nevada, northern Washington Cascades, the Idaho batholith, and portions of the Rocky Mountains in Wyoming and Colorado. However, the percentage of exposed bedrock in a watershed can yield misleading information on the sensitivity of the resources if the bedrock contains even small deposits of calcareous minerals or if physical weathering such as that cased by glaciers causes a high production of base cations within the watershed (Drever and Hurcomb, 1986). Consequently, even though much of the northern Washington Cascades is alpine, many surface waters draining glaciers in this area are moderately alkaline.

Soils in the watersheds of the West, where present, vary considerably from highly alkaline Mollisols to acidic Spodosols. In the volcanic regions of the West, the soils are typically highly siliceous and generate little alkalinity. In alpine areas, soils are absent or poorly developed.

Emissions of SO_2 and NO_x in many areas of the West peaked in the 1970s and decreased through the 1980s (Placet et al., 1990). Sources of SO_2 in the West are more diverse than in the East where the major source is the utility sector. Industrial and other categories constitute major sources of SO_2 in the West, in addition to utilities. Emissions from mobile sources constitute the largest contribution of atmospheric NO_x in the West, accounting for about one-half of the total anthropogenic emissions. Natural sources of NO_x are also regionally important, and represent 17% of the total N emissions (Sullivan and Eilers, 1994).

Atmospheric deposition of both S and N is currently low throughout most portions of the West. Annual wet deposition levels of S and N are generally less than about one-fourth of the levels observed in the high-deposition portions of the northeastern U.S. (Sisterson et al., 1990).

The current chemistry of surface waters in the West is based mostly on synoptic data from the Western Lake Survey (Landers et al., 1987) and a small number of more localized studies. Comprehensive assessments of lake chemistry in the West (cf. Baker et al., 1990a) indicate that there are many low-ANC systems, but few acidic waters. The acidic waters that have been sampled in

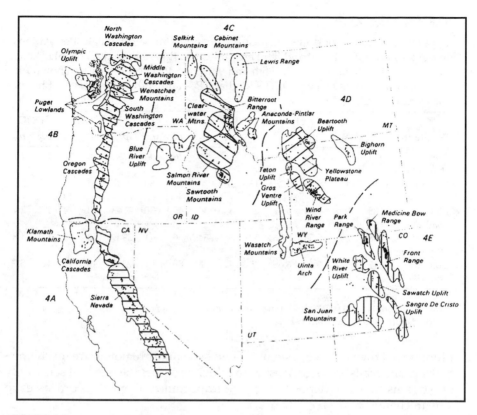

FIGURE 4.6
Major geomorphic units and locations of lakes sampled in the Western Lake Survey (Landers et al., 1987). Those areas known to contain sensitive lake resources are shaded with cross-hatching.

the West (e.g., Fern Lake, WY, Eilers et al., 1987, and West Twin Lake, OR, Eilers and Bernert, 1990) are acidic from natural production of S from watershed sources. It should be noted, however, that the database available to evaluate the occurrence of acidic lakes in the West is fairly limited.

The surface water chemistry in areas of interest delineated for the West shows that the Sierra Nevada and Cascade Mountains constitute the mountain ranges with the greatest number of sensitive resources (Tables 4.4 and 4.5). Sensitive lakes in the West are dilute bicarbonate systems and, unlike many low-ANC lakes in the East, have very low concentrations of DOC. Acid anion concentrations in most western lakes are extremely low in fall samples, but limited analyses of lake chemistry in spring generally show higher concentrations of NO_3^- and SO_4^{2-} (Williams and Melack, 1991a; Melack et al., 1989). The extremely dilute nature of many western lakes raises concerns regarding potential increases in acid anions, derived from acidic deposition, during spring snowmelt. The available data from intensive study sites in the West (e.g., Loch Vale, CO, Emerald Lake Basin, CA, and the Glacier Lakes Watershed, WY) suggest that episodic depression of stream pH is more

TABLE 4.4

Population Estimates of Water Chemistry Percentiles for Selected Lake Populations in the Western U.S. The 1st and 5th Percentiles (P_1, P_5) are Presented for pH, ANC (μeq/L), and C_B (μeq/L) and the 95th and 99th (P_{95}, P_{99}) Percentiles are Shown for SO_4^{2-} (μeq/L) and NO_3^- (μeq/L). The Median (P_{50}) and 90th Percentiles are Shown for DOC (mg/L).

Population	n	$\hat{N}$	pH		ANC		C_B		SO_4^{2-}		NO_3^-		DOC	
			P_1	P_5	P_1	P_5	P_1	P_5	P_{95}	P_{99}	P_{95}	P_{99}	P_{50}	P_{90}
Sierra Nevada	114	2119	5.84	6.31	15	16	21	26	90	386	8	10	0.8	2.7
Cascades	146	1473	5.95	6.25	11	18	20	31	60	97	3	6	1.3	2.6
Idaho Batholith	88	937	6.34	6.42	21	33	30	45	30	43	3	4	1.2	2.4
NW Wyoming[b]	38	648	6.56	6.56	38	38	64	66	41	2909	13	32	1.0	4.8
Colorado Rockies	121	1173	6.02	6.65	25	42	58	80	915	2212	10	13	1.3	5.7

[a] Data from Landers et al. (1987).
[b] Excluding Fern Lake (4D3-017) which is naturally acidic.
Source: Sullivan and Eilers, 1994.

pronounced than for lake systems, yet no systematic regional stream chemistry data are available in the West with which to assess the regional sensitivity of streams to acidic deposition or the importance of episodic processes to stream chemistry.

The Sierra Nevada is particularly sensitive to potential acidic deposition effects because of the predominance of granitic bedrock, thin acidic soils, large amounts of precipitation, coniferous vegetation, and extremely dilute waters (McColl, 1981; Melack et al., 1985; Melack and Stoddard, 1991). Similarly, Cascade and Rocky Mountain lakes are highly sensitive to potential acidic deposition effects (Nelson, 1991; Turk and Spahr, 1991). It appears that chronic acidification has not occurred to any significant degree. It is possible, however, that episodic effects have occurred under current deposition regimes. Unfortunately, the data that would be needed for such a determination have not been collected to a sufficient degree in acid-sensitive areas of the West to permit any regional assessment of episodic acidification.

Concentrations of SO_4^{2-} in western lakes are generally low, but in some cases watershed sources contribute substantial amounts of S to lake waters (Table 4.4). Turk and Spahr (1991) presented a conceptual model for expected SO_4^{2-} distributions in lake waters in the West that can be used as an aid in identifying the proportion of watersheds with significant watershed sources of S. Considering that atmospheric sources can account for generally less than 30 μeq/L of SO_4 in the West (including the Colorado Rockies), it is clear that many lakes, particularly in Colorado, receive major, albeit variable, sources of watershed S (Figure 4.7).

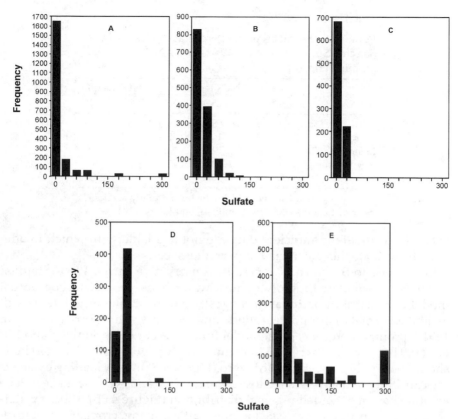

FIGURE 4.7
Sulfate (μeq/L) distribution in lakes in (A) Sierra Nevada, (B) Cascade Mountains, (C) Idaho Batholith, (D) NW Wyoming, and (E) Colorado Rockies. (Data from the Western Lake Survey.)

Nitrate concentrations are virtually undetectable in most western lakes in the fall (Landers et al., 1987). However, in some cases, fall NO_3^- concentrations were surprisingly high (Table 4.4). For example, nearly one-fourth of the lakes in NW Wyoming had NO_3^- greater than 5 μeq/L and almost 9% had NO_3^- greater than 10 μeq/L (Sullivan and Eilers, 1994; Table 4.5). The Sierra Nevada and Colorado Rockies subregions also exhibited many lakes with higher NO_3^- concentrations than would be expected for fall samples. In both areas, about 10% of the lakes had NO_3 concentrations above 5 μeq/L (Table 4.5). Based on existing data, it appears likely that many high-elevation lakes in the West are currently experiencing N deposition sufficiently high to cause chronic NO_3^- leaching, and likely associated chronic and episodic acidification, albeit small in magnitude. Many lakes in Colorado and Wyoming exhibit fall concentrations of NO_3^- in the range of 10 to 30 μeq/L, and likely have substantially higher NO_3^- concentrations during spring. The weight of evidence suggests that episodic acidification associated with N deposition may be occurring to a significant degree in many high-elevation western

Aquatic Effects of Acidic Deposition

TABLE 4.5

Population Estimates of the Percentage of Lakes in
Selected Subregions of the West with ANC and NO_3^-
within Defined Ranges

	ANC (μeq/L)			NO_3^- (μeq/L)	
	<0	<25	<50	>5	>10
Sierra Nevada	0	8.7	39.3	10.6	1.5
Cascades	0	10.2	22.4	1.5	0
Idaho Batholith	0	2.0	23.6	4.6	3.9
NW Wyoming[a]	0	2.3	12.8	22.8	8.9
Colorado Rockies	0	0.9	5.5	9.8	1.8

[a] Excluding Fern Lake (4D3-017) which is naturally acidic.
Source: Landers et al., 1987; Sullivan and Eilers, 1994.

lakes. Unfortunately, sufficient data are not available with which to adequately evaluate this potentially important issue.

In contrast to S, which is generally conservative in the West (Stauffer, 1990), N is assimilated rapidly in most watersheds. Whereas a conceptual model of S distribution in lakes suggests that most lake populations will exhibit a normal distribution around some positive value (Turk and Spahr, 1991), a companion conceptual model for N presented by Sullivan and Eilers (1994) suggests that NO_3^- distributions for undisturbed watersheds should be highly skewed towards zero (Figure 4.8a). As N loading exhausts the capability of the watershed to assimilate NO_3^- and NH_4^+, leakage will be exhibited as an extended regional distribution (Figure 4.8b). Where watershed disturbance is severe (e.g., logging, cattle grazing, cropland, urbanization), NO_3^- concentrations in drainage waters can be substantial. Distributions of NO_3^- concentrations in the western lakes are relatively low (Figure 4.9), but even low to moderate concentrations of NO_3^- may be significant in view of:

1. The low base cation concentrations in many lakes.
2. Potential for continuing N deposition to eventually exhaust natural assimilative capabilities.
3. The fact that these distributions are based on a fall survey.

Limited data collected during snowmelt indicate spring concentrations several times higher than data collected in the fall (e.g., Reuss et al., 1995).

Although it is evident from Tables 4.4 and 4.5 that lakes in the Sierra Nevada, Cascades, and Idaho are generally lower in ANC than those in Northwest Wyoming and Colorado, it also appears that lakes in the central and southern Rockies have considerably higher concentrations of acid anions. Turk et al. (1992) noted that concentrations of SO_4^{2-} and NO_3^- in the snowpack of northern Colorado are several times greater than in snow collected in southern Colorado. Nitrate concentrations in the Wyoming and

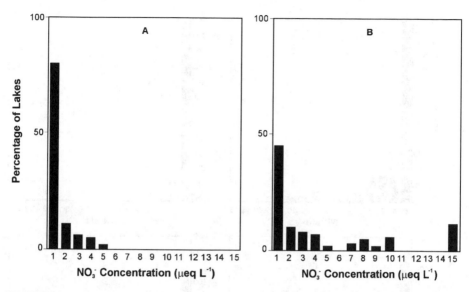

FIGURE 4.8
Conceptual model of regional lake-water nitrate distributions in (A) undisturbed lake populations where the percentage of lakes (Y-axis) with measurable NO_3^- concentrations (X-axis) is
low. As N deposition exceeds the capability for terrestrial and aquatic systems to assimilate
inorganic N, concentrations of NO_3^- in lake populations are expected to show greater dispersion,
as shown in (B). (From Sullivan and Eilers, 1994.)

northern Colorado lakes are also high relative to other sites in the West, and
this is noteworthy for two reasons. First, these population estimates are
derived from fall samples when inorganic N concentrations are expected to
be at minimum levels. Nitrate concentrations in these lakes are expected to
be much higher during spring. Second, there may be an important bias in the
WLS data from Wyoming and Colorado that has caused regional lake-water
NO_3^- concentrations to be underestimated (Sullivan and Eilers, 1994). During
the survey, 49 lakes in the statistical frame were not sampled because the
lakes were frozen when the field crews visited them (Landers et al., 1987).
Most of the frozen lakes were from Wyoming ($n = 21$) and Colorado ($n = 20$).
Because the frozen lakes were generally higher elevation sites, it is likely that
the actual population values for northeast Wyoming and Colorado are higher
in NO_3^- and lower in ANC and base cation concentration than previously
reported. Although the magnitude of this bias may not have been critical to
the primary objective of the WLS (to perform regional characterization of
lake populations), it may be important in view of increasing evidence of elevated N deposition in the Rockies (e.g., Turk et al., 1992) and elevated NO_3^-
concentrations in many high-elevation lakes. It would be very useful to
revisit these unsampled lakes and characterize their chemistry. On the basis
of this characterization, population estimates for Colorado and Wyoming
could be recalculated.

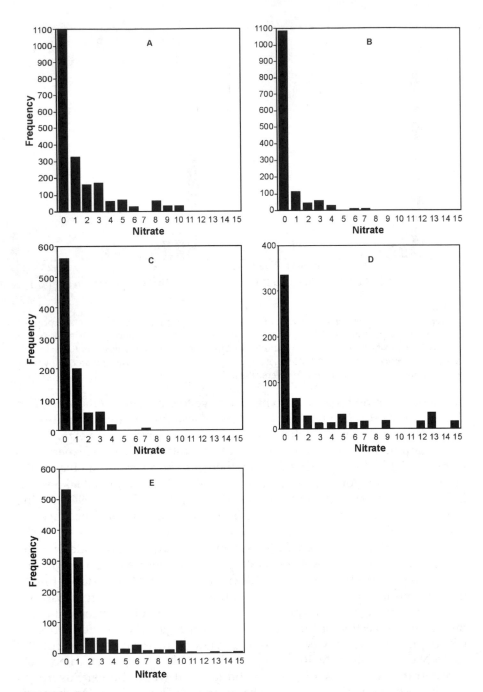

FIGURE 4.9
Nitrate (μeq/L) distribution in lakes in (A) Sierra Nevada, (B) Cascade Mountains, (C) Idaho Batholith, (D) NW Wyoming, and (E) Colorado Rockies. (Data from the Western Lake Survey.)

Musselman et al. (1996) conducted a synoptic survey of surface water chemistry in the mountainous areas east of the Continental Divide throughout the length of Colorado and in southeastern Wyoming that are exposed to increased atmospheric emissions of N and S. High-elevation lakes in catchments with a high percentage of exposed bedrock or glaciated landscape were selected for sampling. A total of 267 lakes were sampled. More than 10% of the lakes had ANC less than or equal to 50 µeq/L. None were acidic (ANC less than 0), although several had ANC less than 10 µeq/L. The lowest pH was 5.4 at the GLEES research site in Medicine Bow National Forest in southeastern Wyoming. Most lakes had pH greater than 6.0. Many of the lakes had high NO_3^- concentrations, especially those sampled during the first half of the field study.

4.5.1 Monitoring Studies

Results of monitoring studies conducted in the western U.S. were summarized in several chapters of the acidic deposition regional case studies book, edited by Charles (1991). Few monitoring data sets were available in any of the western regions. Overall, historical data are mostly limited to the recent past and a number of reconnaissance studies (Turk and Spahr, 1991; Melack and Stoddard, 1991; Nelson, 1991).

Turk et al. (1993) reported the results of 5 years of monitoring for 10 lakes in the Mt. Zirkel and Weminuche Wilderness areas in Colorado. The short period of record and seasonal variability precluded documentation of significant changes in any primary variables. Lake-water SO_4^{2-} concentrations showed the most consistent pattern of change, primarily towards increasing SO_4^{2-} concentrations. Turk et al. (1993) speculated that such increases in SO_4^{2-} concentration may be related to increased atmospheric deposition of S, changes in S release by weathering of sulfide minerals, or changing dilution of SO_4^{2-} by snowmelt. Based on lake concentrations of SO_4^{2-} and Cl^- and on wet deposition concentrations of SO_4^{2-}, NO_3^-, and H^+, Turk and Spahr (1991) concluded that low-ANC lakes have lost no more than 5 µeq/L ANC in the Bitterroot Range of the Northern Rocky Mountains, 12 µeq/L ANC in the Wind River Range of Wyoming, and 10 µeq/L ANC in the Front Range of Colorado. It is likely that the actual ANC losses have been much less than these.

Summit Lake located at 1650 m in the Western Washington Cascades was originally sampled by Landers et al. (1987) and in 1993 by the Mt. Baker-Snoqualmie National Forest (J.M. Eilers, unpublished data). The current lake ANC is 1 µeq/L with a nonmarine SO_4^{2-} concentration of 7 µeq/L. Because the nonmarine base cation concentrations are only 10 µeq/L and can largely be attributed to atmospheric deposition, there appears to be little opportunity for watershed contribution of S. Summit Lake is close to local and regional sources of atmospheric S emissions and it is conceivable that the lake has lost up to 7 µeq/L of ANC in response to atmospheric deposition (Sullivan and Eilers, 1994).

Melack et al. (1998) reported volume-weighted average concentrations of major ions in the Emerald Lake outflow, Sequoia National Park, from 1990 through 1994. The Emerald Lake chemistry data set covered an uninterrupted period from 1983 to 1994. No trend was found in either pH or ANC during this time period, although NO_3^- concentrations declined after about 1990. The cause of the decline in NO_3^- in Emerald Lake was not clear. Peak NO_3^- values in the lake outlet during 1983 through 1987 were above 10 μeq/L in nearly all years, but only 8 to 9 μeq/L during water years 1990 through 1994. Melack et al. (1998) hypothesized that climatic changes, for example the 1987–1992 drought, may have altered biological uptake rates and, therefore, runoff concentrations of NO_3^-.

4.5.2 Paleolimnological Studies

Diatom-inferred pH and ANC were calculated at 24 depth intervals at Emerald Lake, CA, from 1825 to the present (Holmes et al., 1989). Emerald Lake is a very dilute, high-elevation lake in Sequoia National Park. Significant trends were not found for either pH or ANC, and the authors concluded that Emerald Lake has not been affected by acidic deposition. Whiting et al. (1989) completed paleolimnological analyses of three additional lakes in the Sierra Nevada. Eastern Book Lake (pH = 7.06) showed evidence of both long-term alkalization (approximately 0.3 pH units over the past 200 years) and pH fluctuations since 1970. Lake 45 (pH = 5.16) may have acidified slightly (approximately 0.2 pH units) over the last 60 years. Lake Harriet (pH = 6.52) showed no significant change.

Baron et al. (1986) investigated metal stratigraphy, diatom stratigraphy, and inferred pH profiles of four subalpine lakes in Rocky Mountain National Park, CO. The authors of this study also found no evidence of historical influence on pH attributable to atmospheric deposition. Eilers and Dixit (1992) reported a slight dilution or acidification trend in Lake Notasha, OR (ANC equal to 10 μeq/L), based on recent changes in diatom stratigraphy. However, there was no diatom calibration set available to allow quantitative estimates of change in lake pH. A survey of lake-water chemistry and diatom frustules in the surface sediment of Cascade Mountain lakes in Oregon and Washington has recently been completed (Eilers et al., 1998a). A calibration equation for diatom-inferred pH was developed from this database that included about 50 lakes and will be available for future studies. The limited paleolimnological data available for lakes in the western U.S. suggest that widespread chronic acidification probably has not occurred. Some lakes may have experienced recent pH declines, but the magnitude of such changes has likely been small.

4.5.3   Model Simulations

The USDA Forest Service and National Park Service, as part of their programs to evaluate air quality-related values in wilderness areas and national parks, have initiated several projects to assess the sensitivity of wilderness lakes to acidic deposition. The sensitivity of selected lakes in Idaho and Montana was assessed by modeling lake response to simulated increases in S deposition (Eilers et al., 1991). Selected were 15 lakes with alkalinity values ranging from 21 to 835 μeq/L for study from 3 wilderness areas, the Selway-Bitterroot (SBW), Absaroka-Beartooth (ABW), and the Bob Marshall (BMW). The MAGIC model was calibrated to each watershed using area-wide estimates of deposition and soil characteristics. Current wet deposition of S is approximately 2 kg S/ha per year.

The projected chemistry of the study lakes using MAGIC suggested that the lowest alkalinity lakes would remain nonacidic (ANC greater than 0) if the S deposition load were tripled over present values, for example, to about 6 kg S/ha per year. At a 5-fold increase in S deposition (10 kg S/ha per year), 2 of the 6 lakes modeled in the SBW were projected to become chronically acidic. None of the lakes in the ABW or BMW were projected to become acidic at this 5-fold increase in loading, although 1 lake in the ABW exhibited a projected ANC of 4 μeq/L and a pH of 5.65. The lakes in the SBW are largely subalpine systems and generally contain moderate soil cover. Nevertheless, the relatively unreactive granitic parent material, coupled with high rates of precipitation, make these lakes among the most likely in the Northern Rocky Mountains to become acidic under increased deposition loads. Although the watersheds in the ABW had a higher percentage of exposed bedrock than watersheds in the SBW and BMW and showed the largest decreases in modeled alkalinity, the weathering of rocks in the ABW is apparently sufficient to maintain positive alkalinities under moderate to heavy acidic deposition loads. The lakes modeled in the BMW are located on limestone and argilitic sedimentary bedrock. These watersheds have sufficient base cation production to ensure that the lakes will not acidify under any realistic increase in acidic deposition.

The MAGIC model has also been applied recently to estimate the critical load of S and N deposition required for the protection of acid-sensitive watersheds in Colorado and Wyoming (Sullivan et al., 1998). As part of ongoing modeling studies, watersheds in the Cascade Mountains and Sierra Nevada are also being modeled. Results of the completed modeling studies are discussed in Chapter 11.

MAGIC modeling results provide estimates of long-term lake response to sustained increases in S deposition. It is likely, however, that low-ANC lakes in the western U.S. will acidify under episodic conditions before chronic acidification is evident. Estimates of both episodic and chronic acidification potential could be refined by collecting more watershed-specific information on low-ANC lakes. More detailed information is needed on soil properties, N cycling, hydrologic flowpaths, weathering characteristics of bedrock, deposition, and surface water chemistry during snowmelt.

5

Chemical Dose–Response Relationships and Critical Loads

5.1 Quantification of Chemical Dose–Response Relationships

There has been a growing international recognition that air pollution effects, particularly from S and N, may in some cases necessitate emission controls to reduce or limit future increases in atmospheric deposition. Measures to reduce emissions must rely on known or estimated dose–response relationships that reflect the tolerance of natural ecosystems to various inputs of atmospheric pollutants. This need has stimulated interest in evaluating the efficacy of establishing one or more standards for acid deposition. The Clean Air Act Amendments of 1990 (CAAA) also included requirements to assess the effectiveness of the mandated emissions controls via periodic assessments, and to submit an EPA report on the feasibility of adopting one or more acid deposition standards to Congress.

Diverse data are available from a variety of sources with which to quantify the watershed acidification response, as well as recovery from acidification. Such data shed light on the sensitivity of various kinds of watershed systems to changes in acidic deposition. Intercomparisons among the various studies that have been conducted are complicated by different relative watershed sensitivities, S deposition loading rates (and changes in those rates), the relative importance of N leaching and N saturation, temporal considerations, and natural (especially climatic) variability. In addition, these quantitative data have been generated in vastly different ways, including monitoring, space-for-time substitution, whole-watershed or whole-lake acidification, whole-watershed acid exclusion, paleolimnology, and modeling. The only way in which different approaches can be compared on a quantitative basis is by normalizing surface water response as a fraction of the change in SO_4^{2-} concentration (or $SO_4^{2-} + NO_3^-$ concentration where NO_3^- is also important). The principal ions that change in direct response to changes in $(SO_4^{2-} + NO_3)$ concentration are ANC (which can be expressed as $[HCO_3^- - H^+]$), base cations (C_B), inorganic aluminum (Al_i), and organic acid anions (A^-). The proportional changes in (HCO_3^- - H^+), Al_i, C_B, and A^- concentrations should sum to 1.0 in order to satisfy the electroneutrality condition. For aquatic systems that are relatively insensitive to acidic deposition, ΔC_B approximates $\Delta(SO_4^{2-} + NO_3^-)$, and the F factor (Henriksen, 1982) approximately equals 1.0:

$$F = \frac{\Delta[C_B]}{\Delta[SO_4^{2-} + NO_3^-]} \approx 1.0 \qquad (5.1)$$

where brackets indicate concentration in μeq/L and changes in other constituents are insignificant. Where acidification occurs in response to acidic deposition, changes in (HCO_3^- - H^+) and/or Al_i comprise an appreciable percentage of the overall surface water response and, therefore, the F factor is less than 1.0. The F factor is important in evaluating criteria for establishing acid deposition standards because it provides the quantitative linkage between inputs of acid anions (e.g., SO_4^{2-}, NO_3^-) and effects on surface water chemistry. An important limitation of the F factor concept, however, is that the value of F is likely to change as the base cation pools in watershed soils become depleted by acid deposition inputs.

Quantitative dose–response relationships for S have been determined, using a variety of approaches, in a number of regions in North America and Europe. Such studies have included, for example, measured changes in water chemistry during periods when S deposition changed appreciably, regional paleolimnological (e.g., diatom-inferred change in pH and ANC) investigations, whole-catchment manipulation studies, and intensive process modeling. Each type of study has provided quantitative estimates of dose–response that entail different sets of assumptions and limitations. Taken together, they provide a good indication of the range of quantitative acidification response. As a result of these recent studies, we are much better able to quantify acidification and recovery relationships than we were in 1990.

5.1.1   Measured Changes in Acid–Base Chemistry

Measured changes in surface water chemistry in areas that have experienced short-term (less than 20 years) changes in chemical constituents in response to changes in mineral acid inputs are available from a number of sources. Proportional changes in ANC, base cations, and Al_i relative to changes in SO_4^{2-} or (SO_4^{2-} + NO_3^-) concentrations were summarized by Sullivan and Eilers (1994) for lakes and streams in which such changes had been measured. They included lakes in the Sudbury region of Ontario, the Galloway lakes area of Scotland, a stream site at Hubbard Brook, NH, and catchment manipulation experiments in the RAIN project in Norway and Little Rock Lake in Wisconsin. Most of the observed changes were coincident with decreased acidic deposition, and it is unclear to what extent acidification and recovery are symmetrical. F-factors in the range of 0.5 to 0.9 are apparently typical for lakes having low base cation concentrations, although lower values (0.35 to 0.39) were

TABLE 5.1

Measured Short-Term Changes in Surface Water Chemistry Associated with Changes in Mineral Acid Anion Concentrations. (Units in μeq/L.) Proportions are Expressed as Absolute Values.

Site	Location	Period	Type	Initial pH	ΔSO_4^{2-}	$\dfrac{\Delta(HCO_3^--H^+)}{\Delta SO_4^{2-}}$	$\dfrac{\Delta C_B^*}{\Delta SO_4^{2-}}$	$\dfrac{\Delta Al}{\Delta SO_4^{2-}}$	Ref.[d]
Clearwater Lake	Sudbury, Canada	1973–1977 to 1984	Recovery	4.2	-175	0.19	0.66[b]	0.15	1
Swan Lake	Sudbury, Canada	1977 to 1982	Recovery	4.0	-360	0.26	0.67	0.07	1
Baby Lake	Sudbury, Canada	1968–1972 to 1983	Recovery	4.05	-750	0.12	—	—	2
Whitepine Lake	Sudbury, Canada	1980–1988	Recovery	5.4	-42	0.24	—	0.05	6
Laundrie Lake	Sudbury, Canada	1974–1976 to 1979–1983	Recovery	4.7	-58	0.24	—	—	3
Florence Lake	Sudbury, Canada	1974–1976 to 1979–1983	Recovery	4.6	-42	0.22	—	—	3
Average of 37 lakes having pH< 5.5	Sudbury, Canada	1974–1976 to 1979–1983	Recovery	4.7	-42	0.15	—	—	3
Average of 105 trout lakes	Sudbury, Canada	1980–1987	Recovery	—	-45	0.51	—	—	6, 10
Average of 50 lakes	Galloway, Scotland	1979–1988	Recovery	5.4 ± 0.71	-76[a]	0.13	0.84	~0.06	4
Little Rock Lake[e]	Wisconsin	1983–1989	Acid addition	6.6	94	0.44	0.53	—	9
SOG2 catchment	Sogndal, Norway	1984–1987	Acid addition	5.5	28[a]	0.46	0.39	0.11	5
SOG4 catchment	Sogndal, Norway	1984–1987	Acid addition	6.0	20[a]	0.35	0.35	0.15	5
KIM catchment	Risdalsheia, Norway	1984–1987	Acid exclusion	4.1	-139[a,c]	0.09	0.55	0.05	5, 7
Bear Brook	Maine	1987–1992	Acid addition	5.6	62[a]	0.14	0.51	0.20	11
Hubbard Brook	New Hampshire	1969–1979	Recovery	4.8	-30[a]	0.15	0.91	—	8

[a] Also includes NO_3^-.

[b] $\Delta C_B/\Delta SO_4^{2-}$ calculated by difference, assuming that the proportional changes in alkalinity, C_B, and Al sum to 1.0.

[c] Changes in the organic anion contribution to acidity were important at this site where DOC was very high (~ 1250μM).

[d] 1—Dillon et al., 1986; 2—Hutchinson and Havas, 1986; 3—Keller et al., 1986; 4—Wright, 1988b; 5—Wright et al., 1988b; 6—Gunn and Keller, 1990; 7—Wright, 1989; 8—Sullivan, 1990; 9—Sampson et al., 1994; 10—Gunn, personal communication; 11—Norton et al., 1993.

[e] Little Rock Lake experiment involved manipulation of lake only.

TABLE 5.2

Inferred Long-Term Regional Changes in Surface Water Chemistry Associated with Estimated Changes in Mineral Acid Anion Concentrations, Using the Technique of Space-for-Time Substitution

Region	$\dfrac{\Delta ANC}{\Delta SO_4^{2-}}$	$\dfrac{\Delta C_B}{\Delta SO_4^{2-}}$	$\dfrac{\Delta Al}{\Delta SO_4^{2-}}$	Reference	Comments
NE U.S.	0.13	0.54	0.07	Sullivan et al., 1990a	Analysis restricted to lakes having current ANC ≤ 25 µeq/L
S. Norway	0.22	—	—	Brown and Sadler, 1981	Regional data set ($n = 471$)
S. Norway	—	0.82	—	Wright, 1988; Sullivan, 1990	Lakes located across depositional gradient from Bykle to Mandal

observed for the highly sensitive catchments at Sogndal, Norway that are characterized by thin soils and much exposed bedrock, as is common in many areas of southern Norway and the western U.S. The proportional change in ANC relative to the change in ($SO_4^{2-} + NO_3^-$) was variable, within the range of 0.1 to 0.5 (Table 5.1). The proportional change in Al was smaller, ranging up to 0.15. These measured values of acidification and deacidification change in ANC and Al are somewhat smaller than previously anticipated.

Relatively early in the international efforts to quantify the acidification response, Henriksen (1982) proposed that *F* factors for softwater lakes would be in the range 0 to 0.4. More recent research (e.g., Table 5.1) has shown this earlier estimate to be too low in most cases. Based on measured values, only the most sensitive systems, for example at Sogndal, exhibit *F* factors below 0.4.

TABLE 5.3

Diatom-Inferred Long-Term Changes in Lake-water ANC as a Fraction of Estimated Historic Changes in Lake-water SO_4^{2-} Concentration

Region	Number of Lakes	$\dfrac{\Delta ANC}{\Delta SO_4^{2-}}$	References	Comments
Adirondacks, NY	48	0.11	Sullivan et al., 1990a	Statistical sampling
Adirondacks, NY	25	0.18	Sullivan et al., 1990a	Acidic lakes only[a]
Northern New England	12	0.30	Davis et al., 1994	Lakes were selected that were presumed to be acid-sensitive
Florida (Lakes Barco, Suggs)	2	0.27	Sullivan and Eilers, 1994	Seepage lakes

[a] The set of 25 acidic lakes was part of the regional data set of 48 lakes presumed to be acid-sensitive.

TABLE 5.4

Dynamic Model (MAGIC) Estimates of *F*-Factors for Hindcast or Future Forecast
Projections of Acidification or Recovery Responses

Region	Type of Simulation	Number of Lakes or Streams	Median	5th Percentile	Reference
			F-Factor		
Adirondacks	Hindcast	33	0.56	0.25	Sullivan et al., 1996a
Adirondacks	50-year forecast, 50% reduction in S deposition	33	0.73	0.39	Sullivan, unpublished
Wilderness lakes, Western U.S.	Forecasted 3-fold increase in S deposition	15	0.34	0.03	Eilers et al., 1991
Bear Brook, ME	Response to experimental watershed acidification	1	0.85	—	Norton et al., 1992

In addition to the measured acidification and recovery data presented in
Table 5.1, there are several other sources of quantitative or semiquantitative
data with which to evaluate the general applicability of the measured results
that are available. These include the results of space-for-time substitution
(Table 5.2), diatom-inferences of historical acidification (Table 5.3), and results
of process-based model hindcasts or future forecasts (Table 5.4). Each of these
methods has its own assumptions and limitations, and none are as robust as
results of actual field measurements of response. Major advantages of these
alternative sources of quantitative data, however, are that they primarily reflect
acidification, rather than recovery, scenarios, and that they sometimes include
longer periods of response than do the available direct measurements.

5.1.2   Space-for-Time Substitution

Results of space-for-time substitution must be interpreted with caution. This
approach is based on the assumption that changes in chemistry across space,
for example, from low to high levels of acidic deposition, reflect changes that
occurred over time as deposition increased from low to high. It is implicitly
assumed that the waters included in the analysis were initially homogeneous
in their chemistry, and also that potentially important factors other than deposition (e.g., soil characteristics, land use impacts) do not co-vary with deposition. Results should therefore be considered only semiquantitative.
Nevertheless, available data using this method (Table 5.2) appear similar to
results of measured values shown in Table 5.1.

The spatial distributions of lake-water chemical variables across a longitudinal gradient in the upper Midwest for low-ANC groundwater recharge

seepage lakes (Figure 4.5) provides a good example of the use of space-for-time substitution to evaluate acidification dose–response relationships. These distributions also constitute perhaps the best evidence available that many of the most sensitive lakes in the eastern portion of this region have acidified. In the absence of additional paleolimnological data for these systems of most interest, however, it is difficult to substantiate in terms of magnitude much regional acidification in the upper Midwest.

Nitrogen deposition does not appear to be an important issue for sensitive aquatic resources in the upper Midwest. This is likely attributable to the fact that snowmelt is less important to the acid–base chemistry of sensitive (i.e., seepage) lakes in this region, and hydrologic retention times are long. Sulfur deposition appears to be of greater importance, and potential chronic effects are of greater interest than episodic effects because of the nature of the hydrology of sensitive resources in the region. Based largely on the results of space-for-time substitution analyses, Sullivan and Eilers (1994) concluded that current deposition in the eastern portion of the region (approximately 5 kg S/ha per year) is a reasonable approximation of the deposition level required to protect the most sensitive aquatic receptors. Resources in the western portion of the region are less sensitive, however, and an appropriate standard for S deposition would be much higher. Because S deposition has been decreasing in recent years, it does not appear that acidic deposition is an important environmental concern in the upper Midwest at this time.

An S deposition standard has been in effect in Minnesota since 1986. The Minnesota standard was based on the Acid Deposition Control Act, passed by the state legislature in 1982, which required the Minnesota Pollution Control Agency (MPCA) to identify natural resources within the state that were threatened by acid deposition and to develop both an acid deposition standard and an emissions control plan. Small, poorly-buffered lakes in northcentral and northeastern Minnesota were identified as the resources at greatest risk. Based on model simulations, MPCA selected a threshold pH for precipitation of 4.7, below which damage to aquatic biota was thought to occur with prolonged exposure. This threshold pH was correlated with SO_4^{2-} deposition data, and a standard was determined that allowed no more than 11 kg/ha of wet SO_4^{2-} to be deposited during any 52-week period (3.7 kg S/ha per year) (MPCA, 1985). This standard is fairly stringent. In fact, 6 of 12 monitoring sites in Minnesota exceeded the standard in 1992 (Orr, 1993). There appears to be a limited scientific basis for such a standard for protection of aquatic resources in Minnesota.

5.1.3 Paleolimnological Inferences of Dose–Response

Diatom-inferences of change in ANC from pre-industrial times to the present have been reported for a regional population of Adirondack lakes (Sullivan et al., 1990a), and for two lakes in Florida that have shown clear acidification in recent decades (Sweets, 1992). Proportional changes in diatom-inferred

ANC as a fraction of assumed increases in SO_4^{2-} concentration since pre-industrial times show estimates ranging from 0.1 to 0.3 (Table 5.3), in close agreement with measured values (Table 5.1).

Diatom estimates of pH have been compared with measured pH values at numerous lake sites where changes in acid–base status have occurred. Such validations of the diatom approach have been performed for lakes that have been acidified and lakes that have recovered from acidification or have been limed in Canada (e.g., Dixit et al., 1987, 1991, 1992), Sweden (e.g., Renberg and Hultberg, 1992), and Scotland (e.g., Allot et al., 1992). Diatom-inferred pH histories generally agree reasonably well with the timing, trend, and magnitude of known acidification and deacidification periods. In several cases, however, the sedimentary reconstructions were slightly damped in comparison with measured values. That is, the diatom reconstructions did not fully reflect the magnitude of either the water pH decline or subsequent recovery.

For example, Renberg and Hultberg (1992) compared diatom-inferred pH reconstructions with the known pH history for several decades at Lake Lysevatten in southwestern Sweden. The diatom-inferred pH history agreed well with both the acidification period of the 1960s and early 1970s and also the liming that occurred in 1974. The magnitude of pH change inferred from sedimentary reconstructions was slightly smaller, however, than the measured changes in pH for both acidification and deacidification.

Allot et al. (1992) found diatom reconstructions of pH recovery in the deacidifying Round Loch of Glenhead, Scotland to be somewhat smaller than the measured pH recovery since the late 1970s. pH reconstructions from the sediment cores showed an average recovery of 0.05 pH units. Measured increases in pH between 1978–1979 and 1988–1989 averaged 0.23 pH units. The authors attributed this difference to attenuation of the reconstructed pH record owing to sediment mixing processes.

Dixit et al. (1992) analyzed sedimentary diatoms and chrysophytes from Baby Lake (Sudbury, Ontario) to assess trends in lake-water chemistry associated with the operation, and closure in 1972, of the Coniston Smelter. Extremely high S emissions caused the lake to acidify from pH approximately equal to 6.5 in 1940 to a low of 4.2 in 1975. Following closure of the smelter, lake-water pH recovered to pre-industrial levels. The diatom-inferred acidification and subsequent recovery of the lake corresponded with the pattern of measured values. However, the diatom-inferred pH response was more compressed and did not fully express the amplitude of the pH decline or the extent of subsequent recovery.

It is not known why diatom-inferences of pH change are often slightly attenuated relative to measured acidification or deacidification. Possible explanations include the preference of many diatom taxa for benthic habitats where pH changes may be buffered by chemical and biological processes. Alternatively, such an attenuation could be a result of sediment mixing processes.

Some upper Midwestern lakes have acidified since pre-industrial times. However, based on available paleolimnological data, there is little paleolimnological evidence suggesting that widespread acidification has occurred in this region (Kingston et al., 1990; Cook et al., 1990). Land use changes and other human disturbances of upper Midwestern lakes and their watersheds have probably exerted more influence on the acid–base chemistry of lakes than has acidic deposition (Eilers et al., 1989a; Kingston et al., 1990; Sullivan, 1990). This is because acidic deposition has occurred at a much lower level in the upper Midwest than in most areas of the eastern U.S. The portion of the region most likely to have experienced acidification from acidic deposition is the Upper Peninsula of Michigan, where acidic seepage lakes are particularly numerous (Baker et al., 1990b), acidic deposition is highest for the region, and the $[SO_4^{2-}]/[C_B]$ ratio is commonly greater than 1.0 (Figure 4.5). The percentage of acidic lakes in the eastern portion of the Upper Peninsula of Michigan (east of longitude 87°) is 18 to 19% (Schnoor et al., 1986; Eilers et al., 1988b), which is comparable to heavily impacted areas of the Northeast.

Diatom-inferred pH data are available for only two lakes in upper Michigan, McNearney and Andrus Lakes. McNearney Lake was naturally acidic prior to this century and is therefore atypical for the region. Andrus Lake is inferred to have experienced declines in pH and DOC since pre-industrial times that could be related to acidic deposition (Kingston et al., 1990). It is likely that other lakes in this subregion have also experienced recent acidification, although quantitative data are lacking regarding the amount of acidification that occurred in the past or the dose–response relationships of these systems. In addition to the scarcity of paleolimnological data within the portion of the upper Midwest most likely to have experienced widespread historical acidification, there is also a paucity of basic biogeochemical data on the response of the predominant lake type in this region to atmospheric inputs of S and N.

Historical changes in Florida lake-water chemistry, as inferred from diatoms, showed a distinct geographical pattern. All five of the paleolimnological study lakes in the Trail Ridge region showed some evidence of acidification, some strongly linked in timing to both the period of increasing acidic deposition and increased water consumption. Trail Ridge lakes showed diatom-inferred ΔpH ranging from -0.2 (McCloud) to -0.9 (Suggs). No clear evidence of acidification was observed for lakes in the Ocala National Forest (three lakes) or the Panhandle (eight lakes), except Lake Five-O, where gross hydrological change was implicated. It is most likely that several factors have caused the recent acidification of lakes in the Trail Ridge area suggested by the diatom data. Acidic deposition is implicated, but changing lake stage and the linked phenomenon of evapoconcentration may also be important (Sweets et al., 1990).

Diatom-inferred historical changes in pH for all lakes in the Florida Panhandle, except Lake Five-O, were less than -0.10 units. These results appear surprising insofar as the Panhandle seepage lakes are the most dilute lakes in Florida, and have been believed to receive minimal hydrologic in-seepage (ca. 1 to 3% of total hydrologic budget; cf. Baker et al., 1988b). Groundwater monitoring data collected adjacent to Lake Five-O suggested, however, that groundwater may contribute one-third to one-half of the overall hydrologic budget of this lake (Pollman et al., 1991). Calibrated inflows based on Cl⁻ balances for Panhandle lakes also suggested substantial groundwater inflows, ranging from 10 (Moore Lake) to 29% (Lofton Ponds) (Pollman and Sweets, 1990).

Superimposed on the complex heterogeneity of Florida lakes is a high incidence of anthropogenic disturbance. Of the 159 total lakes sampled by ELS-I in Florida, all but 37 were judged by Baker et al. (1988b) to have substantial shoreline or watershed disturbances, mostly related to agriculture. Besides the increased atmospheric deposition in Florida in the 1950s, other changes have also occurred. The human population has increased markedly, as has freshwater withdrawal from the Floridan aquifer (Aucott, 1988). As a result, the potentiometric head has declined substantially in the Trail Ridge area (Healy, 1975; Aucott, 1988; Pollman and Canfield, 1991). The effects of water withdrawal on the acid–base status of lakes is not well understood.

For undeveloped lakes in the northcentral peninsula, lake-water chemistry is consistent with an hypothesis of acidification by acidic deposition (Hendry and Brezonik, 1984; Eilers et al., 1988c; Baker et al., 1986, 1988b). Evaporative concentration of modest amounts of acidic deposition, and in-lake retention of SO_4^{2-} and NO_3^- appear to be important processes. However, Eilers et al. (1988c) concluded it is unlikely that the mechanisms of acidification of clearwater lakes in Florida and the linkages to atmospheric deposition will be satisfactorily understood until the hydrologic pathways are better known. Slight differences in groundwater inputs can have a major influence on base cation supply and lake-water chemistry in these precipitation-dominated seepage systems. Based on limited paleolimnological data, it appears that recent acidification of lakes in Florida may have been restricted to the Trail Ridge district. Furthermore, it is unclear to what extent recent acidification of lakes in the Trail Ridge district may be attributable to acidic deposition, as compared to other anthropogenic activities, especially groundwater withdrawal.

5.1.4   Model Estimates of Dose–Response

Dynamic model estimates of F-factors for watersheds in the northeastern U.S., using the MAGIC model, show reasonably close agreement with measured F-factors for acid-sensitive systems (Tables 5.1 and 5.4). Model-generated median values of the F-factor ranges from 0.56 to 0.85, and values of the 5th percentile of Adirondack lake projections (0.25 to 0.39) were reasonably comparable to the measured values at the highly sensitive Sogndal site (0.35 to 0.39). MAGIC forecasts for western lakes, however, yielded estimated F-factors that were substantially lower (median 0.34, 5th percentile 0.03; Table 5.4). It is not clear how representative these forecasts might be for western lakes, in general, or how accurate the estimates are for the modeled lakes. Nevertheless, these comparative data suggest that western systems are as sensitive, or perhaps more sensitive, than any of the watersheds for which acidification and/or recovery responses have been more rigorously quantified.

5.2   Critical Loads

5.2.1   Background

It has been well documented that acidic deposition has caused environmental degradation of surface waters, soils, and forests in certain areas. Such degradation has been more widespread in Europe than in North America, owing partly to the fact that many regions of Europe have received much higher deposition of S and N for a longer period of time than have comparable North American ecosystems. Recent emissions control efforts have focused on attempts to reduce deposition sufficiently to permit ecosystem recovery, if not to pre-acidification levels, at least to ecologically acceptable levels. The key questions facing scientists and policy-makers, therefore, have to do with the degree in space and time to which S and N emissions will need to be reduced in order to allow ecosystem recovery to proceed (Jenkins et al., 1998).

Public policy measures to reduce emissions must be based upon quantified dose–response relationships that reflect the tolerance of natural ecosystems to various inputs of atmospheric pollutants. This need has given rise to the concepts of critical levels of pollutants and critical loads of deposition (e.g. Bull, 1991, 1992), as well as interest in establishing one or more standards for acid deposition. A critical load can be defined as "a quantitative estimate of an exposure to one or more pollutants below which significant harmful effects on specified sensitive elements of the environment do not occur according to present knowledge" (e.g., Nilsson, 1986; Gundersen, 1992). Such an approach to establishing a standard is intuitively satisfying. However, the assignment of a standard or critical load of S or N for any particular region may be difficult to defend scientifically. A variety of natural processes and anthropogenic activities affect the acid–base chemistry of lakes and streams, in addition to atmospheric deposition of S and N. The loadings of N or S that may be required to protect *the most sensitive* elements of an ecosystem may be unrealistically low in terms of economic or other considerations, and may be difficult to quantify.

The basic concept of critical load is relatively simple, as the threshold concentration of pollutants at which harmful effects on sensitive receptors begin to occur. Implementation of the concept is, however, not at all simple or straightforward. Practical definitions for particular receptors (soils, fresh waters, forests) have not been agreed to easily. Different research groups have employed different definitions and levels of complexity (Bull, 1991, 1992). Constraints on the availability of suitable, high-quality regional data have been considerable.

The acid–base chemistry of surface waters typically exhibits substantial intra- and interannual variability. Seasonal variability in the concentration of key chemical parameters often varies by more than the amount of acidification that might occur in response to acidic deposition. Such variability makes quantification of acidification and recovery responses difficult, and also complicates attempts to evaluate sensitivity to acidification based solely on index chemistry, as is typically collected in synoptic lake or stream surveys. Seasonal variability is particularly problematic in the assessments of standards for N.

5.2.2 Progress in Europe

The United Nations Economic Commission for Europe (UNECE) established the Convention on Long Range Transboundary Air Pollution (LTRAP) in 1979 to promote reductions in the emissions and deposition of S and N throughout Europe. LTRAP has had an enormous impact on European air pollution research and abatement during the last two decades. The convention adopted the First Sulfur Protochol in 1985, which targeted national reductions in SO_2 emissions by 30% compared with 1980 emission levels by the year 1993. Interestingly, the U.K. was widely criticized for failing to sign the First Sulfur Protochol and thereby joining the "30% Club," and yet subsequently agreed in 1994 to an 80% reduction in SO_2 emissions by the year 2010. This is indicative of the fact that enormous political changes have occurred since the 1980s. We scientists like to believe that those political changes have been the direct result of our scientific advancements.

The majority of the critical loads work to date has been conducted in Europe. A number of documents have been prepared in conjunction with the UN/ECE critical loads research efforts over the past decade. These have included documentation of methodologies (e.g., ECE , 1990) and presentation of critical loads maps for portions of Europe. In addition, a number of other background documents have been prepared in conjunction with the ongoing critical loads research efforts in Europe (e.g., Gundersen, 1992; Kämäri et al., 1993; Hessen et al., 1992; Lövblad and Erisman, 1992).

A simplistic and generalized attempt to quantify critical loads for S and N was presented at the Skokloster workshop (Nilsson and Grennfelt, 1988), based on a long-term mass-balance approach. A stable base cation pool was used as the criterion for defining the critical load. This implied an absence of soil acidification, and allowed a connection between the critical loads of S and N. Leaching of both NO_3^- and SO_4^{2-} above the production rate of base cations via weathering will eventually lead to soil acidification. The permissible input of N for designation of the critical load was the amount allocated to forest growth, forest floor accumulation, and an acceptable leaching of 1 to 2 kg N/ha per year. On this basis, Nilsson and Grennfelt (1988) estimated critical loads of N for Europe to be in the range of 3 to 20 kg N/ha per year, depending on forest productivity.

Although some ECE working groups have developed fairly complex, process-based approaches, the severe constraints on data availability generally necessitate creating maps based on the more simplistic steady-state approaches that tend to have more substantial problems. For example, a calculation frequently employed for estimation of the critical load of S to surface waters is based on assumed pre-industrial and current base cation fluxes (Henriksen et al., 1990a,b, 1992). There is significant uncertainty in the estimates of *current* base cation input, especially on a regional basis. It is even more difficult to quantify pre-industrial base cation deposition.

Terminology in this research area can cause some confusion. It has been assumed that it will not be possible to reduce loads below critical values for some sensitive systems in Europe, and also that dynamic watershed processes cause lag periods in the acidification and recovery responses. These problems have given rise to the concept of target loads (e.g., Henriksen and Brakke, 1988) that implies policy relevance, rather than strictly ecological justification. Critical loads and target loads are conceptually different. A critical load is a characteristic of a specific environment that can be estimated by a variety of mechanistic and empirical approaches. A target load can be based on political, economic, or temporal considerations, and implies that the environment will be protected to a specified level (i.e., certain degree of allowable damage) and/or over a specified period of time. For example, a given target load may be sufficiently low as to protect a particular ecosystem from significant environmental degradation over a 10-year period but, in fact, may be substantially higher than would be required for long-term protection of that ecosystem. There has been a rapid acceptance of the concepts of critical and target loads throughout Europe for use in political negotiations concerning air pollution and development of abatement strategies to mitigate environmental damage (e.g., Posch et al., 1997).

Criteria of unacceptable change used in critical loads assessments are typically set in relation to known effects on aquatic and terrestrial organisms. For protection of aquatic organisms, the ANC of runoff water is most commonly used (Nilsson and Grennfelt, 1988; Henriksen and Brakke, 1988; Sverdrup et al., 1990). Critical limits of ANC, that is, concentrations below which ANC should not be permitted to fall, have been set at 0, 20, and 50 μeq/L for various applications (e.g., Kämäri et al., 1992). Designation of an ANC limit is confounded, however, by natural acidification processes that can also reduce ANC to low, or even negative, values.

An ANC limit of 0 has been adopted by the U.K. for the national mapping of critical loads for surface waters (Harriman et al., 1995a). This has been defined as the ANC at which there exists a 50% probability of survival of salmonid fisheries (Sverdrup et al., 1990). However, recent evidence suggests that, for Scottish fisheries, sites with mean surface water ANC less than or equal to zero are currently almost all fishless, although sites having mean ANC greater than zero but periodic fluctuations below zero have relatively healthy populations (Harriman et al., 1995). Jenkins et al. (1997a), therefore, suggested that the critical ANC limit is too low, and should be replaced by a limit of 20 µeq/L (for low TOC waters) that corresponds with significant change in diatom flora in Scottish lakes.

Critical loads are often determined separately for soils and surface waters, and the resulting estimates may differ. In general, surface waters appear to be more sensitive (i.e., have lower critical loads) than soils within a given area. Because the objective of implementing the critical load concept is to protect the entire ecosystem from degradation, the overall critical load for the ecosystem is the lowest critical load observed for the various sensitive receptors. In other words, if the surface waters are protected, the soils also will be protected generally. Critical loads will also differ from site to site depending on the inherent sensitivity of the environment.

The mapping of critical loads throughout Europe was initiated at several international workshops within the UN/ECE. The resulting maps assigned critical load values to discrete geographical areas (grids), and provided the basis for comparison with current or projected atmospheric deposition. A great deal of effort has gone into mapping activities on national and international scales in Europe since 1990. Such maps have been and will continue to be used in developing pollution abatement strategies.

The Protochol on Further Reductions of Sulfur Emissions was signed by 28 countries in 1994. This Second Sulfur Protochol outlined country-specific emissions reductions that were calculated in an effort to protect 95% of the forest ecosystem area from adverse effects. This was the first time that an effects-based strategy (critical loads approach) has been adopted for air pollution effects mitigation (Posch et al., 1997). In the U.K., the Second Sulfur Protochol had as its basis critical load calculations using the Steady State Mass Balance model for soils (Hornung et al., 1995).

Nitrogen-saturation and NO_3^- leaching have been proposed as indicators of ecosystem stability, and as such can be used as criteria for evaluating critical loads for N. The definition of N saturation, and interpretation of N effects on ecosystem stability, require the evaluation of NO_3^- leaching data within the context of data from unaffected areas. This is difficult in Europe because N deposition is elevated throughout most forested regions (Gundersen, 1992). Based on available data, background NO_3^- leaching from coniferous forests has been estimated to be in the range of 1 to 3 kg N/ha per year (e.g., Nilsson and Grennfelt, 1988; Hauhs et al., 1989). Estimates in this range are currently being used in critical loads calculations.

As a forest ecosystem approaches the point of N saturation, NO_3^- leaching will first become pronounced during the dormant season when vegetative uptake is low. The biological control on NO_3^- leaching results in a distinct seasonality in the patterns of NO_3^- leaching from soils and the resulting NO_3^- concentrations in drainage waters. This biological control of NO_3^- leaching, and consequent seasonality in NO_3^- output fluxes, can be eliminated as the ecosystem becomes N saturated. This was emphasized by Hauhs et al. (1989) who showed a progressive reduction of the NO_3^- seasonality at the Lange Bramke and Dicke Bramke sites in Germany. This loss of biological control appears to be a critical factor indicating N saturation (Aber et al., 1989; Stoddard, 1994).

Critical loads can be evaluated on an empirical basis, using input/output budgets. For example, Grennfelt and Hultberg (1986) examined NO_3^- leaching across a gradient of atmospheric N input in Europe, and found increased NO_3^- leaching at a threshold of wet deposition input of about 10 to 15 kg N/ha per year. Such an empirical approach has limitations, however, because other factors besides atmospheric input can regulate the extent of NO_3^- leaching (Skeffington and Wilson, 1988; Gundersen, 1992). Forest decline, in particular, can confound the analysis. Gundersen (1992) emphasized that such empirical analyses can yield useful information, but cautioned that the data should be separated by scale (plot or catchment) and ecosystem type (coniferous or deciduous), and sites with obvious forest decline or N fixation should be excluded.

A variety of model approaches are being used for estimating the long-term (chronic) critical loads of S and N to surface waters. They range from simple empirical calculations to complex dynamic models. Steady-state models can be useful to derive long-term critical loads for S, and potentially for N. They only include processes that influence acid production and consumption over long periods of time, such as mineral weathering and net uptake. An assumption in the application of steady-state models is that dynamic processes are not important for the assessment of long-term critical loads. Dynamic models include evaluation of the time period required to reach critical criteria values. Thus, processes such as cation exchange, N mineralization/ immobilization and SO_4^{2-} adsorption/desorption are often included in the dynamic approaches (deVries and Kros, 1991). Although steady-state models will provide estimates of the final emission or deposition amounts required to achieve a steady state condition over an infinite time period, dynamic models are needed for an assessment of the temporal evolution of the acidification process.

The MAGIC model was applied to 21 upland watersheds involved within the UK Acid Waters Monitoring Network to assess the critical loads of S and the likely future recovery of acidified surface waters in response to the emissions controls agreed upon in the Second Sulfur Protochols (Jenkins et al., 1998). Future estimates of S deposition that would result from lower S emissions were generated with the Hull Acid Rain Model (Metcalfe and Whyatt, 1995), an atmospheric deposition and transport source-receptor lagrangian model that links emissions to deposition for all major point sources of S in the U.K. The MAGIC modeling results suggested that only a limited degree of recovery in surface water chemistry would occur over the next 50 years despite an 80% reduction in emissions from the 1980 baseline. However, the projected recovery was pronounced when compared with model projections that did not consider the emissions reductions of the Second Sulfur

Protochols (Jenkins et al., 1998). The agreed-upon reductions in S deposition are simulated by the model to be insufficient to restore the base saturation at most of the sites. Now that the model has been calibrated to a range of acid-sensitive sites throughout the U.K., it will be somewhat easier to examine the effects of changes in N deposition and other policy-relevant scenarios in the future.

The critical load, as formulated as a science/policy concept in Europe for atmospheric deposition of S or N represents an inherent characteristic of the watershed. Specification of the critical load, or any kind of acid deposition standard, assumes that the ecosystem has reached or will reach steady-state with respect to deposition inputs over some time scale of acidification or recovery response. The environmental consequences of different emissions reductions cannot be fully evaluated using only the empirical and steady-state methods for specifying critical loads (Warfvinge et al., 1992). For example, the long-term critical load for S at the Birkenes site in southern Norway, required to maintain ANC greater than 0 (e.g., ANC criterion equal to 0) is estimated to be approximately 50 meq SO_4^{2-}/m^2 per year (8 kg S/ha per year). However, the time-dependence derived from the MAGIC model illustrates that to obtain ANC greater than 0 within 10 years, the target load would be only 1/4 the critical load (12 meq SO_4^{2-}/m^2 per year); if one could wait 50 years to achieve ANC greater than 0, then the target load would be much greater (41 meq SO_4^{2-}/m^2 per year) and would approach the long-term critical load (Warfvinge et al., 1992). Similarly, the starting point can have a large influence on the model estimate of target load. Starting with pre-acidification conditions, the MAGIC model estimated that the Birkenes watershed could tolerate 270 meq SO_4^{2-}/m^2 per year for 10 years before the stream water would *acidify* to ANC equal to 0. Starting from acidified conditions in 1985, however, MAGIC estimated that the load would have to be reduced by a factor of 22 (to 12 meq SO_4^{2-}/m^2 per year) in order for stream water to *recover* to ANC equal to 0 (Warfvinge et al., 1992).

Thus, model-based analyses suggest that standards for the protection, or restoration, of surface water quality must be specified within a temporal context. Standards suitable for protection of aquatic ecosystems for a short period of time may be less than adequate for long-term protection. Conversely, reductions in deposition that are insufficient for acidified ecosystem restoration in the short term may require additional time, rather than additional emissions reductions, to achieve the desired outcome.

5.2.3   Progress in the U.S. and Canada

In 1990, the Clean Air Act was amended by Congress, in part in an effort to reduce the perceived adverse environmental impacts of acidic deposition. Title IV of the Clean Air Act Amendments of 1990 (CAAA) required a 10 million-ton reduction in annual atmospheric emissions of S dioxide and approximately a 2 million-ton reduction in annual N oxide emissions. The CAAA also included requirements to assess the effectiveness of the mandated emissions controls via periodic assessments. In addition, the EPA was required by Section 404 of the CAAA to submit to Congress a report on the feasibility of adopting one or more acid deposition standards:

> Not later than 36 months after the date of enactment of this Act, the Administrator of the Environmental Protection Agency shall transmit to the Committee on Environment and Public Works of the Senate and the Committee on Energy and Commerce of the House of Representatives a report on the feasibility and effectiveness of an acid deposition standard or standards to protect sensitive and critically sensitive aquatic and terrestrial resources. The study required by this section shall include, but not be limited to, consideration of the following matters:
>
> (1) identification of the sensitive and critically sensitive aquatic and terrestrial resources in the U.S. and Canada which may be affected by the deposition of acidic compounds;
>
> (2) description of the nature and numerical value of a deposition standard or standards that would be sufficient to protect such resources;
>
> (3) description of the use of such standard or standards in other Nations or by any of the several States in acid deposition control programs;
>
> (4) description of the measures that would need to be taken to integrate such standard or standards with the control program required by title IV of the Clean Air Act;
>
> (5) description of the state of knowledge with respect to source-receptor relationships necessary to develop a control program on such standard or standards and the additional research that is ongoing or would be needed to make such a control program feasible; and
>
> (6) description of the impediments to implementation of such control program and the cost-effectiveness of deposition standards compared to other control strategies including ambient air quality standards, new source performance standards and the requirements of title IV of the Clean Air Act.

Technical information required by the EPA for assessing the feasibility of adopting one or more acid deposition standards for the protection of aquatic resources was summarized by Sullivan and Eilers (1994) and Van Sickle and Church (1995). Quantitative model-based analyses were conducted for areas of the U.S. intensively studied in EPA's model forecasting program, the Direct Delayed Response Project (DDRP, Church et al., 1989). The MAGIC model (Cosby et al., 1985a,b) was used to project changes in surface water chemistry for a range of S and N deposition scenarios, assuming a range of N retention efficiencies (Van Sickle and Church, 1995).

A report was prepared for Congress on the feasibility of adopting one or more acid deposition standards (EPA, 1995a). The report concluded that establishment of such standards for S and N deposition in the U.S. was technically feasible, but that two critical areas of uncertainty advised against the setting of standards at that time. First, policy decisions regarding appropriate or desired goals for protecting sensitive systems were needed, especially with respect to the level of protection desired and the costs and benefits of such protection. Second, key scientific unknowns, particularly regarding watershed processes that govern N dynamics, limited the ability to recommend specific standards for N deposition at that time.

Policy decisions regarding appropriate or desired goals for protecting sensitive systems have still, in many cases, not been made. Federal agencies rely upon different approaches to achieve common goals. Nevertheless, all federal land managers (FLMs) are required to make such decisions routinely, and yet still lack a common scientific foundation for those decisions. An effort attempting to coordinate FLM approaches to setting critical loads is currently underway as part of the Federal Land Managers AQRV Group (FLAG).

Prior to and since publication of EPA's Acid Deposition Standards Feasibility Report (EPA, 1995a), considerable research has been conducted on the topics of N dynamics and the effects of atmospheric N deposition (c.f., Sullivan , 1993; Emmett et al., 1997; Jenkins et al., 1997b; Cosby et al., 1997). A variety of dynamic models are now available with which to estimate critical loads for N at the watershed scale. Nitrogen dynamics have recently been added to the MAGIC model (Ferrier et al., 1995; Jenkins et al., 1997b), thus allowing MAGIC to be used for assessment of critical loads for either S or N or a combination of the two.

Critical loads modeling for the 1997 Canadian Acid Rain Assessment (Jeffries, 1997) was conducted for six regional clusters of lakes, four in eastern Canada, one in Alberta, and also the Adirondack Mountains in New York. The Integrated Assessment Model (IAM, Lam et al., 1994) was used to estimate the future steady-state pH of each lake in each region at varying levels of wet SO_4^{2-} deposition over the range 6 to 30 kg SO_4^{2-}/ha per year (2 to 10 kg S/ha per year as wet S). pH was used as the critical load threshold criterion and was evaluated for 3 alternative critical levels (pH 6.0, 5.5, and 5.0). Lakes that were judged to have had pre-industrial pH less than the critical levels (e.g., owing to the presence of organic acidity) were deleted from the analyses. Critical loads of S were specified on the basis of protecting 95% of the regional lake resource from acidity in excess of the designated critical levels. The modeling results suggested critical loads of wet S deposition [converted from units of wet SO_4^{2-} reported by Jeffries (1997)] ranging from less than 2 kg/ha per year for the Kejimkujik, Nova Scotia, Fort McMurray, Alberta, and Adirondack lake clusters in New York to about 5 kg/ha per year at Sudbury, Ontario. There was not a large difference in the estimates of critical load in the various regions in response to varying the critical pH level of protection from 5.0 to 6.0 in most cases.

The adoption of acid deposition standards for the protection of surface water quality in the U.S. from potential adverse effects of S and N deposition is a multifaceted problem. It requires that S and N be treated separately as potentially acidifying agents, and that separate estimates for each be generated for all individual, well-defined regions or subregions of interest.

Appropriate criteria must be selected as being indicative of damaged water quality, for example ANC or pH. Once a criterion has been selected, a critical value must be estimated, below which the criterion should not be permitted to fall. For example, if the selected criterion is surface water ANC, one could specify that ANC should not be permitted to fall below 0, 20, or 50 μeq/L in response to acidic deposition (e.g., Kämäri et al., 1992). Selection of critical values for ANC or pH is confounded by the existence of lakes and streams that are acidic or very low in pH or ANC owing entirely to natural factors, irrespective of acidic deposition (Sullivan, 1990). In particular, low contributions of base cations in solution, owing to low weathering rates and/or minimal contact between drainage waters and mineral soils, and high concentrations of organic acids contribute to naturally low pH and ANC in surface waters. Other factors also can be important in some cases, including the neutral salt effect (cation retention) and watershed sources of S.

Acid deposition standards might be selected on the basis of protecting aquatic systems from chronic acidification; conversely, episodic acidification might also be considered, and would be of obvious importance in regions where hydrology is dominated by spring snowmelt. Thus, selection of appropriate acid deposition standards involves consideration of a matrix of factors, as outlined in Table 5.5.

5.2.4 Establishment of Standards for Sulfur and Nitrogen

Sulfur deposition is a potential concern in all of the acid-sensitive regions of the U.S. Some degree of chronic acidification attributable to S deposition has occurred in the Adirondacks, northern New England, mid-Appalachian Mountains, the eastern portion of the upper Midwest region, and possibly in the Trail Ridge region of northcentral Florida.

MAGIC model projections of change in surface drainage water ANC in response to changes in S deposition have been shown to be relatively consistent from region to region in the eastern U.S. Turner et al. (1992) and Sullivan et al. (1992) presented the results of NAPAP modeling scenarios for 50-year MAGIC simulations for lakes in the Adirondacks, New England, Mid-Atlantic Highlands, and Southern Blue Ridge Province and streams in

TABLE 5.5

Factors that Should be Considered for Selection of Acid Deposition Standards for the Protection of Surface Water Quality

Factors for Consideration	Possible Options
Acidifying agent	Nitrogen or sulfur
Regional delineation	Region- or subregion-specific standards
Temporal response	Chronic or episodic acidification
Damage criterion	ANC or pH
Critical values for criterion	ANC<0, 20, 50 μeq/L
	pH < 5, 5.5, 6

the Mid-Atlantic Highlands. Simulations included changes in S deposition over 1985 values of -50, -30, -20, 0, +20, and +30%. Each kg/ha per year change in S deposition caused approximately a 3.5 µeq/L change in median lake-water ANC for all regions studied. Although the modeled response of individual watersheds to simulated changes in S deposition was more variable, these results demonstrate that the MAGIC model is strongly driven by S deposition input values.

Regional quantification of the amount of acidification that has occurred in the upper Midwest is not possible with existing data. Although more quantitative (paleolimnological) data are available for northcentral Florida and, consequently, historical changes in lake-water pH are better documented, the cause of recent acidification in some Florida lakes cannot be definitively ascribed to acidic deposition. Substantial groundwater withdrawals from local aquifers might explain part, or all, of the historical changes in pH.

In the upper Midwest and Florida, seepage lakes constitute the most sensitive resources of interest. It is difficult to make direct comparisons of deposition and potential impacts between these regions, however. Interpretation of deposition impacts is confounded by the importance of natural marine deposition of SO_4^{2-} and Cl^- in Florida and also by the enhanced importance of evapoconcentration in Florida lakes, which increases the acidity of weakly acidic solutions (e.g., Munson and Gherini, 1991). It is likely that an appropriate S deposition standard for the Upper Peninsula of Michigan would be somewhat less than peak deposition values recorded in the 1970s, although it is not possible to quantify how much less, based on available data. Furthermore, S deposition in this region has been declining steadily in recent years, and will, therefore, likely be of less concern in the future than in many other regions of the country. As an interim guideline, Sullivan and Eilers (1994) suggested the use of a standard for S in the range of 5 kg S/ha per year that approximates current deposition in the eastern portion of the region.

Based on analysis of available S dose–response data for sensitive watersheds worldwide (Tables 5.1 to 5.4), it is clear that proportional changes in ANC and base cations in drainage waters in response to changes in S inputs are highly variable. Documented F-factors are generally above 0.5, although lower values have been found. Perhaps the best available estimate of an appropriate F-factor for highly sensitive watersheds, such as are found throughout the western U.S., would be based on the experimental values obtained at Sogndal, in western Norway (near 0.4). This alpine watershed exhibits substantial areas of exposed bedrock, and contains shallow acidic soils. As such, it appears to be a reasonable surrogate for sensitive watersheds in the West. Although MAGIC model projections for western lakes (e.g., Eilers et al., 1991; Sullivan et al., 1998) suggest that some watersheds may exhibit values for the F-factor lower than 0.4, assessments using multiple approaches have concluded that MAGIC projections may represent upper bounds for watershed acidification response (NAPAP, 1991; Sullivan et al., 1992). Sullivan and Eilers (1994), therefore, recommended a value for F of 0.4 as most likely representative for highly sensitive aquatic systems in the western U.S. As a worst case scenario, a value as low as perhaps 0.2 may not be unreasonable for extreme cases of acid sensitivity. Assuming such a high level of sensitivity ($F = 0.2$) would certainly not be appropriate for watersheds in the northeastern U.S., based on all available information. It must be recognized, however, that surface waters in the western U.S. probably are among the most sensitive in the world to inputs of acidic deposition (Eilers et al., 1990; Melack and Stoddard, 1991).

The first and fifth percentiles of measured ANC for acid-sensitive subregions of the West are presented in Table 5.6. Also provided are calculated estimates of the amount of increase in lake-water SO_4^{2-} that would be required to acidity the first and fifth percentile lake of the subregional ANC distributions from current values to ANC equal to zero. It was assumed for these calculations that 40% of the increased SO_4^{2-} concentration is neutralized by base cation release ($F = 0.4$) and the remainder causes a stoichiometric decrease in ANC. If a lower value of F is assumed (e.g., $F = 0.2$), then the estimates of SO_4^{2-} change provided in Table 5.6 would decrease by 25%. These calculations suggest that relatively minor increases in lake-water SO_4^{2-} concentration would lead to chronic acidity (ANC less than zero) in the Sierra Nevada and Cascade Mountain ranges. An estimated 5% of the lakes in these subregions would become acidic with increased SO_4^{2-} concentration of only 27 to 30 µeq/L. This would occur under S deposition loadings of about four times current levels, based on current SO_4^{2-} concentrations. Although uncertainties are large in current estimates of S deposition in these regions, total S deposition is likely in the range of 0.5 to 2 kg S/ha per year (Sisterson et al., 1990). Thus, a reasonable standard for preventing 5% of the lakes in the Sierra Nevada and Cascade Mountains from becoming chronically acidic owing to S deposition is approximately 2 to 8 kg S/ha per year. In other subregions of the West, the required SO_4^{2-} increase estimated to cause 5% of the lakes to become acidic is somewhat higher (55 to 70 µeq/L), but still low compared to SO_4^{2-} concentrations currently found throughout the eastern U.S. Total S

TABLE 5.6

First and Fifth Percentiles of the Regional Lake ANC Distributions for Subregions of Interest in the Western U.S. Having Large Numbers of Acid-Sensitive Lakes, and Estimates of the Increase in Lake-water SO_4^{2-} Concentration that Would be Required to Drive Chronic ANC to Zero (Units are in µeq/L.)

Subregion	Current Lake ANC		ΔSO_4^{2-} to drive ANC to O[a]	
	1st Percentile	5th Percentile	1st Percentile	5th Percentile
Sierra Nevada	15	16	25	27
Cascade Mountains	11	18	18	30
Idaho Batholith	21	33	35	55
Wyoming	38	39	63	65
Colorado Rocky Mountains	25	42	42	70

[a] Calculation based on an assumed F-factor equal to 0.4.
Source: Sullivan and Eilers, 1994.

deposition levels approximately in the range of 3 times (Colorado) to 5 times (Idaho) current deposition would be required to chronically acidify 5% of the lakes in these other western regions. These estimates equate to acid deposition standards equal to approximately 5 to 10 kg S/ha per year. If this analysis is based on the lowest percentile lake in the subregional ANC distribution, increased SO_4^{2-} concentrations of 35 to 63 µeq/L would cause chronic acidity in the Idaho Batholith, Wyoming, and Colorado subregions, assuming F = 0.4. There are unquantifiable uncertainties associated with such approximations, although the results are generally consistent with calculations for sensitive watersheds in the Northeast and in Europe. These uncertainties could be substantially reduced by conducting MAGIC simulations (or other models of acid–base chemistry) in a suite of watersheds in the western subregions identified as potentially highly sensitive to acidic deposition inputs. Such modeling work has only been conducted for a limited number of watersheds.

The estimates of increased SO_4^{2-} concentration required to acidify western lakes within the lower percentiles of acid-sensitivity, presented previously, are based on fall chemistry and chronic acidification processes. It is likely, however, that sensitive watersheds in the western U.S. would experience episodic acidification (especially during snowmelt) at S deposition levels lower than those that would cause chronic acidification. In most cases, episodic pH and ANC depressions during snowmelt are driven by natural processes (mainly base cation dilution) and NO_3^- enrichment (cf., Wigington et al., 1990, 1993). Where pulses of increased SO_4^{2-} are found during hydrological episodes, they are usually attributable to S storage and release in streamside wetlands. More often, lake- and stream-water concentrations of SO_4^{2-} decrease or remain stable during snowmelt. This is probably attributable to the observation, based on ratios of naturally occurring isotopes, that most stream flow during episodes is derived from pre-event water. Water stored in watershed soils is forced into streams and lakes by infiltration of meltwater via the piston effect. This is not necessarily the case for high-elevation watersheds in the West, however. Such watersheds often have large snowpack accumulations and little soil cover. Selective elution of ions in snowpack can, therefore, result in relatively large pulses of both NO_3^- and SO_4^{2-} in drainage water early in the snowmelt. It appears likely that S deposition will contribute to episodic acidification of sensitive western surface waters at deposition levels below those that would cause chronic acidification. Episodes have been so little studied within the region, however, that it is not possible to provide quantitative estimates of episodic S standards for the western subregions of concern.

Webb et al. (1994) estimated F factors for the long-term VTSSS sampling sites in the various watershed response classes identified for western Virginia. They assumed that there should be a similar ratio between SiO_2 and that part of the base cations associated with primary mineral weathering. Thus, stream-water SiO_2 concentrations provided a theoretical basis for discriminating between the background (pre-acidic deposition) base cation concentrations and the increase in base cation concentrations in response to strong acid anions from acidic deposition. They applied regression analysis to estimate stream-water base cation concentrations as a function of SiO_2 and $(SO_4^{2-} + NO_3^-)$ concentration, whereby the coefficient for SiO_2 represents the primary mineral weathering ratio and the coefficient for $(SO_4^{2-} + NO_3^-)$ represents an instantaneous estimate of the F factor. The resulting regressions were highly significant ($p \leq 0.01$) and suggested that the mean F factor for siliclastic watersheds in the Blue Ridge Mountains was 0.69, with a standard error of 0.14. Results for siliclastic watersheds in the Allegheny Ridges suggested slightly greater acid sensitivity with a mean estimated F factor of 0.39 (se, 0.11). Estimated F-factors were higher, as expected, for the minor carbonate watersheds (0.88, se = 0.20) and the basaltic watersheds (1.14, se = 0.17).

Stream-water concentrations of NO_3^- are typically below about 5 µeq/L in boreal forested regions, and such a concentration is considered to have no harmful effect on the biota of freshwater and near-coastal aquatic systems. Therefore, 5 µeq/L has been suggested as a reasonable critical concentration for surface waters to protect against significant harmful effects (Rosén et al., 1992). The relationship between measured wet deposition of N and stream-water output of NO_3^- was evaluated by Driscoll et al. (1989a) for sites in North America (mostly eastern areas), and augmented by Stoddard (1994). The resulting data showed a pattern of N leaching at wet inputs greater than approximately 5.6 kg N/ha.

Stoddard (1994) presented a geographical analysis of patterns of watershed loss of N throughout the northeastern U.S. He identified approximately 100 surface water sites in the region with sufficiently intensive data to determine their N status. Sites were coded according to their presumed stage of N retention, and sites ranged from Stage 0 through Stage 2 (see additional discussion in Chapter 7). The geographic pattern in watershed N retention depicted by Stoddard (1994) followed the geographic pattern of N deposition. Sites in the Adirondack and Catskill Mountains, where N deposition is about 11 to 13 kg/ha per year, were typically identified as Stage 1 or Stage 2. Sites in Maine, where N deposition is about one-half as high, were nearly all Stage 0. Sites in New Hampshire and Vermont that receive intermediate levels of N deposition were identified as primarily Stage 0, with some Stage 1 sites. Based on this analysis, a reasonable threshold of N deposition for transforming a northeastern site from the natural Stage 0 condition to Stage 1 would correspond to the deposition levels found throughout New Hampshire and Vermont, approximately 8 kg N/ha per year. This agrees with Driscoll et al.'s (1989a) interpretation that suggested N leaching at wet inputs above about 5.6 kg N/ha per year would correspond to total N inputs near 7 to 8 kg N/ha per year. This is likely the approximate level at which episodic aquatic effects of N deposition would become apparent in some watersheds of the northeastern U.S. Wet deposition of N was reported by Stoddard and Kellog (1993) for two monitoring stations in Vermont (Bennington and Underhill), based on 1987 data from the National Atmospheric Deposition Program (NADP). Total wet N deposition at the NADP sites in Vermont ranged from 4.8 kg/ha (Bennington) to

6.0 kg/ha (Underhill), of which NO_3-N contributed approximately two-thirds. These wet deposition values are intermediate between estimates for the Adirondacks (8.6 kg/ha; Pollack et al., 1989) and both the Bear Brook site in Maine (4.3 kg/ha; Kahl et al., 1993a) and Hubbard Brook in New Hampshire (4.2 kg/ha; Stoddard and Kellog, 1993).

Lake-water concentrations of NO_3^- were surprisingly high in many high-elevation sites included in the Western Lake Survey, despite the possible bias caused by the failure to collect samples at many of the highest elevation areas owing to frozen lake conditions at the time of sampling. Based on existing data, some high-elevation lakes in the West are currently experiencing N deposition sufficiently high to cause chronic NO_3^- leaching, and likely associated chronic acidification. Furthermore, it is also likely that many of these sites that exhibit fall concentrations of NO_3^- in the range of 10 to 30 μeq/L have substantially higher concentrations during the spring. Thus, the weight of evidence suggests that episodic acidification associated with N deposition may already be occurring to a significant degree in many high-elevation western lakes. Unfortunately, sufficient data are not available with which to adequately evaluate this potentially important issue.

Specification of numerical standards for S and N deposition is dependent on a host of both scientific and policy decisions. These include, for example

- Scientific determination of the extent to which water chemistry will change in its acid–base character in response to various deposition loading rates (chemical dose–response relationship).
- Scientific estimation of the biological responses associated with given changes in water chemistry (biological dose–response relationship).
- Policy determination of the percent of sensitive resources within a given region that one wishes to protect against adverse changes.
- Policy determination of what biological changes must be protected against.

Such decisions are not made easily, nor should they be. More progress has been made in the U.S. in dealing with the scientific decisions than with the policy decisions. It is now fairly straightforward to estimate the dose–response functions for a given watershed or group of watersheds within a region, although this does entail a moderate level of uncertainty (e.g., Turner et al., 1992; van Sickle and Church, 1995; Sullivan and Eilers, 1994). Furthermore, there are generally well-accepted criteria for specifying biological response functions, both chronically and episodically (e.g., Baker et al., 1990c; Wigington et al., 1993) and episodic excursions from measured chronic chemistry and a general knowledge of regional hydrology (e.g., Eshleman, 1988; Webb et al., 1994). The policy decisions are somewhat more difficult, and for the most part have not been adequately addressed (EPA, 1995a). For example, one may be willing to accept the damage of 15 or 20% of the lakes in Adirondack Park, NY (as estimated currently), but not be willing to accept the damages of 1% of the lakes in Rocky Mountain National Park. This is because the latter are *expected* to be pristine. FLMs are required to protect sensitive resources in Class I areas from any harmful effects, whereas in some cases extremely low levels of air pollution may damage *the most sensitive receptor* without compromising the ecological integrity of the ecosystem at large. Despite such difficulties, some progress has been made.

The West is the most susceptible region in the U.S. to potential acidification from acidic deposition. Because of the paucity of dose–response data for the region, it is unclear what level of deposition of either S or N would be appropriate for the protection of aquatic resources from adverse effects. Based upon the weight of evidence, Sullivan and Eilers (1994) concluded that an appropriate standard for S deposition would be less than 10 kg S/ha per year to protect against chronic acidification in large areas of the West. A standard sufficient to protect against episodic acidification may be much lower than that, perhaps in the range of 5 kg S/ha per year. Furthermore, in the most sensitive portions of the West (e.g., Sierra Nevada and Cascade Mountains), an appropriate standard for protecting the most sensitive aquatic resources against chronic and episodic acidification is probably below 5 kg S/ha per year (Sullivan and Eilers, 1994). Such estimates are highly subjective, however, and should be considered as "best guesses" at this time.

6

Episodic Acidification

6.1 Background and Characteristics of Sensitive Systems

The acid–base chemistry of surface waters typically exhibits substantial intra- and inter-annual variability. Seasonal variability in the concentration of key chemical parameters often varies by more than the amount of acidification that might occur in response to acidic deposition. Such variability makes quantification of acidification and recovery responses difficult, and also complicates attempts to evaluate sensitivity to acidification based solely on index chemistry. The latter term is applied to chemical characterization data that correspond with periods when the chemistry is expected to be relatively stable. These are typically summer or fall for lakes and spring baseflow for streams. Lakes and streams exhibit short-term episodic decreases in ANC, and often also pH, usually in response to hydrological events, such as snowmelt or rainfall. Periods of episodic acidification may last for hours to weeks, and sometimes result in depletion of ANC to negative values with concurrent increases in potentially toxic inorganic Al in solution.

Precipitation inputs to a watershed typically pass through the soil profile prior to reaching stream channels. The typical soil profile in acid-sensitive watersheds has lowest pH (approximately 4) in upper organic soil horizons, increasing down the profile to pH greater than 6 at depth (Norton et al., in press). Drainage water chemistry is generally somewhat reflective of conditions in the lower soil horizons and, therefore, generally has pH greater than 6. During high discharge snowmelt or rainfall events, however, flow routing favors water flowpaths through upper horizons. During such events, drainage water chemistry, therefore, typically reflects the lower pH, higher organic content, and lower ANC of these upper soil horizons. This is one of the major reasons why many surface waters are lower in pH and ANC during hydrological episodes.

Many of the same characteristics that predispose aquatic systems to chronic acidification from acidic deposition (discussed in Chapter 3) also predispose aquatic systems to episodic acidification. Geology and soils characteristics are important in this regard. However, the single most important factor governing the sensitivity of a given watershed to episodic acidification is hydrology. The pathways followed by snowmelt and stormflow water through the watershed and into streams or lakes largely determine the extent of acid neutralization provided by the soils and bedrock in that watershed. High-elevation watersheds with steep topography, extensive areas of exposed bedrock, deep snowpack accumulation, and shallow, base-poor soils are most sensitive. Such systems are common throughout the mountainous West and in portions of the Northeast and Appalachian Mountains.

Episodes are generally accompanied by changes in at least two or more of the following chemical parameters: ANC, pH, base cations, SO_4^{2-}, NO_3^-, Al^{n+}, organic acid anions, and DOC. These changes in chemistry can adversely impact biota, particularly when changes involve pH, Al_i, and/or Ca^{2+} (Baker et al., 1990c). Aquatic biota vary greatly in their sensitivity to episodic decreases in pH and increases in Al_i in waters having low Ca^{2+} concentration, and it is difficult to classify chemical episodes according to potential biological effects. Baker et al. (1990c) concluded, however, that episodes are most likely to impact biota if the episode occurs in waters with baseline (pre-episode) pH above 5.5 and minimum pH during the episode of less than 5.0. In addition, for episodes that occur in systems that are chronically acidic or nearly so, the increase in acidity during the episode may be biologically significant, particularly when it is accompanied by increased concentrations of Al_i (Baker et al., 1990c).

Episodic acidification is nearly ubiquitous in drainage waters. Lakes and streams that have been studied throughout the U.S., Canada, and Europe nearly all experience loss of ANC during hydrologic events (Wigington et al., 1990). Chemical changes during episodes are controlled by a number of natural processes, including dilution of base cation concentrations, nitrification, flushing of organic acids from terrestrial to aquatic systems, and the neutral salt effect.* Acidic deposition can also contribute to episodic acidification, particularly via enhanced NO_3^- leaching. Under some conditions, episodes can also be partially caused by increased SO_4^{2-} concentration, although S-driven episodes appear to be less common than N-driven episodes. There is also the likelihood that chronic acidification by acidic deposition can precondition a watershed, thereby increasing the severity of episodic acidification.

Since preparation of the NAPAP 1990 Integrated Assessment, the EPA has completed the Episodic Response Project (ERP), an integrated evaluation of episodic acidification of surface waters during high-discharge periods (e.g., storms, snowmelt) in portions of the eastern U.S. (Wigington et al., 1993). This research provided important confirmatory evidence regarding the chemical and biological effects of episodic pH depressions in lakes and streams in parts of this country. The ERP clearly demonstrated that episodic processes are mostly natural, that SO_4^{2-} and, especially, NO_3^- attributable to

---

* The neutral salt effect is a process whereby addition of a neutral salt (e.g., NaCl) to base-poor soils can cause acidification of drainage water that passes through that soil. The mechanism involves ion exchange between H^+ from the soil ion exchange complex and the neutral salt cation (e.g., Na^+) in solution.

atmospheric deposition play important roles in the episodic acidification of some surface waters, and that the chemical response that has the greatest impact on biota is increased Al concentration. Similar findings have been reported elsewhere, especially in Europe, but the ERP helped to clarify the extent, causes, and magnitude of episodic acidification in portions of the U.S.

Short-term pulses of increased NO_3^- concentration have been identified as the primary factor contributing to episodic depressions of pH and ANC during snowmelt in many acid-sensitive Adirondack lakes and streams (Driscoll and Schafran, 1984; Driscoll et al., 1987a,b; Stoddard, 1994). The magnitude of episodic acidification is strongly regulated by the base cation, and therefore also ANC, concentration in lake waters. High-ANC Adirondack lakes experience episodes driven primarily by dilution of base cations during snowmelt, whereas low-ANC lakes often experience episodes driven by NO_3^- increases (Schaefer et al., 1990). The source of the N released during snowmelt in Adirondack watersheds includes nitrified snowpack N and also likely mineralized N from soil organic matter (Schaefer and Driscoll, 1993).

Nitrogen has been experimentally added to a small pristine alpine catchment in Norway at deposition levels similar to those received by some Adirondack watersheds. Since 1993, 7 kg N/ha per year have been applied to the Sogndal minicatchment as part of the RAIN and NITREX projects to augment the ambient loading of 2 kg/ha per year (Wright and Tietema, 1995). Runoff contained high concentrations of NO_3^- only during events of high flow, however, during 9 years of treatment (Wright and Tietema, 1995). These findings suggest that during low-flow periods, the flow routing of drainage water and its contact with watershed soils and terrestrial biota allow for efficient utilization of essentially all of the incoming N. In contrast, during high-flow periods, a portion of the increased N is not utilized, mainly because drainage water containing relatively high concentrations of NO_3^- moves too quickly through the soil reservoir to allow efficient N utilization.

The EPA's National Lake Survey (NLS), conducted in 1984 and 1985, provided the most comprehensive database on the acid–base chemistry of lake waters in areas of the U.S. potentially susceptible to the effects of acidic deposition. This synoptic survey was conducted during the autumn index period, during which time lake-water chemistry typically exhibits low temporal and spatial variability. Although autumn is an ideal time for surveying lake-water chemistry in terms of minimizing variability, lake-water samples collected during autumn provide little relevant data on the dynamics or importance of N in most aquatic systems. Nitrate concentrations in lake water are elevated during the autumn season only in lakes having watersheds that exhibit fairly advanced symptoms of N saturation (e.g., Figure 7.6; Stoddard, 1994). It is, therefore, not surprising that results of both the Eastern and Western Lake Surveys, both of which were conducted during the fall season, suggested that NO_3^- is of only minor importance compared to SO_4^{2-} as an acid anion in lake waters in this country. For example, the median value of the ratio of lake water NO_3^- to $(SO_4^{2-} + NO_3^-)$ concentration in Florida, the upper Midwest, and the West were very low and varied from about 0.01 to

0.06 (Landers et al., 1987; Stoddard, 1994). Survey data with which to evaluate the (largely episodic) effects that might be associated with N deposition were not collected in these surveys.

Most research on episodic processes has been conducted on stream systems that are generally more susceptible to such effects than are lakes. Spatial variability can be considerable in lakes, particularly during snowmelt episodes. Strong vertical and horizontal gradients in lake-water chemistry often preclude quantification of the magnitude of the effects in lake systems (Gubala et al., 1991). Because of the logistical difficulties and expense associated with sampling lake-water chemistry during episodic events in a manner sufficient to characterize these spatial gradients, few data are available for lakes in the areas of concern.

A great deal of the research on episodic processes, both before and after 1990, has focused at least in part on the identification of source areas of storm flow within the watershed. To a large degree, the results of this research have been less than satisfying. It is clear that water flowpaths are enormously important in the regulation of drainage water chemistry. It is also clear that watershed hydrology is complex and varies with different flow regimes. This complicates the task of attributing chemistry to particular watershed soil horizons or other source areas. Efforts to "explain" stream-water chemistry on the basis of soil water chemistry at various points within the watershed have only been useful in the extent to which the results of such efforts have communicated to watershed scientists that we do not know everything. In fact, when it comes to drainage water flow routing in upland catchments, we seem to know little.

It is very difficult to determine the immediate source of solutes in drainage water. It is intuitive that the soil water chemistry of the predominant soil/vegetation types within the watershed should correspond approximately to the chemistry of drainage waters. In practice, however, this turns out not to be the case. Data from the Bear Brook watershed manipulation project in Maine provide a good example. Flow separation calculations using the ratio of ^{18}O to ^{16}O chemical isotopes suggested that the majority of the stream flow during high discharge events was derived from "old water" (Kendall et al., 1995), a finding common to most ^{18}O studies. However, none of the soil lysimeter sites showed soil water chemistry comparable to stream chemistry (Norton et al., 1999). The conclusion of Norton and co-workers was that storm water chemistry at Bear Brook is governed by a mixture of soil solution and deep groundwater, new water (precipitation and snowmelt), and micropore water that is not well-sampled by tension lysimeters. The observed constancy or only slight dilution of base cation concentrations during high discharge periods suggested water sources deep within the soil profile, whereas the observed episodic increases in NO_3^- and DOC concentrations suggested sources in shallow soil areas. Partly in response to these recognized uncertainties in the routing of drainage waters within the watersheds, a hydrologic analysis was conducted for the Bear Brook watershed manipulation site by Chen and Beschta (in press).

To simulate the dynamic hydrological processes at the Bear Brook watershed manipulation site, Chen and Beschta (in press) used the Object Watershed Link System (OWLS) model of Chen (1996), a physically-based, three-dimensional distributed watershed hydrologic model. The OWLS model attempts to represent the major hydrologic processes within the watershed and also allows dynamic three-dimensional visualization of flow separation processes and variable source areas. Results of the flow separations suggested that surface flow from riparian areas was the predominant component of the flood rising limb, whereas macropore flow from riparian areas dominated during the falling limb of the hydrograph. Downstream riparian areas appeared to be the major contributing areas for peak flow. Because the ^{18}O results suggested that most of the high-flow discharge was "old water," it must be assumed that deep groundwater in the uplands re-emerges as near-surface flow in the lower riparian areas. More specific linkages between the simulated flow routing of drainage water and the observed soil water chemistry may further refine our understanding of these complex interactions.

6.2 Causes

Episodic acidification can be caused by several factors, including base cation dilution and organic acid enrichment, both of which are natural components of the hydrological response. Other potentially important factors include S and N enrichment that can be either natural and/or result from acidic deposition. The relative importance of these various factors and the extent to which they contribute to episodic acidification vary by region and individual watershed.

6.2.1 Natural Processes

The most important cause of episodic acidification of surface waters is base cation dilution. It is a completely natural process and typically accounts for a sizable fraction (often more than one-half) of the overall acidification response during snowmelt or rainfall events. Because hydrological episodes entail rapid water flow routing through upper soil horizons, base cations are contributed to drainage waters to a lesser extent than during periods of low flow. The additional large influx of water in the form of rain or meltwater, some of which makes only limited contact with watershed soils, further contributes to the observed dilution of base cation concentrations in stream water during high-flow events.

The altered hydrological flow-routing during episodes that contributes to lower base cation concentrations in stream water also causes increased concentrations of organic acid anions. This is because upper soil horizons tend to be relatively rich in organic C. Some of the organic acidity of the upper soils is transported to streams during high-flow events. The fraction of the episodic acidification that is caused by organic acid enrichment varies from watershed to watershed. In some cases, mainly in wetland-influenced high-DOC streams, the organic acid component of episodic acidification can dominate the episodic response. In other cases, episodic organic acidity is negligible compared with other components of the episodic response.

For the most part, base cation dilution and organic acid enrichment account for the ubiquitous nature of episodes. These processes operate with or without acidic deposition and can account for episodic loss of a few μeq/L of ANC to losses of 50 μeq/L or more during snowmelt or rainstorms.

Episodic acidification owing to S or N enrichment can also be a natural process in some areas, but both are typically associated with anthropogenic effects of acidic deposition. In particular, N-driven episodic acidification is frequently associated with high levels of N deposition.

6.2.2   Anthropogenic Effects

Nitrate in snowmelt runoff has been recognized for some time as an important component of biological damage resulting from atmospheric deposition (c.f., Wigington et al., 1990). Nitrate is the principal acid anion in snowmelt in many areas of northern Europe and the northeastern U.S. Selective elution of NO_3^- from the snowpack can result in early spring runoff having concentrations substantially greater than the average snowpack concentrations. The biological response to acidic runoff is similar, regardless of whether the predominant acid anion is NO_3^- or SO_4^{2-}, assuming concentrations of other ions, including Al_i, are the same.

Nitrate concentrations in surface waters exhibit a strong seasonality; NO_3^- is typically elevated during late winter and spring, particularly during periods of snowmelt, and reduced to low or nondetectable levels throughout summer and fall. This can be attributed to seasonal growth patterns of forest vegetation. Vegetation growth is reduced or stopped entirely during winter months, and microbial assimilation of N is also reduced during this season. Spring snowmelt can act to flush N into lakes and streams that was deposited in the snowpack from atmospheric deposition or N mineralized within the forest floor or soil during winter.

Except in cases of excessive N saturation, the effects of N deposition on surface waters are expected to be primarily episodic in nature. Unfortunately, data required to make regional assessments of episodic effects are generally not available. Such data need to be collected on an intensive schedule and must include sample periods during late winter and early spring when snowmelt often causes the most severe N-driven episodes of surface water acidification. Sampling during this time of year is more difficult and expensive than during the more common summer/fall sampling seasons. Sampling during snowmelt can be particularly difficult in the high mountains of the West, when study sites are often inaccessible, and when motorized transport (e.g., via snowmobile) is often not allowed owing to wilderness restrictions.

Aluminum concentration in drainage water is also greatly affected by hydrological variations. For example, Wigington et al. (1993) sampled 4 Adirondack streams in New York, 3 of which contained maximum concentrations of Al_i greater than or equal to 485 µg/L and maximum weekly average concentrations of Al_i greater than or equal to 264 µg/L. These are high concentrations of Al_i by any standards, and would be toxic to many species of fish. Nevertheless, all of these streams had minimum weekly average concentrations of 0 µg/L and during 25% of the weeks during the course of the year, the average weekly Al_i concentration was less than or equal to 94 µg/L. In other words, an assessment of potential Al toxicity conducted during the times of year when Al concentrations were at their lowest (summer and fall) would likely conclude that Al was of minor or negligible importance in these streams. Such an assessment would of course be in sharp contrast to the extremely high (and toxic) levels of Al_i achieved during winter and, especially, spring.

During periods of high discharge, especially during snowmelt, it has been frequently observed that increasing NO_3^- concentration contributes greatly to seasonal or episodic chemistry of streams, and to a lesser extent lakes. The observed NO_3^- pulse is often accompanied by a large increase in the concentration of Al_i. This has led to speculation that NO_3^- may be a more effective mobilizer of Al_i than SO_4^{2-} (Driscoll et al., 1991). However, the concentration of NO_3^- *per se* is not necessarily related to the concentration of Al_i in surface waters. For example, Wigington et al. (1993) reported data from three streams in the Catskill Mountains, NY, that had very similar maximum NO_3^- concentrations (129, 108, 106 µeq/L) and maximum weekly average NO_3^- concentrations (68, 72, 67 µeq/L). Despite these similarities in peak NO_3^- concentrations, however, peak Al_i concentrations differed by a factor of 4 (159, 72, 505 µg/L) and maximum weekly average Al_i concentrations differed by a factor of 7 (92, 49, 380 µg/L) in these 3 streams. Similarly, the 1 Adirondack stream (Biscuit Brook) sampled by Wigington et al. (1993) that had low concentrations of Al_i (75th percentile of weekly average concentration equal to 20 µg/L) had similar NO_3^- concentrations to the three Adirondack streams that exhibited much higher concentrations of Al_i.

The strongly elevated concentrations of Al_i in drainage waters that are often observed during winter and spring in general, and snowmelt in particular, suggest that assessments of acidification effects based on fall index, or synoptic survey, data are totally inadequate with respect to evaluating the dynamics and potential toxicity of Al. The general inadequacy of the hydrological component of most acidification assessments is widely acknowledged, and is perhaps most problematic when considering Al (Sullivan, 1994).

In near coastal environments, the neutral salt effect can have a large influence on episodic chemistry. For example, mechanisms of episodic acidification at Bear Brook watershed in Maine were at least partially controlled by the volume and chemistry of the precipitation event, especially the contribu-tion of ions from seaspray (Kahl et al., 1992; Norton and Kahl, in press). Sim-ilar results were found at Lake Skjervatjern in western Norway (Gjessing, 1994). It was well-known previously that antecedent soil moisture is an important determinant of episodic chemical response. It now also seems clear that, for near-coastal watersheds, the ionic make-up of the rainfall event is also important.

6.3   Extent and Magnitude

The ERP data for Adirondack streams (Figure 6.1) showed that as stream-water ANC decreased during hydrological episodes, the relative importance of NO_3^- vs. SO_4^{2-} acidity increased. In the acidic (ANC $\leq$ 0) samples from the ERP, NO_3^- concentrations were often of the same or similar magnitude as SO_4^{2-} concentrations. These are the chemical conditions (high concentrations of H^+ and Al_i) that are most toxic to fish. In samples that had positive ANC values, NO_3^- was of much less importance relative to SO_4^{2-} concentrations (Figure 6.1), but such conditions are not typically associated with adverse biological effects (Wigington et al., 1993).

All four of the Adirondack ERP study sites were located in the southwest-ern highlands area of the Adirondack Park, with maximum elevations of 710 to 775 m. Buck Creek, Bald Mountain Brook, and Seventh Lake Inlet are typical first- or second-order Adirondack streams. The Buck Creek catch-ment is characterized by steep terrain with numerous rock ledges and thin soils. The Seventh Lake Inlet watershed is a combination of moderately sloping terrain with deeper soils and bedrock outcrops. Bald Mountain Brook contains a higher percentage of deeper soils (Wigington et al., 1993). Buck Creek and Seventh Lake Inlet showed similar relationships between the ratio of NO_3^- to (SO_4^{2-} + NO_3^-) concentration and ANC, with increasing importance of NO_3^- as a strong acid anion at lower ANC values ($P < 0.0001$, Figure 6.1). Both had ANC that generally remained below about 20 μeq/L. The watershed with steepest slopes and shallowest soils (Buck Creek) had the lowest ANC. Bald Mountain Brook and Fly Pond Outlet showed much wider ranges of ANC values, and ANC was less related to the relative importance of NO_3^- as a strong acid anion in these streams. The stream gra-dient (9m/km) of Fly Pond outlet was the lowest of the 4 Adirondack ERP study sites. This stream had circumneutral water chemistry and constituted the reference stream for ERP biological studies in the Adirondacks (Wiging-ton et al., 1993).

Episodic acidification of streams in Shenandoah and Great Smoky Moun-tains National Parks has been demonstrated in several recent studies (e.g., Hyer et al., 1995; Eshleman et al., 1995; Nodvin et al., 1995; Webb et al., 1995). Streams with chronic ANC less than about 25 μeq/L, in particular, have been

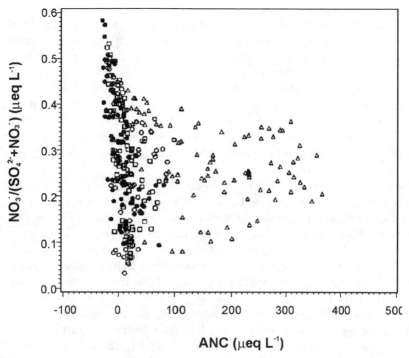

FIGURE 6.1

Ratio of NO_3^- : $(SO_4^{2-} + NO_3^-)$ concentration vs. ANC in stream-water samples collected during hydrological episodes in the four streams included in the Adirondack region of the ERP. Site identifications: Buck Creek, •; Bald Mountain Brook, °; Seventh Lake Inlet, ▫; Fly Pond Outlet, Δ. (Source: *Water, Air, Soil Pollut.*, Vol. 95, 1997, p. 330, Increasing role of nitrogen in the acidification of surface waters in the Adirondack Mountains, New York, Sullivan, T.J., J.M. Eilers, B.J. Cosby, and K.B. Vaché, Figure 8, copyright 1997. Reprinted with kind permission from Kluwer Academic Publishers.)

found to be subject to substantial ANC declines during precipitation events. Gypsy moth defoliation has also apparently contributed to episodic acidification at White Oak Run in Shenandoah National Park owing to increased NO_3^- leaching (Eshleman et al., 1995). Mean episodic ANC depressions increased by about a factor of two after the onset of defoliation.

Webb et al. (1994) developed an approach to calibration of an extreme event (episodic acidification) model for VTSSS long-term monitoring streams in western Virginia that was based on the regression method described by Eshleman (1988). Median, spring quarter ANC concentrations from 1988 to 1993 were used to represent chronic ANC, from which episodic ANC was predicted. Regression results were very similar for the four lowest ANC watershed classes, and they were, therefore, combined to yield a single regression model to predict the minimum measured ANC from the chronic ANC. Extreme ANC values were about 20% lower than chronic values, based on the regression equation:

$$\text{ANC}_{min} = 0.79\text{ANC}_{chronic} - 5.88 \ (r^2 = 0.97; \text{ se of slope} = 0.02, \ p \leq 0.001) (6.1)$$

Because the model was based on estimation of the minimum ANC measured in the quarterly sampling program, it is likely that the true minimum ANC values were actually somewhat lower than 20% below the measured chronic ANC. Nevertheless, regression approaches for estimation of the minimum episodic ANC of surface waters, such as was employed by Webb et al. (1994) for western Virginia, provide a basis for predicting future episodic acidification. A model such as MAGIC can be used effectively to derive estimates of future chronic ANC under various assumptions of atmospheric deposition. Episodic ANC can then be estimated by applying the regression approach.

Simple mixing models have also been used successfully elsewhere to simulate the magnitude of episodic acidification (c.f., Hendershot et al., 1992; Hooper and Christopherson, 1992; Schaefer and Driscoll, 1993). In the Sierra Episodes Study (Melack et al., 1998), minimum lake-water ANC was related to chronic lake-water ANC in a manner that was very similar to that found by Webb et al. (1994) for Virginia streams. For Sierra Nevada lakes, minimum ANC during snowmelt was about 12% lower than chronic ANC (Melack et al., 1998).

Episodes of surface water acidification involving increases in NO_3^- concentration have been reported for a few sites in the western U.S. (e.g., Loranger et al., 1986; Loranger and Brakke, 1988; Melack and Stoddard, 1991). Maximum NO_3^- concentrations in western lakes and streams reported during episodes tend to be low, however, typically less than 15 µeq/L (Stoddard, 1994). Although such episodic concentrations are quite low in comparison with many sites in the eastern U.S., increases in NO_3^- concentration during episodes at Emerald Lake in the Sierra Nevada have apparently been sufficiently high, when coupled with natural base cation dilution, to drive lake-water ANC to zero on at least two occasions (Williams and Melack, 1991a,b; Stoddard, 1994).

Episodic acidification is an important issue for surface waters throughout high-elevation areas of the West. A number of factors predispose western systems to potential episodic effects. There is an abundance of dilute to ultradilute lakes (i.e., those having extremely low concentrations of dissolved solutes), exhibiting very low concentrations of base cations, and therefore ANC, throughout the year. Large snowpack accumulations occur at the high elevation sites, thus causing substantial episodic acidification via the natural process of base cation dilution. Many of the high-elevation drainage lakes exhibit short hydraulic retention times, thus enabling snowmelt to rapidly flush lake basins with highly dilute meltwater. The hydrology, physical characteristics (e.g., bedrock geology, lake morphometry), and climate throughout high elevation areas of the West provide justification for considering potential episodic acidification to be an important concern. In addition, the few studies that have been conducted to date confirm the general sensitivity of western lakes to episodic processes.

Lakes and streams in the Sierra Nevada, Cascade, and Rocky Mountains are highly sensitive to potential acidic deposition effects because of the predominance of granitic bedrock, thin acidic soils, large amounts of precipitation, coniferous vegetation, and extremely dilute waters. There are no data to suggest, however, that lakes in these areas have experienced chronic acidification to ANC values less than zero to date, and based on examination of current chemistry, it appears that chronic acidification has not occurred to any significant degree. It is possible, however, that episodic effects are occurring under current deposition regimes, and that NO_3^- concentrations have caused a small loss of ANC on a chronic basis at some high-elevation sites. Unfortunately, the data that would be needed for such determinations have not been collected to a sufficient degree in acid-sensitive areas of the West to permit any regional assessment of either episodic or chronic N-driven acidification.

In the Rocky Mountain region, Loch Vale has been the subject of intensive research on hydrochemical responses to snowmelt. Loch Vale is located in Rocky Mountain National Park, CO, at an elevation of 3000 to 4000 m. The watershed is comprised primarily (81%) of exposed bedrock and talus (Baron and Bricker, 1987). The snowmelt is characterized by three general periods:

1. Flushing of concentrated lake water out of the lake during early snowmelt, and release of the winter's accumulated atmospheric deposition from the snowpack. The minimum lake-water pH (approximately 5.7) and highest NO_3^- concentrations occur at this time.

2. Dilution phase during which ionic concentrations in lake water are diluted by the continuing melt.

3. Decreasing hydrologic flows, yielding very dilute lake water during the summer months (Baron and Bricker, 1987; Wigington et al., 1990).

High concentrations of NO_3^- and SO_4^{2-} in the early phases of snowmelt have also been documented for other high-elevation sites in Colorado (Reddy and Caine, 1988) and Wyoming (Clow et al., 1988).

Vertucci and Corn (in press) collected weekly water samples during the snowmelt season at three sites in the Front Range, CO, ranging in elevation from 2800 to 3050 m. They also examined data from sites in the Elk Mountains, CO, and West Glacier Lake, WY. Snowmelt adjacent to West Glacier Lake was acidic (ANC = -20 µeq/L), although water sampled from the lake outlet maintained ANC greater than 0 on nearly all sampling occasions. Patterns of ANC at the Elk Mountain site were related to changes in base cation concentrations owing to dilution. Vertucci and Corn (in press) found no evidence of chronic acidic conditions at any of the three Front Range sites, and minimum ANC values were associated with low base cation concentrations, rather than elevated acid anion concentrations.

In the North Cascade Mountains, Loranger and Brakke (1988) observed a 50% decline in the ANC of Bagley Lake to a low of 67 µeq/L following snowmelt. Because the anion concentrations were so low in the snowpack, most of the ANC loss was attributed to dilution of base cations by meltwater. The annual average ANC of Bagley Lake was 100 µeq/L. From August through October, the ANC in the lake increased by nearly 30 µeq/L. The shallow depth of the lake promoted rapid mixing and flushing (short residence time). Thus, most of the NO_3^- in the snowpack passed through Bagley Lake (elevation 1800 m) during the period of most rapid snowmelt; by mid-June NO_3^- concentrations in the lake had decreased from over 5 µeq/L to less than 2 µeq/L. Loranger and Brakke (1988) concluded that NO_3^- and SO_4^{2-} concentrations in the snowpack were too low to result in significant pH depression during snowmelt. In contrast, Eilers et al. (1994b) observed virtually no change in ANC in a small lake in the southern Cascade Mountains, OR, during snowmelt. Lake Notasha, a seepage lake in southern Oregon also situated at 1800 m, varied only a few microequivalents from an average ANC of 10 µeq/L. Despite its extremely low base cation concentrations (approximately 20 µeq/L; Eilers et al., 1990), the lack of surface inlets to Lake Notasha results in a long hydraulic residence time (5 to 10 years) which provides a buffer against even substantial increases in acidic inputs. Because of high anion retention, seepage lakes in the southern Cascades such as Lake Notasha are likely far less sensitive to potential increases in acidic deposition than lakes with similar C_B elsewhere in the West.

In contrast to lakes in the southern Cascades, lakes in the Sierra Nevada receive both greater acidic deposition and higher rates of runoff. In the Emerald Lake watershed (elevation 2800 to 3416 m), NO_3^- increases of about 120% have been observed in the streams during snowmelt (Melack and Stoddard, 1991; Williams and Melack, 1991b). Concentrations of NO_3^- in the Emerald Lake outlet increased from 2 to 3 µeq/L in the fall to 10 to 13 µeq/L during spring runoff. The increase in NO_3^- and SO_4^{2-} (approximately 50%) was attributed to preferential elution from the snowpack and low retention rates in the watershed. Reduction of NO_3^- and SO_4^{2-} within Emerald Lake was relatively small, and most of the acid anions passed through the lake outlet. This increase of both SO_4^{2-} and NO_3^- during snowmelt in Emerald Lake in the Sierra Nevada contrasts with observations in the Adirondacks where snowmelt runoff diluted SO_4^{2-} as well as base cation concentrations (Schaefer et al., 1990).

Melack et al. (1993) reported two years of intensive research at seven high-elevation lakes in the Sierra Nevada. The lake elevations varied from 2475 to 3425 m, and the catchments spanned the length of the high-elevation Sierra Nevada. Solute concentrations, particularly ANC and base cation concentrations, were greatest during winter, declined to minima during snowmelt, and gradually increased during summer and autumn. Sulfate concentrations varied most in lakes with lowest volumes, and were generally less than 10 µeq/L in 5 of the 7 lakes. In Speuller and Topaz lakes, however, SO_4^{2-} concentrations generally ranged from 10 to 30 µeq/L, and from 30 to 60 µeq/L, respectively. These were the 2 shallowest lakes (mean depth 1.6 and 1.5 m) and were relatively high in elevation (3121 and 3219 m). Nitrate concentrations were generally low in all lakes (less than 4 µeq/L), although NO_3^- concentrations increased during snowmelt in most lakes owing to inputs of stream water enriched with NO_3^-. In the first few days of snowmelt, the concentration of NO_3^- in High Lake increased by a larger amount than did the concentration of base cations, causing ANC to decline to values below zero (Stoddard, 1995). Zooplankton species known to be intolerant of acidification were found in all seven lakes, and Melack et al. (1993) concluded that their presence is evidence that Sierra Nevada lakes are not currently showing chronic biological effects of acidic deposition.

6.4 Biological Impacts

The ERP (Wigington et al., 1993) included chemical and toxicological analyses of 13 streams in the northeastern U.S. during multiple seasons. *In situ* bioassays were conducted for brook trout, sculpins, and blacknose dace. At all of the streams, some bioassays resulted in low mortality (less than 10%), illustrating that toxic conditions do not occur at all times in a given stream throughout the year. Chemical conditions toxic to fish occurred at some time during the year in all of the ERP study streams, except the reference streams that were comparatively stable in their chemistry and generally had pH greater than 5.5. A key finding of the ERP study was that the single most important variable for predicting episodic fish mortality was the episodic concentration of Al_i. Of lesser importance, but also significant predictors of mortality, were Ca^{2+}, pH, and DOC. Furthermore, although fish exposed to high Al_i for longer periods of time had higher mortality, the time-weighted median Al_i concentration usually provided as good or better predictions of fish mortality than did more complex expressions of integrated chemical exposure (Wigington et al., 1993).

Episodic Response Project results clearly demonstrated that episodic acidification can have long-term adverse effects on fish populations. Streams with suitable chemistry during low flow, but low pH and high Al levels during high flow, had substantially lower numbers and biomass of brook trout than were found in nonacidic streams. Streams having acidic episodes showed significant mortality of fish. Some brook trout avoided exposure to stressful chemical conditions during episodes by moving downstream or into areas with higher pH and lower Al. This movement of brook trout only partially mitigated the adverse effects of episodic acidification, however, and was not sufficient to sustain fish biomass or species composition at levels that would be expected in the absence of acidic episodes. These findings suggested that stream assessments based solely on chemical measurements during low flow conditions will not accurately predict the status of fish populations and communities in small mountain streams (Baker et al., 1990c; Baker et al., 1996; Wigington et al., 1996).

The Shenandoah National Park Fish in Sensitive Habitats Project (FISH) is an ongoing effort to assess fish community responses to stream acidification in Shenandoah National Park, VA (Bulger et al., 1995). The aim is to assess potential impacts and to predict likely future effects based on current relationships between water chemistry and fish population responses. The study design includes high-frequency measurement of stream discharge and water chemistry (especially episodic chemistry) in three streams, fish community and habitat surveys, and bioassays with brook trout and blacknose dace. Additional streams are surveyed less intensively to address issues of spatial variability. Results of the FISH project will provide a wealth of biological response data, and will greatly improve our predictive capabilities in the area of biological effects of acidification. The results of the episodic aspects of this project will be prepared for publication in the near future (Art Bulger, University of Virginia, personal communication.).

The potential role of acid deposition in the observed recent declines in amphibian populations has been the subject of several recent studies in the western U.S. For example, the boreal toad (*Bufo boreas*) has experienced recent widespread decline throughout the southern Rocky Mountains since about 1975 (Corn et al., 1989; Carey, 1993). Leopard frogs (*Rana pipiens*) have also declined in Colorado and Wyoming (Corn et al., 1989). Harte and Hoffman (1989) hypothesized that episodic acidification was the principal cause of the decline of at least one amphibian species in Colorado, the tiger salamander (*Ambystoma tigrinum*). If episodic acidification is an important factor with respect to amphibian decline in the Rocky Mountains, then two conditions must be met: episodic acidification to toxic levels (of pH, Al, etc.) must occur, and sensitive life stages of the amphibian species must also be present at the time of the episodic acidification. Contrary to the situation in eastern North America, where spring and summer rainstorms are the dominant hydrological events that influence the chemistry of amphibian breeding habitats, episodic acidification events in the Rocky Mountains occur primarily during early snowmelt in the spring. The life history strategies of most amphibian species make them unlikely to be exposed to acidification during snowmelt (Corn and Vertucci, 1992; Vertucci and Corn, 1996). Direct mortality of *Bufo boreas* embryos from exposure to low pH is unlikely because most snowmelt and associated pH depression occurs prior to egg deposition (Vertucci and Corn, 1996). In addition, *Rana pipiens* generally occupy lakes at lower elevations that tend to be insensitive to episodic acidification (Corn and Vertucci, 1992). Vertucci and Corn (1996) concluded that there is no evidence that episodic acidification has led to acidic conditions in the Rocky Mountains or that sensitive amphibian embryos are present during the initial phases of snowmelt when episodic acidification might occur.

Quite different conclusions were reached by Kiesecker (1991) and Turk and Campbell (1997). In the area around Dumat Lake, just south of the Mt. Zirkel Wilderness Area in Colorado, 60 to 70% of tiger salamander eggs were dead or unviable in ponds at about pH 5.0 or less, about 40% in ponds at pH between 5.0 and 6.0, and about 20% at about pH 6.0 or greater (Kiesecker, 1991). Turk and Campbell (1997) used bulk snowpack acidity data from Buffalo Pass, adjacent to the Mt. Zirkel Wilderness Area, and measured amplification factors from an acid pulse during snowmelt at Loch Vale in Rocky Mountain National Park to predict the acidity of meltwater at Buffalo Pass. Their estimates of snowmelt acidity were high (e.g., pH less than 5.0) for a large portion of the snowmelt. It is important to note, however, that the greatest concentration of acidity in snowpack within the Rocky Mountains appears to occur in and near the Mt. Zirkel Wilderness Area (Turk and Campbell, 1997) where this study was conducted.

In the Sierra Nevada, no relationship was found linking current deposition chemistry and the status of amphibian populations (Bradford et al., 1992; Soiseth, 1992). We have insufficient data at this time with which to fully evaluate the role of episodic acidification as a causal factor in amphibian decline. Episodic acidification probably plays a role in the areas with the most acidic snowpacks, but it is unlikely to be regionally important throughout the high-elevation areas of the West. The extent to which acidic snowmelt influences the pH of amphibian breeding habitat in Rocky Mountain National Park and other Class I areas in Colorado remains uncertain. We would not expect many locations to exhibit such high levels of acidity pulses as were estimated by Turk and Campbell (1997) for Buffalo Pass simply because the bulk snowpack at most locations seems to be higher in pH (Peterson and Sullivan, 1998). In the absence of more detailed surveys of the chemistry of known or suspected amphibian breeding habitat within potentially sensitive areas, it is not possible to draw any firm conclusions at this time regarding potential effects on amphibians.

7
Nitrogen Dynamics

7.1 Nitrogen Cycle

Nitrogen is a critical element that controls species composition, biological diversity, and ecosystem functioning in a variety of ecosystem types, including forests, grasslands, fresh waters, estuarine, and near-coastal environments. Where N is limiting, many species are adapted to low levels of available N and can be adversely impacted when the N supply is increased. Fossil fuel combustion, agriculture, fertilizer production, and other human activities have greatly increased the availability and mobility of N over large areas, and in the process have substantially altered the global N cycle (Vitousek et al., 1997).

Nitrogen is added to watersheds in several forms (Figure 7.1). In areas not subjected to air pollution, the most important external source of N is N fixation, whereby atmospheric N_2 is converted into organic N that is incorporated into biomass. A fraction of this organic N is recycled each year through animal manure and biomass decomposition and mineralization, thereby constituting an internal input of NH_4^+ to the watershed soils. Anthropogenic inputs can include atmospheric deposition (wet, dry, occult), fertilizer application, and livestock manure, and can be in the form of NO_3^-, NH_4^+, or organic N. Internal ecosystem cycling results in biologically mediated transformations that mineralize organic N through decomposition processes, thereby converting organic N into NH_4^+. Ammonium can then be taken up by biota or converted to NO_3^- through the process of nitrification. A portion (usually small) of this NO_3^- can be lost to the atmosphere via denitrification. Nitrogen lost to leaching is mostly in the form of NO_3^- and dissolved organic N (DON). Tree or crop harvesting and livestock removal can also constitute important N losses from the watershed in some cases (Figure 7.1).

The N cycle and effects of excess N deposition on aquatic and forested terrestrial ecosystems are now reasonably well understood in general terms, owing in large part to research programs conducted within the last decade. Nitrogen is an essential nutrient for both aquatic and terrestrial organisms, and is a growth-limiting nutrient in most ecosystems. Thus, N inputs to

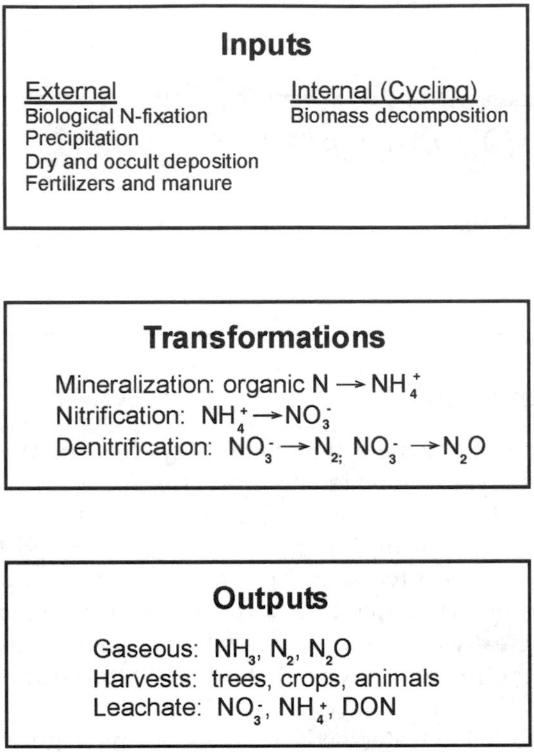

Inputs

External
Biological N-fixation
Precipitation
Dry and occult deposition
Fertilizers and manure

Internal (Cycling)
Biomass decomposition

Transformations

Mineralization: organic N $\rightarrow$ NH$_4^+$

Nitrification: NH$_4^+ \rightarrow$ NO$_3^-$

Denitrification: NO$_3^- \rightarrow$ N$_2$; NO$_3^- \rightarrow$ N$_2$O

Outputs

Gaseous: NH$_3$, N$_2$, N$_2$O
Harvests: trees, crops, animals
Leachate: NO$_3^-$, NH$_4^+$, DON

FIGURE 7.1
Major components of the nitrogen cycle.

natural systems are not necessarily harmful. For each ecosystem, there is an optimum N level that will maximize ecosystem productivity without causing significant changes in species distribution or abundance. Above the optimum level, harmful effects can occur in both aquatic and terrestrial ecosystem compartments (Gunderson, 1992).

Nitrogen compounds are found in the atmosphere in reduced (NH$_3$, NH$_4^+$) and oxidized (NO, NO$_2$, HNO$_2$, HNO$_3$, PAN) forms. Whereas S emissions in North America and Europe increased to maximum levels in the 1970s or 1980s and have subsequently been declining, N emissions have remained stable in recent years or in some areas have been increasing. Emissions into the atmosphere of N oxides (NO$_x$) are mainly derived from fossil fuel combustion. Important sources include motor vehicles, power plants, biomass burning, and industry. Reduced N is mainly emitted into the atmosphere from agricultural sources, especially animal production and the production and application of fertilizers.

Many forested areas in Europe currently receive N deposition in excess of 20 kg N/ha per year. This elevated deposition of N to European forests is a chronic addition to the natural background flux of mineral N from net mineralization. N deposition levels in North America tend to be much lower, only exceeding 10 to 12 kg N/ha per year in limited areas. Recent studies have also indicated that some watersheds in Japan are N saturated (Ohrui and Mitchell, 1997; Mitchell et al., 1997). Forest decline has also been reported at some Japanese sites that receive high levels of N deposition (Katoh et al., 1990). Recently, Vitousek et al. (1997) discussed alterations to the global N cycle caused by human activities, and Fenn et al. (1998) provided an overview of the effects of excess N deposition on sensitive North American ecosystems.

The biogeochemical cycling of S and its role in watershed acidification has been better understood for a longer period of time than is the case for N. The N cycle is extremely complex and controlled by many factors besides atmospheric emissions and consequent deposition. Also, N inputs that may be beneficial to some species or ecosystems may be harmful to others. Increased atmospheric deposition of N does not necessarily cause adverse environmental impacts. In most areas, added N is taken up by terrestrial biota and the most significant effect seems to be an increase in forest productivity (Kauppi et al., 1992). However, under certain circumstances, atmospherically deposited N can exceed the capacity of forest ecosystems to take up N. In some areas, especially at high elevation, terrestrial ecosystems have become N saturated* and high levels of deposition have caused elevated levels of NO_3^- in drainage waters (Aber et al., 1989, 1991; Stoddard, 1994). This enhanced leaching of NO_3^- causes depletion of Ca^{2+} and other base cations from forest soils and can cause acidification of soils and drainage waters in areas of base-poor soils.

An international conference on N and its environmental effects was convened in The Netherlands in March 1998, under the Convention on Long-Range Transboundary Air Pollution of the United Nations Economic Commission for Europe (UN/ECE). Conclusions and recommendations from the conference included the following (Erisman et al., 1998):

- Increased growth of trees in European forests have been owing in part to increased atmospheric deposition of N. Tree growth increases as N deposition increases until the ecosystem becomes N-saturated, and then growth may decline.

- Increased N deposition can cause nutrient imbalances in forest vegetation and loss of biodiversity.

- An integrated approach to N-pollution abatement is needed, with consideration of acidification, eutrophication, human health, and climate change issues.

N inputs to forested and alpine ecosystems include atmospheric deposition of NO_3^-, NH_4^+, and organic N, as well as N fixation, and in some cases can also

---

* The term nitrogen saturation has been defined in a variety of ways, all reflecting a condition whereby the input of nitrogen (e.g., as nitrate, ammonium) to the ecosytem exceeds the requirements of terrestrial biota and a substantial fraction of the incoming nitrogen leaches out of the ecosystem as NO_3^- in groundwater and surface water.

include fertilization. N fixation provides variable quantities of N to the forest and can be carried out by bacteria associated with plant roots, soil microbes, and lichens found in the forest canopy (Bormann et al., 1993; Sollins et al., 1980). In rare instances, geologic N can be an important contributor to the N flux through forests (Dahlgren, 1994). Most of the N contributed to the forest from the mix of potential N sources described previously is subsequently retained to a significant extent within the watershed, largely by plant uptake, microbial assimilation, and abiotic incorporation of N into soil humus.

Variation in the percent of N inputs that is retained in watershed soils and biota is generally rather small (c.f., Kahl et al., in press); retention is typically in the range of 80 to 100% at sites that receive low to moderate levels of atmospheric N deposition (i.e., less than 20 kg N/ha per year). For example, estimates of the retention of NO_3^- and NH_4^+ at Hubbard Brook Experimental Forest, New Hampshire were 85 and 84%, respectively, and 85% N retention was estimated for Arbutus watershed in the Adirondack Mountains of New York (Mitchell et al., 1996). At the experimental West Bear Brook catchment, N retention has consistently been about 82% in all except the first year of acidification (Kahl et al., in press). At the NITREX reference sites in Europe, only those sites that receive fairly high levels of N deposition (greater than 15 kg N/ha per year) leaked significant percentages of input N. The percent N retention at the treated Sogndal catchment in Norway (SOG4) was identical (88%) to that of the untreated reference catchment (SOG1) over a 9-year period of record (Wright and Tietema, 1995). At the Gårdsjön NITREX site in Sweden, percent N retention remained very high during the first 2 years of experimental addition of N even though the total N deposition to the site (ambient plus experimental loading) was greater than 40 kg N/ha per year. The percent watershed retention of N was only 1% lower at the treatment catchment (98.9%) than it was at the reference catchment (99.9%, Moldan et al., 1995) after 2 years. After 5 years of experimental treatment, the catchment was still retaining about 95% of the N input (Moldan and Wright 1998a).

Fertilization of a mixed hardwood forest plot at Harvard Forest, MA, with very high levels of N over an 8-year period (greater than 900 kg N/ha) resulted in virtually no net loss of NO_3^- (Aber et al., 1995). The observed large N-retention capacity of this forest is believed to have been caused in part by intensive land management during previous decades.

Although there are certainly exceptions to this pattern, the percent N retention by forested ecosystems under depositional regimes that can reasonably be expected to occur in the U.S. should generally be greater than 80%. For alpine ecosystems that lack extensive soil coverage, it would not be unreasonable to expect that the percent watershed retention of N could be much less. We have insufficient data on alpine systems that receive more than about 5 kg N/ha per year to form a judgement at this time.

N saturation of watershed soils, and associated high levels of NO_3^- leaching in soil waters and surface waters can cause a wide range of environmental problems in a wide array of ecosystems and ecosystem compartments. This is a result of the critical importance of N for life processes (e.g., protein synthesis) and the fact that N is poorly stored in soils in a form that is readily available to biota.

Even though soils store very large quantities of N and, in fact, constitute by far the largest ecosystem pool for N in forested ecosystems, biologically available N pools in the soil are generally very small relative to vegetative and microbial demand. Thus, the N cycle of forest ecosystems is usually very tight, turning over several times per year. Most biocycling involves NH_4^+, that is produced by the mineralization (decomposition) of organic materials. NH_4^+ is readily taken up by plant roots and microbes and converted into organic N that is recycled back into the soil system through litterfall, death, and decomposition. A relatively small amount of the soil NH_4^+ is converted to NO_3^- by the process of nitrification, and only a small amount of that NO_3^- is typically lost from the ecosystem as NO_3^- leaching or gaseous losses (e.g., N_2O) via denitrification.

Nadelhoffer et al. (in press) conducted an ^{15}N tracer study at Bear Brook watershed to characterize N cycling processes and identify sinks for experimental NH_4-N additions to the watershed. Changes in the ^{15}N content of plant tissues, soils, and stream water after adding the isotopic tracer illustrated that soils were the dominant sink for the added NH_4^+. Although the $(NH_4)_2SO_4$ addition caused increased NO_3^- leaching, the ^{15}N data suggested that only 15% of the NO_3-N exported from the watershed during 2 years of tracer addition was derived from the 42 kg/ha of labeled NH_4-N additions. Thus, most of the exported NO_3^- was derived from watershed N pools with residence times greater than 2 years, and not directly from nitrification of the recent deposition (Nadelhoffer et al. in press).

In areas of the U.S. heavily influenced by photochemical smog, such as the Los Angeles Basin, deposition of oxidized N compounds can be quite high, in some cases higher than 20 kg N/ha (Fenn and Bytnerowicz, 1993). Nitrogen deposition in these highly exposed areas has caused N saturation of chaparral and mixed conifer stands and consequent high NO_3^- concentrations in stream water and soil water (Fenn and Bytnerowicz, 1993; Riggan et al., 1985, 1994; Bytnerowicz and Fenn, 1996).

Dry deposition of N is of greater magnitude than wet deposition in many parts of California owing to the arid climate, and this pattern is magnified in areas that experience frequent temperature inversions. Bytnerowicz and Fenn (1996) reviewed atmospheric concentrations and deposition of N compounds and their biological effects in California forests that receive much higher deposition of N than S.

7.2   Environmental Effects

Excess N can affect the ecosystem at many levels. At high concentration, NO_3^- contaminates drinking water and can be directly toxic to aquatic life. NO_3^- is relatively efficient at mobilizing and transporting Al from soils to soil waters and surface waters. Dissolved Al and associated acidity can deplete base cations (e.g., Ca^{2+}, Mg^{2+}) from the soil cation exchange complex and lead to toxic responses in aquatic biota and plant roots. Because N is frequently limiting for algal growth in aquatic ecosystems, eutrophication can result from excess N, especially in estuarine systems. In pristine alpine and subalpine terrestrial ecosystems, the limiting factor for primary production is often N supply that is largely determined by the ability of soil microbes to fix atmospheric N_2 and to mineralize organic N. Most terrestrial ecosystems are considered N limited (Friedland et al., 1991; Bowman et al., 1993). Inputs of anthropogenic atmospheric N to plant communities have the potential to alter plant community structure and increase sensitivity to water stress, frost, and herbivory (Bowman et al., 1993), as well as to contribute NO_3^- to drainage waters.

The end results of N saturation can include forest decline, reduced forest growth, increased forest susceptibility to disease and insect infestation, eutrophication of estuaries and near-shore oceans, fresh water and soil acidification, loss of fish and other aquatic life, and changes in terrestrial and aquatic biodiversity. Fortunately, atmospheric N inputs to most forests are not high enough to cause such problems. Because of the severity of the potential effects, however, it is important that we understand the N cycle and the extent to which it is being perturbed by atmospheric emissions.

The complexities of the N cycle make development of such understanding challenging, to say the least. These complexities are also what makes it so interesting to study environmental N effects. Study of the N cycle encompasses a huge diversity of disciplines, from atmospheric physics and hydrology to chemistry and biology. All levels of life are directly tied to and/or affected by the cycle, from microbes and mycorrhizal fungi to plants and animals. All major ecosystem compartments are involved in the cycling of N through the system: foliage, roots, soil, microbial communities, soil water, stream water, algal communities, and so forth.

To further complicate the situation, N cycling is also regulated to a significant degree by climate, disturbance, and land management. Such factors are believed to have both short- and relatively long-lasting (i.e., decadal to century) effects on the response of forest ecosystems to atmospheric N deposition (Mitchell et al., 1996; Aber et al., 1989, 1995b, 1998; Fenn et al., 1998). Thus, the extent to which the land was logged, burned, or used for agricultural production in the past, perhaps even during the previous century, can profoundly affect the N status of the soils and, therefore, the extent to which N deposition will or will not cause environmental degradation. Many scientists, policymakers, and concerned citizens long for simple environmental cause/effect relationships. Increase N deposition and bad things happen. Decrease N deposition and good things happen. This is clearly not how it works.

The concentration of NO_3^- in runoff at the NITREX experimental N-addition site at Gårdsjön, Sweden, showed a pattern of higher N loss during winter and lower N loss during summer. Moldan and Wright (1998a) demonstrated a strong nonlinear inverse relationship between mean air

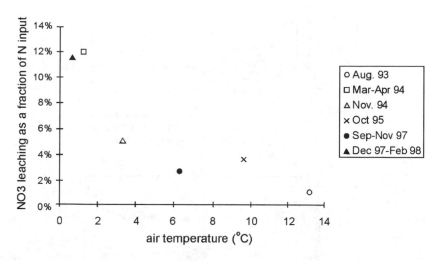

FIGURE 7.2

Observed relationship between NO_3^- leaching loss in runoff and mean air temperature at the G2 NITREX site at Gårdsjön, Sweden. Each point represents an average of 14 to 90 days. (Reprinted from Forest Ecology and Management, Vol. 101, Moldan, F. and R.F. Wright, Changes in runoff chemistry after five years of N addition to a forested catchment at Gårdsjön, Sweden, p. 442, Copyright 1998. With permission from Elsevier Science.)

temperature and NO_3^- leaching, with a threshold between about 2 and 5°C, below which NO_3^- leaching losses greatly accelerated (Figure 7.2). Moldan and Wright (1998a) speculated that the rates of most N uptake processes, for example by microbes, fine roots, and mycorrhiza, are strongly reduced below such a temperature threshold.

Interestingly, Murdoch et al. (1998) found pretty much the opposite effect at Biscuit Brook, a headwater stream in the Catskill Mountains, NY. Volume-weighted mean stream NO_3^- concentration and both annual (Figure 7.3a) and seasonal (data not shown) average air temperature were positively correlated. Water year (WY) 1990 was an outlier in the observed relationship, and this was attributed to higher N deposition and an unusually severe cold weather period with little snow cover during the 1989–1990 winter. Similarly, Murdoch et al. (1998) found that the concentration of NO_3^- in stream water during the late summer base flow period was highly correlated with average annual air temperature (Figure 7.3b). They attributed these results to the temperature-dependence of nitrification, whereby nitrification is greater at higher temperatures (Figure 7.3c).

We do not know why the results at Gärdsjön and Biscuit Brook are opposite, but it seems that nitrification and uptake processes are affected by temperature in opposite directions. Where NO_3^- leaching is limited by nitrification, a positive relationship, such as was found at Biscuit Brook, might be expected. Perhaps in other situations, NO_3^- leaching is more limited by N uptake, which is enhanced at warmer temperatures.

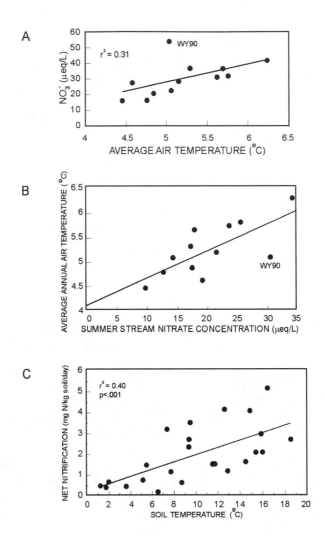

FIGURE 7.3

Relationships between nitrogen and temperature: (A) average annual stream NO_3^- concentration and average annual air temperature; (B) average annual air temperature and late summer baseflow stream NO_3^- concentration; and (C) net nitrification and soil temperature. Stream data are from Biscuit Brook in the Catskill Mountains, NY, from 1984 to 1995; nitrification data are from a nearby deciduous forest stand, July 1993 to July 1996. Water year 1990 was identified as an outlier owing to unusually high dormant season N deposition and extremely cold temperatures during December of that year. (Source: Reprinted with permission from Murdoch et al., 1998, Environmental Science & Technology, Vol. 32, p. 1644-1646, Figures 3A, 6, 7B, Copyright 1998, American Chemical Society.)

It has recently been hypothesized that prior land use history, extending back 100 years or more, can have a major effect on forest response to N deposition (Aber and Driscoll, 1997; Foster et al., 1997; Aber et al., 1997, 1998). The greater the previous extraction of N from the site by agricultural conversion, fire, logging, or other disturbance, the more N the forest will be able to absorb without becoming N saturated. Aber et al. (1998) contended that previous land use is more important than either current or total accumulated N deposition as a controlling factor for N saturation in the northeastern U.S.

N cycling operates on multiple time scales. Assimilation of N by microbes and consequent mineralization can be very rapid (e.g., days) whereas N turnover and cycling between plants and soils can occur over much longer periods of time (e.g., year or longer; Fenn et al., 1998). Aber et al. (1989) provided a conceptual model of the changes that occur within the terrestrial system under increasing loads of atmospheric N. Stoddard (1994) described the aquatic equivalents of the stages identified by Aber et al. (1989), and outlined key characteristics of those stages as they influence seasonal and long-term aquatic N dynamics. In a recent review for North American ecosystems, Fenn et al. (1998) described the geographic extent of known N saturation and the factors predisposing terrestrial ecosystems to N saturation.

Although some high-quality research has been conducted in the U.S. on the environmental effects of atmospheric N deposition, such research has been conducted to a far greater extent in Europe. The number, and quality, of European N studies have increased tremendously since the 1980's (Sullivan, 1993). As new research initiatives are developed in the U.S. that include N, much can be gained from examining recent findings and research priorities developed overseas. This information is critical to assure that new research priorities, monitoring efforts, modeling studies, and process-level research on N are fully integrated with, and complementary to, studies already conducted or underway in Europe.

The European scientists have concluded that it is important to study N questions as large multidisciplinary, multi-investigator research teams. This is because of

1. The complexities of the N cycle.
2. The multitude of scientific disciplines involved in its study.
3. The importance of expensive, large-scale, whole-system manipulations as a tool for studying N effects (see further discussion in Chapter 8).

A high degree of international and interinstitutional cooperation has developed in recent years within Europe. This spirit of cooperation has been evident in several international umbrella projects on N effects, especially NITREX and EXMAN (Wright and van Breemen, 1995; Beier and Rasmussen, 1993; Tietema and Beier, 1995).

NITREX (Nitrogen Saturation Experiments) is a consortium of experiments in which N deposition has been drastically changed for whole catchments or forest stands at eight sites spanning the present-day gradient of N deposition across Europe (Dise and Wright, 1992; Wright and van Breemen, 1995). At sites receiving low to moderate N deposition (3 to 20 kg N/ha per year), N has been experimentally added to precipitation in an effort to induce N saturation. At sites with high N deposition (20 to 54 kg N/ha per year) and significant leaching losses of NO_3^-, N is removed from precipitation by means of roofs and ion-exchange systems. A variety of ecosystem processes are investigated at each of the sites in an effort to quantify the factors that lead to enhanced NO_3^- leaching. EXMAN (Experimental Manipulation of Forest Ecosystems in Europe) has involved a similar approach; N inputs, water availability, and nutrient inputs have been manipulated to varying degrees (Beier and Rasmussen, 1993). These experimental programs are discussed in greater detail in Chapter 8.

Emmett et al. (1998) summarized data on the N status of the forested NITREX sites and ecosystem responses to experimental N additions and exclusions. Nitrogen leaching losses were highly variable, ranging from less than 5 to about 80% of inputs, and this seemed to depend on the initial N status of the site and the form of deposited N (as NO_3^- or NH_4^+). At low N-status stands, such as Gårdsjön and Klosterhede, both NO_3^- and NH_4^+ were strongly retained and, therefore, leaching losses of N were low in response to both ambient and enhanced N deposition (Figure 7.4). At high N-status

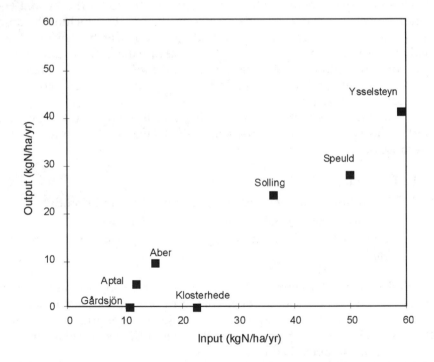

FIGURE 7.4

Ambient inputs in throughfall and leaching losses in streams or soil water of inorganic N at forested NITREX sites. (Source: Ecosystems, Predicting the effects of atmospheric nitrogen deposition in conifer stands: evidence from the NITREX ecosystem-scale experiments, Emmett, B.A., D. Boxman, M. Bredemeier, P. Gundersen, O.J. Kjønaas, F. Moldan, P. Schleppi, A. Tietema, and R.F. Wright, Vol. 1, p. 354, Figure 2, Copyright 1998, Springer-Verlag. With permission.)

stands, N retention was lower, because N supply was somewhat greater than demand, and the magnitude of leaching loss was apparently determined by the availability of NO_3^- for leaching. Nitrate is made available for leaching either directly through NO_3^- deposition or is generated internally by nitrification of available NH_4^+ in the soil. If the nitrification rate at the site is low, leaching losses are much greater if N is deposited as NO_3^- rather than NH_4^+. This situation was evident at the Aber site where NO_3^- leaching losses were double when N was applied as NO_3^- relative to application of 50% NO_3^- and 50% NH_4^+. Where the nitrification rate is high, such as Ysselsteyn, leaching losses of N can be higher than the input of NO_3^--N. This is because a sizable component of the NH_4^+ input is nitrified and subsequently leached (Emmett et al., 1998).

Terrestrial effects have not been pronounced at most of the NITREX sites in response to the experimental treatments (Emmett et al., 1998). Measurements included nutritional status of the trees, wood accumulation, mycorrhizal diversity, fine root biomass, deposition processes, and invertebrate populations. However, at the highly N-saturated Ysselsteyn site that received 45 kg N/ha per year of input, tree growth increased markedly following reduction in N deposition (Boxman et al., 1998).

The C : N ratio of the forest floor was strongly related to the fraction of N input leached from the European sites as NO_3^- (Emmett et al., 1998; Figure 7.5). This has been attributed to the controlling influence of the forest floor C : N ratio on soil nitrification (Gundersen et al., 1998). The NITREX results suggested a threshold forest floor C : N below which NO_3^- leaching is enhanced. As illustrated by the data shown in Figure 7.5, it would be highly desirable to obtain additional data on NO_3^- leaching at sites that exhibit C : N between about 23 and 30.

A number of factors can be involved in controlling the loss of N from a forested watershed to drainage waters, including atmospheric inputs, forest stand age and condition, the size of soil N and C pools, and flowpaths of percolation and melt-water within the catchment. The stages of N loss from the watershed, as described by Stoddard (1994), are depicted in Figure 7.6. Changes in both the seasonal and long-term patterns in surface water NO_3^- concentrations reflect changes that occur within the watershed in N cycling and the degree of N saturation. At Stage 0, NO_3^- concentrations in drainage waters are very low throughout most of the year, and increase to measurable concentrations typically only during snowmelt or spring rainfall hydrological events (Figure 7.6a). The loss of N in runoff is short-lived and small in magnitude. This was viewed by Stoddard (1994) as the "natural" pattern. At Stage 1, that pattern is amplified; spring concentrations of NO_3^- in surface waters reach relatively high concentrations and the seasonal onset of N limitation is delayed (Figure 7.6b). In Stage 2, N begins to percolate beneath the rooting zone of the soil, resulting in elevated groundwater concentrations of NO_3^-. Seasonality is damped because baseflow concentrations of NO_3^- are high (Figure 7.6c). In Stage 3, the watershed becomes a source, rather than a sink, for atmospheric N. The combined inputs of N from deposition, miner-

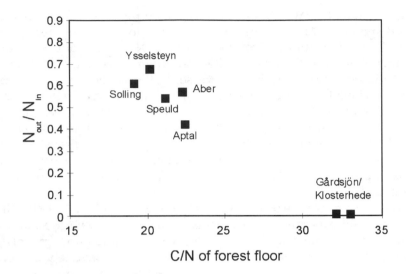

FIGURE 7.5

Observed relationship between the proportion of N inputs leached (N_{out}/N_{in}) and the C/N ratio of the forest floor at forested NITREX sites. (Source: Ecosystems, Predicting the effects of atmospheric nitrogen deposition in conifer stands: evidence from the NITREX ecosystem-scale experiments, Emmett, B.A., D. Boxman, M. Bredemeier, P. Gundersen, O.J. Kjønaas, F. Moldan, P. Schleppi, A. Tietema, and R.F. Wright, Vol. 1, p. 355, Figure 4B, Copyright 1998, Springer-Verlag. With permission.)

alization, and nitrification produce concentrations of NO_3^- in drainage water of Stage 3 watersheds that can be higher than deposition (Figure 7.6d, Stoddard, 1994).

A variety of other symptoms, in addition to NO_3^- leaching, are suggestive or indicative of N overfertilization of forest ecosystems. Most have to do with measurement of the ratio of N to one or more other elements in one or more ecosystems compartments. In foliage, the N : Mg and N : P ratios are commonly employed. Soil C : N is also a common indicator of N saturation (Fenn et al., 1998).

Although most forests retain the majority of N inputs that they receive, some forested ecosystems leach significant amounts of NO_3^- to drainage water, and this occurs under a range of N deposition input levels. At some sites in the U.S., relatively high levels of N deposition (10 to 30 kg N/ha per year) have been shown to result in high N leaching losses (7 to 26 kg N/ha per year). Good examples include watersheds in Great Smoky Mountains National Park in Tennessee (Johnson et al., 1991) and the San Bernadino Mountains in southern California (Fenn et al., 1996). At Fernow, WV, high N loading (15 to 20 kg N/ha per year) has caused moderate N leaching losses (approximately 6 kg N/ha per year; Gilliam et al., 1996; Peterjohn et al., 1996). Despite the observed variability in ecosystem response, a number of generalizations can be made. Coniferous forests and alpine ecosystems seem to be more prone to N saturation than deciduous forests (Aber et al., 1995a, 1998;

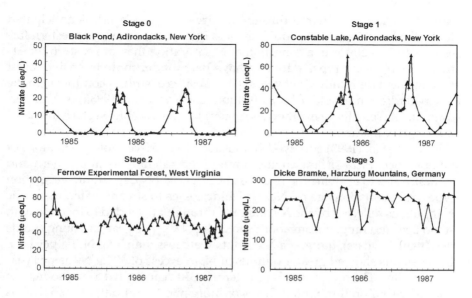

FIGURE 7.6

Example patterns of NO$_3^-$ concentration in surface water at various stages of watershed N saturation. Stage 0, Black Pond in the Adirondack Mountains, NY. Nitrogen transformations are dominated by plant uptake and microbial assimilation. Only small amounts of NO$_3^-$ leach out of the watershed, primarily during snowmelt. Stage 1, Constable Pond in the Adirondack Mountains, NY. As compared to Stage 0, this stage is characterized by delayed onset of N limitation during the spring and larger peaks of NO$_3^-$ in runoff. Stage 2, Watershed #4 at Fernow Experimental Forest, WV. Uptake of N by plants and microbes is reduced, as compared with Stage 1, resulting in loss of NO$_3^-$ to streams during winter and spring and to groundwater during the growing season. Stage 3, Dicke Bramke, Harzburg Mountains, Germany. Nitrate concentrations are chronically high throughout the year; the watershed can become a source, rather than a sink, for atmospheric N. From Stoddard (1994).

Williams et al., 1996a). Young and successional forests, with their greater nutrient demand, are less prone to N saturation than mature stands (Peet, 1992). Climate, soil N pool size, and land use dramatically alter biological N-demand (Cole et al., 1992; van Miegroet et al., 1992; Feger, 1992; Magill et al., 1996; Fenn et al., 1998). It is not difficult to understand such observed differences. Because N is so critical to ecosystem primary productivity, other factors that influence plant and microbial growth (e.g., water availability, temperature) or the abundance or availability of other nutrients (e.g., Ca, Mg, P) will have significant effects on the ability of plants and microbes to utilize N and, therefore, the extent to which N will leach in drainage waters.

Forest stand age has been found to be associated with N retention by the forest system. Watersheds having older trees seem more likely to leach N to a higher degree than forests having younger trees (e.g., Vitousek, 1977; Elwood et al., 1991; Emmett et al., 1993). In general, forests in the eastern U.S. have been logged on many more occasions and subjected to more intensive forest management practices than forests in the western U.S. In addition, stands of old growth trees that are generally absent from the eastern U.S., can still be found scattered throughout areas of the West. It is anticipated that, all other things being equal, the less intensively managed western forests should be closer to a condition of N saturation than are eastern forests that have been cut repeatedly and have been under more intensive forest management for hundreds of years. Thus, results of studies conducted in the east regarding thresholds of N saturation (e.g. Kahl et al., 1993a) may underestimate the sensitivity of otherwise generally comparable western forests to increased N loading.

Emmett et al. (1993) surveyed 20 forested (10 to 55-year-old plantations of Sitka spruce, *Picea sitchensis*) and 5 moorland catchments in northern and central Wales to evaluate the relative importance of N deposition, vegetation type, and plantation age on NO_3^- leaching losses to stream water. Inorganic N inputs, as reflected by the N content of throughfall, tended to increase with stand age, and ranged from about 9 to 25 kg N/ha per year. In young stands (less than 30 years), inorganic N outputs were less than 5 kg N/ha per year. Older stands showed greater outputs of N, in excess of 30 kg N/ha per year at some sites. The oldest stands (greater than 50 years) acted as net sources of total N, based on throughfall fluxes of inorganic N and outputs of DON as well as NO_3^-.

Tietma and Beier (1995) integrated preliminary results from long-term ecosystem manipulation experiments at 12 sites in northwest and central Europe. The sites were included in the EXMAN (Rasmussen et al., 1990) and/or NITREX (Dise and Wright, 1992) research programs. The integration focused on site properties indicative of aspects of N cycling, such as ecosystem fluxes and elemental concentrations in the various ecosystem compartments at the unmanipulated NITREX control plots. All of the sites are located in coniferous forests except Sogndal, Norway, which is alpine. Data were collected to describe the inorganic N input (as reflected in the N concentrations in precipitation and throughfall) and output (in soil water or drainage water) fluxes, stand age, and various parameters that characterize internal N cycling. Regression analyses were conducted to ascertain if N fluxes in precipitation and throughfall were correlated, if N concentrations in ecosystem compartments were related to N inputs in deposition, and the extent to which N output fluxes could be predicted on the basis of inputs or N concentrations in various ecosystem compartments. Such issues are critical to development of an understanding of N dynamics at the plot or catchment scale and also for developing a foundation for predicting changes in N cycling processes and process rates under changing levels of atmospheric inputs.

At the 8 forested sites characterized by high NH_4^+ inputs (greater than 50% of total inorganic N), much of which occurred in the form of dry deposition, Tietema and Beier (1995) found the total N flux in precipitation was highly correlated ($r^2 = 0.93$, $p < 0.001$) with NH_4^+ fluxes in throughfall. All of these sites are located in areas of moderate to intensive agricultural activity. At the other three forested sites located in less polluted areas (Switzerland, Wales, Sweden), atmospheric inputs of N primarily occurred as NO_3^-. The analyses supported the conclusion that increased N concentrations in the various eco-system compartments is caused by increased N inputs (McNulty et al., 1991; Heinsdorf, 1993; Tietema, 1993). Total N wet deposition (kg N/ha per year) was significantly correlated with N concentration in current year needles (r^2 = 0.76, p < 0.05), needle litterfall (r^2 = 0.80, p < 0.05), and the ectorganic forest floor layer (r^2 = 0.82, p < 0.005). Somewhat weaker but statistically significant relationships were also found between NH_4^+ deposition in throughfall and the N concentration in those ecosystem compartments.

Excluding 1 site in The Netherlands that exhibited an extremely high N output flux (greater than 40 kg N/ha per year), total N output flux was highly correlated with wet N deposition (r^2 = 0.83, p < 0.001) and the N concentration in each of the ecosystem compartments investigated. Multiple regressions using N concentrations in current-year needles, needle litterfall, and in the ectorganic layer as independent variables explained 96% of the variation in N output. Thus, the N concentration in these ecosystem compartments provides a useful index of N-saturation status at the plot or catchment scale (Tietema and Beier, 1995). These results, and those from additional study areas have been, and are being, used as the basis for modeling N saturation.

A threshold of NO_3^- leaching at deposition levels above about 10 kg N/ha per year was found; no significant leaching was observed at deposition levels below this threshold, in agreement with results of a survey of 65 forested European plots and catchments by Dise and Wright (1995).

The conceptual model of N saturation that was presented by Aber et al. (1989) was revised and updated in response to these and other more recent research results. Aber et al. (1998) summarized the results of field experiments in 4 forest stands in Massachusetts, Vermont, and Maine and also a survey of 161 spruce–fir forest stands situated along an N depositional gradient (4 to 13 kg N/ha per year) across New York and New England. The manipulated plots included two coniferous stands (Harvard Forest pine stand, MA, and Mt. Ascutney spruce–fir stand, VT) and two mixed hardwood stands (Harvard Forest and Bear Brook, ME). Results of these studies and those of the European NITREX experiments were evaluated with respect to their consistency with the ecosystem responses hypothesized earlier by Aber et al. (1989). In most cases, the earlier hypotheses were supported by the more recent data. Experimental sites showed increased net N mineralization, as did the stands in the midpoint of the spruce–fir transect, relative to the stands receiving the lowest N deposition. With one exception, net nitrification was induced or increased, resulting in increased mobility and leaching losses of NO_3^-, as predicted by the earlier conceptual model. All sites showed increases in foliar N concentrations, as hypothesized. Both of the experimentally manipulated conifer stands and the regional survey showed declining tree growth or increased tree mortality (McNulty et al., 1996; Aber et al., 1998). There were also a number of surprises, however, that caused Aber et al. (1998) to revise their conceptual model in several ways (Figure 7.7). First, although net N mineralization increased initially, longer-term responses in all except the Harvard Forest hardwood stand showed subsequent decreases

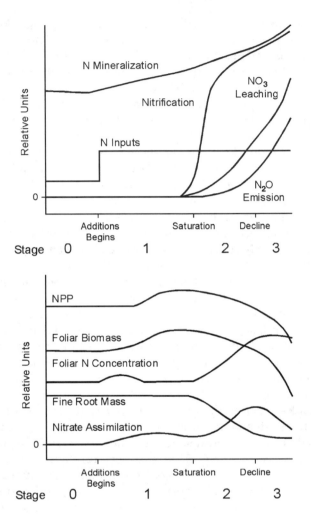

FIGURE 7.7
Hypothesized major responses of temperate forest ecosystems to long-term chronic N addition. (Source: Aber et al., 1998. Reprinted with permission from BioScience, Vol. 48, p. 922, Figure 1, 1998, © 1998 American Institute of Biological Sciences.)

in net N mineralization from early peak values (that were 1.2 to 2.4 times higher than low deposition or control values) down to near or below control values. Gunderson et al. (1998) and Tietema (1998) also found that net N mineralization peaked at intermediate N deposition in Europe. Second, the observed increases in NO_3^- cycling and leaching were very small, relative to N inputs. Nitrogen retention efficiencies were much higher than expected, ranging from 85% at the high-N pine stand at Harvard Forest to values between 93 and 99% at the other 8 treatments and sites. Third, increased foliar N concentrations were accompanied by decreased foliar Mg : N and Ca : Al ratios in all except one site. This observation may provide a link between experimental N addition and forest health.

Aber et al. (1998) found that the deciduous forest stands did not progress toward N saturation as rapidly or as far as the coniferous stands. Growth declined or mortality increased at all of the coniferous stands, but not all the deciduous stands.

7.3 Nitrogen in Surface Waters

Atmospheric deposition of N (as NO_3^- and as NH_4^+ that can be quickly nitrified to NO_3^-) in some cases causes increased concentrations of NO_3^- in drainage waters. An increase in the concentration of NO_3^- will generally result in a number of additional changes in water chemistry that are analogous to those caused by SO_4^{2-}. These can include:

- Increased concentration of base cations (Ca^{2+}, Mg^{2+}, K^+, Na^+).
- Decreased acid neutralizing capacity (ANC).
- Increased concentration of hydrogen ion (decreased pH).
- Increased concentration of dissolved Al.

Increased concentrations of H^+ and/or Al^{n+} occur mostly in response to higher concentrations of SO_4^{2-} or NO_3^- when ANC has decreased to near or below zero. At higher ANC values, increases in SO_4^{2-} or NO_3^- concentrations are mainly balanced by increasing base cation concentrations and some decrease in ANC in some cases. High concentrations of H^+ or Al^{n+} can be toxic to fish and other aquatic biota.

If NO_3^- leaches into stream or lake water as a result of increased N deposition, the result can be eutrophication or acidification. If N is limiting for aquatic primary production, the added NO_3^- will generally result in increased algal productivity that can cause disruption of aquatic community dynamics. If N is not limiting (e.g., P or some other nutrient can be limiting), then the added NO_3^- will remain in solution, possibly leading to acidification.

Except in cases of N saturation, the effects of N deposition on surface waters are expected to be primarily episodic in nature. Unfortunately, data required to make regional assessments of episodic effects generally are not available. Such data need to be collected on an intensive schedule and must include sample periods during late winter and early spring when snowmelt often causes the most severe N-driven episodes of surface water acidification. Sampling during this time of year is more difficult, dangerous, and expensive than during the more common summer/fall sampling seasons. Sampling during snowmelt can be particularly difficult in the high mountains of the West when study sites are often inaccessible and where motorized transport (e.g., via snowmobile) is not allowed in some places owing to wilderness restrictions.

Most lakes receive the majority of their hydrologic input from water that has previously passed through the terrestrial catchment. As long as N retention in the terrestrial system remains high, as is generally the case for forested ecosystems, N concentrations in lakes will remain low in the absence of contributions from land use (e.g., agriculture) or other pollution sources. However, if N retention in the catchment is low and the lake has not yet acidified, N deposition can in some cases increase primary production. This is most likely to happen in groundwater recharge lakes where nutrient inputs are derived largely from deposition to the lake surface. Lakes that are most likely to be low in base cations (therefore potentially sensitive to acid deposition) and also N limited are often systems overlaying volcanic bedrock (these rocks are often high in P).

The relationship between measured wet deposition of N and stream-water output of NO_3^- was evaluated by Driscoll et al. (1989a) for sites in North America (mostly eastern areas), and augmented by Stoddard (1994). The resulting data showed a pattern of N leaching at wet-inputs greater than approximately 400 eq/ha (5.6 kg N/ha).

Stoddard (1994) presented a geographical analysis of patterns of watershed loss of N throughout the northeastern U.S. He identified approximately 100 surface water sites in the region with sufficiently intensive data to determine their N status. Sites were coded according to their presumed stage of N retention, and sites ranged from Stage 0 through Stage 2. The geographic pattern in watershed N retention depicted by Stoddard (1994) followed the geographic pattern of N deposition. Sites in the Adirondack and Catskill Mountains in New York, where N deposition is about 11 to 13 kg/ha per year, were typically identified as Stage 1 or Stage 2. Sites in Maine, where N deposition is about one-half as high, were nearly all Stage 0. Sites in New Hampshire and Vermont that receive intermediate levels of N deposition were identified as primarily Stage 0, with some Stage 1 sites. Based on this analysis, a reasonable threshold of N deposition for transforming a northeastern site from the natural Stage 0 condition to Stage 1 would correspond to the deposition levels found throughout New Hampshire and Vermont, approximately 8 kg N/ha per year. This agreed with Driscoll et al.'s (1989a) interpretation that suggested N leaching at wet inputs above about 5.6 kg N/ha per year that would correspond to total N inputs near 7 to 8 kg N/ha per year. This is likely the approximate level at which episodic aquatic effects of N deposition would become apparent in many watersheds of the northeastern U.S.

A survey of N outputs from 65 forested plots and catchments throughout Europe was conducted by Dise and Wright (1995). Below the throughfall inputs of about 10 kg N/ha per year, there was very little N leaching at any of the study sites. At throughfall inputs greater than 25 kg N/ha per year, the study catchments consistently leached high concentrations of inorganic N. At intermediate deposition values (10 to 25 kg N/ha per year), Dise and Wright (1995) observed a broad range of watershed responses (Figure 7.8). Nitrogen output was most highly correlated with input N ($r^2 = 0.69$), but also significantly correlated with input S, soil pH, percent slope, bedrock

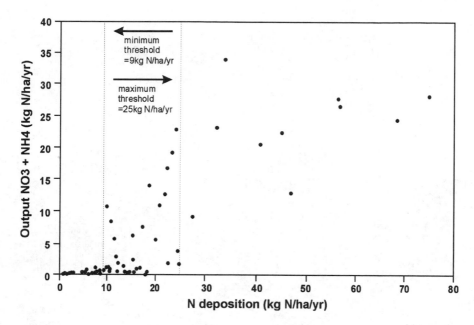

FIGURE 7.8

Nitrogen outputs in soil water and stream water vs. N deposition inputs throughout Europe. Thresholds were identified below and above which forested plots and catchments did not leach inorganic N and consistently leached large quantities of inorganic N, respectively. (Reprinted from Forest Ecology and Management, Vol. 71, Dise, N.B. and R.F. Wright, Nitrogen leaching from European forests in relation to nitrogen deposition, p. 157, Copyright 1995. With permission from Elsevier Science.)

type, and latitude. A combination of input N (positive correlation) and soil pH (negative correlation) explained 87% of the variation in output N at 20 sites (Dise and Wright, 1995).

The issue of recovery of water quality subsequent to emissions reduction was examined by Emmett et al. (1998) using data from the roof exclusion experiments in The Netherlands (Ysselsteyn and Speuld) and at Solling in Germany. Nitrogen deposition was experimentally reduced at these sites by 60 to 95%. In response, the N flux in drainage water decreased substantially within a few years (Bredemeier et al., 1998a; Emmett et al., 1998). Thus, a continuing supply of N from the atmosphere appears to be important to maintaining high N leaching losses at these high N status sites. Improvement in water quality may be expected at such sites when deposition levels are decreased.

8

Experimental Manipulation Studies

The scientific and political value of experimental field studies of acidification processes have been well recognized for some time (Wright, 1991). Other sources of quantitative information, including survey results, monitoring, laboratory studies, and modeling are insufficient, on their own, as a foundation for understanding and predicting acidification and recovery responses. The results of surveys of water quality in areas impacted by acidic deposition, as well as areas not impacted by acidic deposition (c.f., Sullivan, 1990) have been used for two decades as evidence of acidification effects. Interpretation of such data is always compromised, however, by differences between the impacted and unimpacted areas that are independent of acidic deposition. Such differences may include aspects of soils, geology, climate, land use, and hydrology that in some cases can overwhelm the effects of S or N deposition.

Acidification of aquatic and terrestrial ecosystems operates on time scales of many years to many decades. There are few time series of monitoring data available with long enough period of record to confirm the validity of our understanding of key acidification processes. Furthermore, interpretation of time series data is often uncertain because a variety of mechanisms can provide plausible explanations of observed responses. Concurrent changes in climate, land use, disturbance, or other factors confound the interpretation of monitoring results.

There has been a large increase during the past decade in the amount of experimental research being conducted on the environmental effects of atmospheric deposition, especially of N. This research has been initiated mostly in Europe; little comparable work has been conducted in the U.S. The experimental approach has shifted heavily into the area of whole-ecosystem experimental manipulations that have been and are being conducted across gradients of atmospheric deposition and other environmental factors throughout northern Europe (Sullivan, 1993). Individual investigators have, in many cases, been working at a variety of sites, thus enhancing the comparability of the resulting databases. Manipulations have focused primarily on coniferous forest ecosystems, and have involved

1. Increasing deposition of S and/or N.
2. Excluding the previously existing deposition via construction of roofs over entire forested plots or mini-catchments.
3. Manipulating climatic factors, especially water availability.

This research has been highly interdisciplinary, and experiments were designed to continue for relatively long periods of time (i.e., 5 to 10 years). Manipulation studies conducted thus far have clearly demonstrated that a long-term commitment is an essential component of whole-ecosystem research. With the notable exception of the watershed manipulation projects at Bear Brook and Coweeta and several smaller scale projects elsewhere, research of this type and scope has been generally lacking in the U.S.

The whole-ecosystem manipulation experiments in Europe have been augmented by a number of detailed, process-level studies at the various manipulation sites. Key aspects included stable isotope (^{15}N) tracer studies to quantify the partitioning of N into various ecosystem pools (i.e., soil, litter, trees, ground vegetation) and to measure changes in the quantities of stored N in these pools. Other studies focused on quantifying the rates of important ecosystem processes, including the N conversion processes of denitrification and mineralization.

Results of both the broad-scale and detailed studies have been used to build, test, and validate mathematical models that simulate N processing, nutrient cycling, and water regulation in coniferous forest ecosystems under varying depositional and climatic regimes. Ultimately, these models will be used to predict N saturation, estimate the critical loads of N for European forests, and to specify emission controls needed to protect European forests from the detrimental effects of excess N deposition.

Such large-scale, controlled whole-ecosystem experiments have become an increasingly important tool in environmental research regarding the effects of atmospheric pollutants. It is now realized that all parts of the ecosystem are involved in the response to an environmental perturbation such as atmospheric N or S input. Key processes must be evaluated in the broader context of whole-ecosystem structure and function. It is not possible to understand environmental impacts on the basis of isolated process studies alone. A holistic approach is required. In addition, whole-ecosystem experimental manipulations are needed across gradients of atmospheric deposition, climate, and other important factors.

It has also become increasingly evident in recent years that it is not possible to separate research on ecosystem effects attributable to acidic deposition from the effects of other ecosystem stressors. Climatic fluctuations, especially precipitation input and its effects on water availability, act synergistically with a variety of indirect effects of acidic deposition. The obvious linkages between short-term climatic fluctuation and anthropogenic inputs of N and S were incorporated into the experimental approach followed by the European EXMAN program (Beier and Rasmussen, 1993). Both drought and also N and S inputs were evaluated alone and in combination under a variety of conditions. The linkage with climatic change was taken further still in the European CLIMEX project that entailed simultaneous whole-ecosystem manipulation of temperature, atmospheric CO_2, and acidic deposition (Jenkins et al., 1992). Thus, not only short-term climatic fluctuations, but also long-term climatic trends (hypothesized global climate change) have been under investigation as they relate to ecosystem responses to acidic deposition.

Some of the most important contributions to the state of scientific understanding of the aquatic effects of acidic deposition during the past decade have been made in the area of N effects. Much of this research has been conducted in Europe and has involved experimental manipulation of atmospheric inputs to small catchments or forested plots. Although it is well beyond the scope of this review to attempt to cover all of the important research findings of European studies in recent years, it is helpful to summarize some of the key elements of the experimental ecosystem manipulation research.

The European scientists have concluded that it is important to study N questions as large multidisciplinary, multi-investigator research teams. This is because of

1. The complexities of the N cycle.
2. The multitude of scientific disciplines involved in its study.
3. The emerging importance of very expensive, large-scale, whole-system manipulations as a tool for studying N effects.

A high degree of international and inter-institutional cooperation has developed during the last decade within Europe. This spirit of cooperation has been evident in several recent international umbrella projects on N effects, especially NITREX and EXMAN (Wright and van Breemen, 1995; Rasmussen, 1990; Tietema and Beier, 1995).

The NITRogen Saturation EXperiments (NITREX) project was a large, international, interdisciplinary research program that focused on the impacts of NO_3^- and NH_4^+ on forest ecosystems (Wright and Van Breemen, 1995). NITREX included 11 separate large-scale N addition or removal experiments at 9 sites that span the European gradient in N deposition, from less than 5 kg N/ha per year in western Norway to greater than 50 kg N/ha per year in The Netherlands (Figure 8.1). In general, the same team of investigators and the same techniques were used across sites.

At each site, precipitation, throughfall, soil, soil solution, and runoff (catchments only) were monitored, before and after initiation of the experimental manipulations. Nutrient status and nutrient cycling were studied by periodically examining litterfall, needle composition, soil organic matter composition and mass, and fine root biomass. Nitrogen-15 tracer studies were conducted at several sites to follow the fate of added N through the forest

FIGURE 8.1
Location of NITREX research sites as described by Emmett et al. (1998).

ecosystems. The overall objective of the program was to obtain direct experimental data at the ecosystem level of N-saturation from atmospheric deposition, and subsequent ecosystem recovery. Questions regarding the threshold for N saturation, critical loads, and reversibility were addressed by means of a matrix of experimental manipulations that included increasing N inputs to some forest ecosystems and excluding ambient N (and S) inputs to other forest ecosystems.

The NITREX sites clearly separate into those that receive less than 150 meq/m^2 per year of N (approximately 30 kg N/ha per year) input and do not have NO_3^- leaching (Sogndal, Aber, Gårdsjön, Klosterhede) and those that receive greater than 200 meq/m^2 per year (approximately 40 kg N/ha per year) and leach significant amounts of NO_3^- (Speuld, Solling, Ysselsteyn; Dise and Wright, 1995). Other aspects of ecosystem acidification also illustrate the gradient in the NITREX sites. For example, leaching of Ca^{2+} and Al^{n+} are low at Sogndal and Aber, intermediate at Klosterhede and Gårdsjön, and high at Solling, Speuld, and Ysselsteyn. A potential complication in interpreting these data, however, is that S deposition in Europe follows approximately the same gradient as N deposition. It is unclear as to what extent the N data may be confounded by the effects of S.

The EXperimental MANipulations of Forest Ecosystems in Europe (EXMAN) project conducted experimental manipulations of five forest sites in Denmark, Germany, and The Netherlands, with an unmanipulated control site in Ireland. Major objectives of the program were quantification of element biogeochemical cycling, biomass turnover, and the effects of atmospheric deposition on forest ecosystems. Comparable manipulations have been conducted within similar forest types across a range of atmospheric inputs. Experimental approaches and methods were generally standardized. Treatments included simulated summer drought, irrigation, optimal nutrition and water, fertilization, liming, and exclusion of ambient atmospheric deposition via roof emplacement (Rasmussen, 1990).

8.1 Whole-System Nitrogen and/or Sulfur Enrichment Experimental Manipulations

There are a number of research sites where ambient N, and in some cases also S, deposition has been augmented. Typically, the experimental approach involved acid application during rainfall events by means of sprinkler systems, using chemically altered water from a nearby lake or spring as a carrier for the acid or acid precursor addition. In some cases, ammonium sulfate was applied periodically by helicopter to the experimental watershed. Several of these studies are highlighted below.

8.1.1 Gårdsjön, Sweden

At the Gårdsjön experimental manipulation catchment included within NITREX (Catchment G2), about 35 kg N/ha per year was added to the ambient deposition (12 kg N/ha per year) as NH_4NO_3. The sum of the experimentally added N plus the ambient deposition in this 0.52-ha catchment was in the range of deposition received by damaged forest ecosystems in central Europe, but much higher than the deposition levels in sensitive areas of North America. Data have been collected since 1988 and the treatment began in April 1991. Data were routinely collected of meteorology, deposition, throughfall, soil solution, hydrology (tensiometers), runoff, soil chemistry, root vitality, rhizosphere soil chemistry, micorrhizal fungi, litter, vegetation, foliage chemistry, and rates of mineralization, nitrification, and denitrification. Fish toxicity studies were conducted with runoff from the manipulated catchment.

The forest is a mixture of Norway spruce (*Picea abies*) and some Scots pine (*Pinus sylvestris*) less than 100 years old. Soils are mostly acidic silty and sandy loams to an average depth of 38 cm.

Moldan et al. (1995) presented results from input-output measurements at the Gårdsjön manipulation and reference sites for the first 2 years of treatment. During year 1, slightly elevated levels of NO_3^- in discharge were found during the first 2 weeks of treatment in April 1991 and again during late fall and winter. Loss of NO_3^- continued during the second year of treatment, including increased losses during the growing season. However, the watershed retention of deposited N during year 2 was still quite high ($98.9 \pm 0.1\%$). In the untreated reference catchment, N retention was about 99.9% of the total inorganic N inputs (Moldan et al., 1995). The monthly mean volume-weighted concentrations of NO_3^- in runoff increased from near zero during the 2-year pre-treatment period to values typically in the range of 5 to 17 µeq/L during year 2. Moldan et al. (1995) also conducted intensive sampling for a 2-week period during which three experimental NH_4NO_3 additions occurred. They found that NO_3^- concentrations in runoff consistently exhibited a sharp increase to concentrations in the range of 15 to 35 µeq/L immediately following N addition, followed by a recession to below the NO_3^- detection limit (0.4 µeq/L) within 48 h. Thus, the N retention capacity was only exceeded for short periods of time associated with the experimental treatments, and even then only by a relatively small amount. This happened despite the very high N inputs. After 2 years, NO_3^- appeared in soil solution at shallow depth, and after 4 years at all soil depths (Moldan and Wright, 1998a). Nevertheless, the annual loss of inorganic N was only about 5% of the incoming N. The cumulative effect of the N addition was apparent when NO_3^- concentration was plotted by Moldan and Wright (1998a) as a function of stream discharge during the autumn periods of 1994, 1995, and 1997. Nitrate concentrations reached higher values at a given discharge as the experimental acidification proceeded. Discharge rate was the most important factor influencing NO_3^- leaching loss. Peak NO_3^- concentrations in discharge (approximately 20 to 100 µeq/L) corresponded temporally with either times of experimental NH_4NO_3 addition or high discharge (Moldan and Wright, 1998a).

During the first 3 years of experimental N addition, NO_3^- concentrations in discharge were only high during winter. During the fourth and fifth years, however, elevated concentrations also were observed during summer months. The inorganic N lost in discharge, as a percentage of input, was 0.6%, 1.1, 5.0, 5.7, and 4.5%, respectively, during the 5 years of treatment (Moldan and Wright, 1998b). The somewhat reduced N loss during year 5 was attributed to drought and consequent low runoff during that year.

Total N deposition at the experimental Gårdsjön site is about 3 to 5 times higher than ambient N deposition in the high-deposition areas of the north-eastern U.S. It may be, however, that additional time, rather than higher N dose, may be required before NO_3^- leaching shows a more dramatic response at this site.

One of the first biological changes attributed to the experimental treatment was a change in the ectomycorrhizal fungus flora. Brandrud (1995) reported that the micorrhizal fruit body production was reduced after 1 1/2 years of treatment, especially for the dominant genera *Cortinarious* and *Russula*. A decrease in the amount of fine roots, especially in the upper soil horizons, was also observed for the *Vaccinium*-dominated portion of the study area (Clemensson-Lindell and Persson, 1995).

8.1.2 Sogndal, Norway

The Sogndal site in western Norway was part of the Reversing Acidification in Norway (RAIN) project (Wright et al., 1993). One of the small catchments (SOG4) received a 1 : 1 mixture of sulfuric and nitric acid (50 meq/m^2 per year each) additions since 1984. The region receives only 4 kg S and 2.5 kg N/ha per year of ambient atmospheric deposition.

Located at 900 m elevation in western Norway, the Sogndal site has gneissic bedrock, thin and patchy soils averaging about 30 cm depth, and alpine vegetation. Shrub vegetation is dominated by birch (*Betula verrucosa, B. nana*), juniper (*Juniperus communis*), and willow (*Salix hastata*), with a ground cover of *Calluna vulgaris, Empetrum nigrum*, several species of *Vaccinium*, grasses, mosses, and lichens.

Addition of H_2SO_4 and HNO_3 at SOG4 has caused large changes in runoff chemistry. During the first 5 years of treatment, NO_3^- concentrations in runoff at SOG4 were elevated above concentrations at the control sites (SOG1 and SOG3) only immediately after acid applications. Since 1989, however, the NO_3^- concentration in runoff has been chronically high. Alkalinity and pH decreased in parallel fashion at the H_2SO_4 treated catchment (SOG2) and the H_2SO_4 + HNO_3 treated catchment (SOG4; Wright et al., 1994).

The RAIN project ended in 1991, and the site was at that point included within NITREX. Sogndal represents the catchment receiving lowest deposition within the NITREX framework, and is also the only nonforested (alpine) catchment in the project. The experiment at Sogndal represents the only long-term study of chronic N addition to an alpine site.

Results of 9 years of N deposition at a level of 9 kg N/ha per year were summarized by Wright and Tietema (1995). As was found by Moldan et al. (1995) at Gårdsjön, the general pattern of NO_3^- concentration in runoff was one of sharp peaks during and immediately after each acid addition, followed by a rapid decline to concentrations near zero. It was only during the last few years of treatment that the decline in NO_3^- concentration in runoff following experimental N additions proceeded more slowly, and runoff between additions also contained elevated concentrations of NO_3^-. Ecosystem N saturation was not achieved after 9 years of N loading at a rate of 9 kg N/ha per year. More data of this type are needed to improve our understanding of the susceptibility of alpine soils and vegetation communities to N saturation. What is particularly noteworthy about the Sogndal study is the fact that the total N deposition (7 kg/ha per year experimental plus 2 kg N/ha per year ambient) is in the range of deposition found in many parts of the U.S. The results of the Sogndal research are, therefore, perhaps more relevant to the situation in the U.S. than are the results at many of the other NITREX sites.

The input–output budget for N at the treatment catchment (SOG4), summarized over the 9-year period indicated that 88% of the total N input of 72 kg N/ha was retained in the catchment. The percent N retention at the untreated reference catchment SOG1 was identical (88%), although the total N input was much lower, only about 20 kg N/ha.

Wright and Tietema (1995) concluded that there was little evidence that the 9 years of N deposition at a level of about 9 kg N/ha per year had induced N-saturation. Most of the NO_3^- leaching occurred during the early phases of snowmelt and immediately during or following experimental N addition. They attributed the increased leaching loss to insufficient time or capacity to immobilize the NO_3^- flux during times of high flow and high input concentrations and emphasized that the total N deposition at Sogndal was near the 10 kg N/ha per year apparent threshold for N saturation proposed by Grennfelt and Hultberg (1986) and Dise and Wright (1995).

8.1.3   Lake Skjervatjern, Norway

The Humic Lake Acidification Experiment (HUMEX) was initiated by the Norwegian Institute for Water Research (NIVA) in 1987. The principal goals of HUMEX were to evaluate the role of humic substances in the acidification of surface waters and the effects of S and N deposition on the properties of humic substances in watershed soils and surface waters (Gjessing, 1992).

HUMEX is an investigation of the interaction between acid deposition and natural organic acids by means of acid addition to the entire catchment of a pristine humic lake in western Norway. Skjervatjern is a small (2.4 ha), pristine, naturally acidic, humic lake located near Førde, western Norway. The lake has pH 4.6 with average concentrations of TOC of about 9 mg C/L, nonmarine base cations of about 30 μeq/L, and Al_i of less than 50 μg/L.

Lake Skjervatjern is located in an area of western Norway that receives low levels of anthropogenically derived atmospheric deposition of S and N. The 6.5 ha catchment is underlain by granitic bedrock, covered by histosols in the lower portions and podsols developed on thin glacial till in the upland areas. Annual precipitation at the site is about 2 m.

The lake was divided in 1988 by a plastic curtain that effectively separated the lake and its drainage basin into two systems, a manipulated side

(Basin A) and control side (Basin B). A 105 m plastic curtain was installed from the middle of the natural outlet to the opposite shore. The bottom edge of the curtain was pressed into the soft upper lake sediments by sand bags that minimized water movement between the two lake halves. Water quality was monitored for two years prior to treatment and three years during which N and S were applied to one-half of the lake and its respective drainage basin.

Artificial acidification of the treatment side of the catchment (Basin A) was initiated 2 years after installation of the dividing curtain. A sprinkling system consisting of 50 sprinklers (15 m sprinkling radius) was mounted at the top of the taller trees throughout the treatment basin. A combination of H_2SO_4 and NH_4NO_3 was applied at pH 3.0 to 3.2 weekly, in a volume equivalent to approximately 10% of ambient precipitation, using water pumped from nearby Lake Åsvatn. Annual target loadings for SO_4^{2-} and total N were 63 to 66 and 17 to 32 kg/ha, respectively. Water chemistries in the treatment and reference sides of the lake were monitored weekly, for 2 years prior to initiation of the artificial acid additions and for 5 years during which N and S were applied to one-half of the lake and its respective drainage basin (Gjessing, 1994; Lydersen et al., 1996).

The physical division of Lake Skjervatjern into two basins had some effects on the water chemistry of the lake, likely due to small differences in the terrestrial catchments that drain into the two lake halves. Lake water in the treatment side had equivalent or lower concentrations of all ions and lower electrical conductivity than did lake water in the reference side. The most pronounced differences prior to chemical manipulation were for Na^+ (-4 µeq/L), Cl^- (-2 µeq/L), SO_4^{2-} (-1 µeq/L), and K^+ (-1 µeq/L). Lake water pH was slightly higher on the experimental side (approximately 0.03 pH units) and TOC was 0.67 mg C/L lower (Gjessing, 1992, 1994).

During the first 2 years of treatment, 8.5 g m^{-2} of H_2SO_4 and 6.7 g m^{-2} of NH_4NO_3 were applied to the catchment and lake surface of the experimental side (A) (Gjessing, 1992). About 4% of the total chemicals were sprayed directly on the lake surface, and the balance would have received some contact with the terrestrial catchment prior to entering the lake. The majority of the increased lake water SO_4^{2-} concentration (greater than 80%) was attributable to SO_4^{2-} that had made some contact with the catchment. Nevertheless, a considerable amount of the added SO_4^{2-} was apparently retained in the terrestrial system.

The amount of SO_4^{2-} applied to the experimental catchment during the first 2 years of treatment should have caused an increase in lake-water SO_4^{2-} concentration of about 44 µeq/L above the premanipulation concentrations, assuming steady-state conditions and average annual runoff of about 1950 mm. The observed increase in lake-water SO_4^{2-} concentration was 15 µeq/L (Gjessing, 1992), suggesting that about two-thirds of the S added during the first 2 years were retained in the watershed. Over the 5-year treatment period reported by Lydersen et al. (1996), the mean SO_4^{2-} concentration in the treatment catchment increased by 16 µeq/L.

The increase in lake-water NO_3^- concentration was fairly small at Skjer-vatjern, only about 3 µeq/L. A substantial portion of that increase can be attributed to NO_3^- added directly to the lake surface, even without assuming that some of the NH_4^+ applied to the lake surface would have been converted to NO_3^- in the lake water. Total N also increased in the treatment side of the lake, however, and by an amount considerably greater than the total N applied to the lake surface (Gjessing, 1994). This suggested that at least some of the N applied to the terrestrial portion of the catchment also reached the lake. Nevertheless, about 90% of the added N was retained in the terrestrial system, lake sediments, and/or biota, and did not contribute to increased concentrations of NO_3^- and NH_4^+ in lake water.

Lydersen et al. (1996) used randomized intervention analysis (RIA) to test for differences between runoff chemistry from the two basins before and after the artificial acidification treatment. Significantly higher concentrations were found of SO_4^{2-}, NO_3^-, Ca^{2+}, Mg^{2+}, H^+, NH_4^+, and Al_i in Basin A after treatment compared with the control basin. The average ANC increased in the control basin during the course of the study, and this was attributed by Lydersen et al. (1996) to the long-lasting effect of Na^+ leakage after storms having high inputs of sea salts. During a hurricane in January 1993, the concentration of Cl^- in rainfall exceeded 400 µeq/L at the nearby weather station. During that event, the lowest runoff pH (4.25) and ANC (-62 µeq/L) values were recorded in the control basin. ANC remained unchanged in Basin A. Acidification of Basin A was observed as a gradual change in the difference in ANC between the two basins.

Highest concentrations of SO_4^{2-} were observed during summer, likely related to low flow conditions and consequent reoxidation of S stored in wetland soils. One of the most dramatic results of the acidification experiment was the observed decrease in the anion deficit (an estimate of the organic acid anion concentration) in Basin A. The difference in average anion deficit between the 2-year pre-acidification period and the 5-year post-acidification period in Basin A was nearly as large as the corresponding change in base cation concentrations (Lydersen et al., 1996). Thus, organic acid anions became more protonated in the treatment basin compared with the control basin as a consequence of the experimental treatment.

8.1.4   Aber, Wales

Most forests in the U.K. are thought to immobilize a high proportion of incoming atmospheric N. One known exception is the Beddgelert forest in North Wales, where N outputs are higher than inputs in bulk precipitation. The Aber forest study component of NITREX was designed to examine how a forest in this region would process additional loadings of N. The experimental site is located at an elevation of 300 m, 10 km from the North Wales coast. Originally moorland, the site was planted in Sitka spruce (*Picea sitchensis*) in 1960. Ambient deposition includes about 17 kg N/ha per year.

The experimental design included three replicate plots that received each of five treatments in a randomized block design. The treatments began in 1990 and included control, added water, added sodium nitrate (35 and 75 kg N/ha per year), and ammonium nitrate (35 kg N/ha per year). Emmett et al. (1995) reported the results after 2 1/2 years of experimental treatment. Soil water NO_3^- leaching losses increased in parallel with NO_3^- additions, although NH_4^+ additions were virtually completely retained. Little or none of the applied NO_3-N appeared to be taken up by the vegetation, retained in the soil or microbial community, or lost to denitrification (Emmett et al., 1995).

8.1.5 Klosterhede, Denmark

Several studies have been conducted at the Klosterhede Research Station in West Jutland, Denmark, to elucidate key aspects of forest ecosystem response to changing levels of acidic deposition. In 1991, the ongoing ion balance and acid-exclusion studies were expanded to include an N addition experiment as part of the NITREX project. The forest is a 74-year old, second generation Norway spruce (*Picea abies*) plantation on flat terrain, 27 m above sea level. The site was heathland before the plantation was established. Total atmospheric N deposition is about 23 kg N/ha per year, about 55% of which is NH_4^+-N. At the N addition plot (75 m × 75 m), an additional 35 kg N/ha per year was added as NH_4NO_3 beginning in February 1992. Pretreatment data and results after 1 year were reported by Gunderson and Rasmussen (1995). Results after 4 years of treatment were summarized by Gundersen (1998).

As was found by Emmett et al. (1995) at Aber, Wales, virtually all of the added NH_4^+ was taken up by the soil, vegetation, and microbial community (Gunderson and Rasmussen, 1995). There was an immediate response to the added NO_3^-, however, at all soil depths. Nitrate leaching accounted for 13% of the added NO_3^-, and the forest plot retained 92% of the total N input. Similar results were obtained by Aber et al. (1989) at an N-limited pine stand at Harvard Forest, MA, exposed to a comparable NH_4NO_3 application rate (50 kg N/ha per year), where complete N retention was observed after 3 years of experimental N addition.

Soil solution N chemistry changed immediately in response to the N additions, with NO_3^- concentration increasing at all depths. The total NO_3^- leaching increased from less than 0.3 to 4.2 kg N/ha per year by the third year. Ammonium concentrations increased to 15 cm depth, but NH_4^+ did not leach out of the system. Changes were not observed in the concentrations of other ions (Gundersen, 1998).

Based on these results and the results of a nationwide survey of soil water beneath the rooting zone, Gundersen and Rasmussen (1995) concluded that nearly total retention of NH_4^+ inputs was quite common for coniferous forests planted on sandy, former heathland soils in western Jutland. The high C : N molar ratio (28 : 35) of the organic layer of the soils may explain, at least in part, the low rates of nitrification observed (Gundersen and Rasmussen, 1990, 1995).

8.1.6   Bear Brook, ME

The Bear Brook watershed covers the top 210 m of the southeast slope of Lead Mountain (475 m) in eastern Maine. Ambient loading of inorganic N in deposition is about 9 kg/ha per year (Kahl et al., in press). East and West Bear Brooks drain the contiguous approximate 10 ha watersheds. The former has served as a reference catchment and the latter has been experimentally altered since 1990. The forest is mixed northern hardwoods (*Fagus grandifolia*, *Acer* spp., and *Betula* spp.), with spruce, fir, and hemlock (*Picea rubens*, *Abies balsamea*, and *Tsuga canadensis*) at the higher elevations. Coarse, loamy soils, averaging about 0.9 m depth, overlay quartzite and metapelite intruded by granite.

The watershed manipulation of West Bear Brook has included a 2 1/2-year calibration period (1987–1989), 9 years of chemical addition of $(NH_4)_2SO_4$, and will soon be followed by a recovery period. Chemical additions of 1800 eq of SO_4^{2-} and NH_4^+ per hectare per year effectively increased total atmospheric loading about 200% for S and 300% for N (Norton et al., 1999). Prior to the manipulation, stream-water chemistry of both the East and West Bear Brook catchments showed a volume–weighted annual mean pH of about 5.4, ANC 0 to 4 µeq/L, base cation concentrations about 184 µeq/L, and SO_4^{2-} concentration slightly over 100 µeq/L. DOC values were generally low (less than 3 mg/L) and NO_3^- concentrations varied seasonally between about 0 and 30 µeq/L (Norton et al., 1999).

The response of West Bear Brook stream-water chemistry to the experimental manipulation to date has been summarized by Norton et al. (1994, 1999) and Kahl et al. (in press). The major responses of the stream-water chemistry have included increased concentrations of SO_4^{2-}, NO_3^-, base cations, Al^{n+}, and H^+. ANC and organic acid anion concentrations have decreased. After 3 years of chemical manipulation, the volume–weighted mean annual concentration of $(SO_4^{2-} + NO_3^-)$ had increased by 72 µeq/L. This change was compensated primarily (approximately 80%) by increased base cation concentrations. The remaining portion of the change in mineral acid anion concentrations was mostly compensated by decreased ANC, followed by increased Al concentrations (Norton et al., 1994). By 1995, the proportion of the increase in stream-water concentrations of $(SO_4^{2-} + NO_3^-)$ that was charge-balanced by increased base cation concentrations had decreased to about 50% ($F = 0.5$) and the remainder of the response was approximately evenly split between decreased ANC and increased Al concentrations. Base cation concentrations have continued to decrease since 1995, resulting in progressively lower estimated F-factors (Kahl, personal communication.). Minimum pH values achieved during high-flow periods decreased from about 5.3 to below 4.7 through 1995, representing an increase in episodic H^+ concentration of about 15 µeq/L (Norton et al., 1999).

During the first year of treatment, 94% of the added N was retained by the Bear Brook watershed. Percent retention subsequently decreased to about 82% for the next 7 years of treatment (Kahl et al., 1993a, in press).

Both the immediate nature of the N response and the magnitude of the increase in NO_3^- flux from the treated West Bear catchment were unexpected (Kahl et al., 1993a).

The concentrations of SO_4^{2-} in stream water progressively increased throughout the 7 years of experimental acidification. By 1995, the volume-weighted mean SO_4^{2-} concentration reached 185 µeq/L and the maximum concentrations during high-flow periods exceeded 220 µeq/L, approaching the expected steady-state concentration that would be achieved if outputs equaled inputs (approximately 285 µeq/L). The concentration of SO_4^{2-} was flow-dependent, with lowest concentrations at low flow. This suggests a progressive saturation of the soil profile with added S (Kahl et al., in press). Deeper soils that contribute proportionately more discharge during low-flow periods are apparently less saturated than are the more shallow soils that contribute proportionately more discharge during high-flow periods.

Prior to experimental acidification, stream-water concentrations of Al^{n+} were typically below 10 µM, and have increased with experimental acidification, approaching episodic concentrations of 60 µM by 1995. Kahl et al. (in press) found that the majority of the Al^{n+} was divalent, as evidenced by the observed linear empirical relationship between Al^{n+} and the square of the H^+ concentration (in units of µM). Thus, Al^{n+} concentrations approached and exceeded 100 µeq/L during high-flow events, nearly as high as the Ca^{2+} concentrations. Kahl et al. (in press) further speculated that the Al/Ca molar ratio in soil solution would soon exceed the hypothesized 50% threshold (Al/Ca greater than 2) for damage to tree roots (Shortle et al., 1997) in some parts of the watershed.

An important finding of the Bear Brook watershed research has been the observation that Ca^{2+} and Mg^{2+} concentrations have declined at high flow during the period 1993 to 1995. This implies that the base cation supply in the upper soils is becoming depleted, which will lead to further acidification and mobilization of Al (Kahl et al., in press).

8.2   Whole-System Nitrogen Exclusion (Roof) Studies

An important tool that has developed in recent years for the study of ecosystem processes and the impacts of atmospheric deposition at the ecosystem level is the construction of transparent roofs over entire mini catchments or forested plots. The roof emplacement technique was pioneered at Risdalsheia, Norway, in 1983 in the RAIN project. A number of additional roof studies were constructed in Europe during the past decade, with roofs ranging in size from about 300 m² to the extremely impressive 0.6 ha roof at Gårdsjön in Sweden. In most cases, the roofs are constructed below the canopy in well-developed forests. Trees protrude through holes that are often sealed to prevent throughfall and stemflow from passing through the holes. Runoff from the roof is collected, chemically altered, and reapplied beneath via sprinklers. The technique allows simulation of drought and decreased atmospheric deposition to entire terrestrial systems.

8.2.1   Gårdsjön, Sweden

Whole-catchment exclusion of incoming ambient atmospheric pollution has reached its apex with the construction of the large roof at the Gårdsjön site in Sweden in 1990. The clear plastic roof intercepts atmospheric deposition at a height of 2 to 4 m above the ground. Approximately 350 Norway spruce trees protrude through the roof. The roof experiment excludes about 20 to 30 kg S/ha and 15 to 20 kg N/ha ambient atmospheric deposition.

Approximately 20 different and coordinated subprojects have been conducted at Gårdsjön, including:

- Input–output budgets for all major ions (including Hg and other heavy metals).
- Water pathways through the soil.
- Vegetation and fine root effects.
- Chemistry of soil, soil water, and groundwater.
- Soil processes (e.g., S retention, weathering).
- Isotope studies for partitioning of S, N, Hg.
- Wet, dry, and fog deposition processes for S, N, Hg.
- Trace gas emissions.
- Geochemical modeling and model testing.

Results of some of this work, conducted through 1995, were recently summarized by Hultberg and Skeffington (1998).

Anthropogenic S and N deposition to the experimental site were effectively reduced by more than 95%. Concurrently, ambient S deposition to the control watershed also declined by nearly 50% in response to emissions reductions (Ferm and Hultberg, 1998). The effects of these reductions on input fluxes and concentrations of major ions were described by Hultberg et al. (1998).

During the first 5 years of experimental treatment, the accumulated output fluxes of key elements decreased as follows:

SO_4^{2-}	(nonmarine)	45%
NO_3^-		60%
Al_i		29%
Ca^{2+}		20%
Mg^{2+}		28%

The calculated F factor was about 0.5 after 5 years. Sulfate concentrations in discharge remained high, however, about 215 µeq/L, in response to desorption and mineralization of S stored in the watershed soils (Torssander and Mörth, 1998; Gobran et al., 1998). The pH has changed more slowly than the other ions. Most of the reduction in acidity has been expressed as declining Al_i concentrations (Skeffington and Hultberg, 1998). Although there has been a steady improvement in the quality of the output water at Gärdsjön in response to the experimental treatment, the realized reductions in the concentrations of H^+ and Al_i have not been sufficient to mitigate the toxicity of the water to fish (Hultberg et al., 1998). Further recovery will depend largely on the rate of release of stored S and the future supply of base cations from watershed soils.

8.2.2 Ysselsteyn and Speuld, Netherlands

The Netherlands receives extremely high levels of N deposition (50 to 100 kg N/ha per year), resulting in N saturation of forest soils. NITREX sites were established in 1988 at Ysselsteyn and Speuld in which roofs were constructed over 10 m × 10 m forest plots (Boxman et al., 1994). The major objective of the studies at Ysselsteyn and Speuld was to investigate the potential reversibility of the existing N saturation. Nitrate is currently being exported in large quantities from the soils into the groundwater at these sites.

Ysselsteyn, in southeastern Netherlands, is a Scots pine (*Pinus sylvestris*) forest that has experienced significant needle loss. The site at Speuld, in central Netherlands, is a Douglas fir (*Pseudotsuga menziesii*) forest. Throughout the 4-year study period reported by Boxman et al. (1995), N deposition in throughfall ranged between 45 and 60 kg N/ha per year.

Transparent roofs, 2 to 3 m above the ground, were constructed at each of these sites during the winter of 1988–1989. At each site, 3 (10 m × 10 m) experimental plots received

- "Clean" throughfall reapplied beneath the roof to which all nutrients were added in the same amount as in the throughfall water except N and S.
- Ambient (polluted) throughfall reapplied unaltered beneath the roof.
- Ambient throughfall outside the roof.

Initially, the covered plots were watered weekly, using precipitation collected by the roofs.

It was found that watering the trees once per week at a high rate appeared to have major disadvantages. Much of the water probably flowed away from the site via a few preferential pathways. As a consequence, less water was available to the trees, and the soils dried out between watering events, thereby hindering nutrient uptake and microbial processes. In response to this problem, the watering regime was shifted to real-time watering in 1992.

The reduced input of N and S to the "clean" roof plot at Ysselsteyn resulted in reduced NH_4^+, NO_3^-, and SO_4^{2-} concentrations in soil solution. Between 1990 and 1992, the NO_3^- concentration was reduced 45% and NH_4^+ concentration 80% compared with the roof control plot. The total S flux was reduced 70%. At Speuld, the N flux was reduced by 80% and the S flux by 20 to 60%, depending on depth in the soil profile. Boxman et al. (1998) summarized the major results for the Scots pine stand at Ysselsteyn after 5 years of treatment. The mean flux of N in drainage water at 90 cm depth from 1990 to 1995 was reduced to 16 kg N/ha per year under the clean roof, as compared with 36 kg N/ha per year under the control roof and 69 kg N/ha per year under the open-air control plot. Vegetation response showed a more pronounced lag period, although some signs of ecosystem recovery were evident after 5 years. The concentration of N in needles decreased, but was still high. The treatment significantly decreased the arginine-N concentration in the needles, however, and this response was seen after 1 year. The diameter growth of the dominant trees in the clean roof plot significantly improved and was inversely related to the arginine-N concentration in the needles ($r^2 = 0.92$, $p < 0.001$).

In the Scots pine stand, fine root biomass and the number of root tips increased as N deposition decreased, suggesting an increased nutrient uptake capacity. K^+ and Mg^{2+} concentrations in needles increased and N concentrations decreased (Boxman et al., 1995).

8.2.3   Klosterhede, Denmark

The Klosterhede Plantation in western Denmark, a 73-year old even-aged Norway spruce (*Picea abies*) stand, was included within both the NITREX and EXMAN projects. In the Klosterhede Plantation, 3 study plots (each containing about 25 spruce trees) were established beneath the roof and another plot outside the roof served as a control. The treatments included

- Summer drought
- Irrigation with optimal amount of water and removal of incoming acid deposition.
- Irrigation with optimal amount of water plus optimal nutrition with macro and micronutrients (fertigation).

Significant biological response was observed in the treatments. Biomass increment, photosynthetic activity, needle element content, cone production, root development, ground vegetation, and microbial activity have all changed. Results to date show large variations for individual trees, core samples, and annual diameter growth. Despite the variability, however, higher tree growth rates have been observed in the fertilized and irrigated plots since 1988.

8.2.4   Solling, Germany

The Solling research site in central Germany has been under study for about 30 years, and as such, is one of the longest ecosystem studies in the world. Manipulation experiments began in 1989 when four plots were established to investigate ecosystem recovery under clean precipitation and the effects of drought. Solling was included in both the NITREX and EXMAN networks.

Roofs of 300 m² area were constructed in the summer of 1991 over three plots

- The NITREX clean precipitation site.
- Drought manipulation site.
- Roofed control.

The fourth plot served as an open-air control.

The site is located in central Germany at 500 m elevation in a 60-year old Norway spruce (*Picea abies*) plantation. The roofs are underneath the canopy at a height of about 3 m, with the tree trunks passing through preformed holes in the roofs and fitted with plastic collars. Ambient N deposition is very high (approximately 38 kg/ha per year). The "clean" rain roof simulates a 90% reduction of wet N input to the soil.

Bredemeier et al. (1995) reported results after 1 1/2 years of treatment. Nitrogen levels in soil water were reduced dramatically. Within the rooting zone, NH_4^+ and NO_3^- concentrations in soil water declined to near zero. Dramatic declines were also observed for SO_4^{2-} and Al, and soil water pH increased from 3.67 in 1989 to 4.03 in the first quarter of 1993 (Bredemeier et al., 1995).

Subsequent to the rapid decline in NO_3^- during the first 2 years after roof construction, a seasonal pattern was established of higher concentrations during winter and spring and near zero concentrations during the growing season (Bredemeier et al., 1998b). The amount of living fine roots increased at all depths in the rooted zone of the mineral soil under the roof. The total increase was 30 to 40% after 5 years compared with pre-experimental conditions. In the main rooting zone, Al concentrations decreased by about a factor of two and this may be one reason for the increase in fine root biomass (Bredemeier et al., 1998b). Changes were also observed in the concentration of N and Mg in current year and 1-year old needles, and the N/Mg ratio in needles decreased by about a factor of two. The cumulative diameter increment of the trees at breast height was highest at the clean rain roof site, but this difference existed, at least partially, prior to roof construction. Bredemeier et al. (1998b) could not ascribe a growth effect to clean rain manipulation with any certainty. They suggested that growth and other above-ground processes were the slowest components to react to the treatment in a temporal cascade of soil solution leading to fine roots leading to above-ground stand.

8.2.5   Risdalsheia, Norway

The Risdalsheia site in southernmost Norway receives a high loading of acidic deposition (SO_4^{2-} wet + dry loading, 18 kg S/ha per year), and is characterized by exposed granitic bedrock (30 to 50% of surface) and thin, organic-rich, truncated podzolic soils (Wright et al., 1986). Acid exclusion at the KIM catchment is accomplished by a 1200 m^2 transparent roof that completely covers the 860 m^2 catchment. Incoming precipitation is collected from the roof and pumped through a filter and ion exchange system. Seawater salts are added back at ambient concentrations and the clean precipitation is automatically applied beneath the roof by a sprinkler system. During winter, artificial snow is applied beneath the roof using commercial snow-making equipment. Controls include a mini catchment with a roof (EGIL) and one without (ROLF), both of which receive acidic rain and snow. Acid exclusion at KIM has resulted in substantially lower concentrations of SO_4^{2-} and NO_3^- that have been compensated mainly by a decrease in base cation concentrations and an increase in ANC.

Organic acids played a major role in the acid–base chemistry of runoff at the site and in moderating pH change following reduction in acid deposition. This site is particularly important because of its long period of record (greater than 10 years) and large change in the concentrations of SO_4^{2-} and NO_3^- in both the deposition and drainage water.

Major results of the acidic deposition exclusion experiment at Risdalsheia have been discussed in detail by Wright and co-workers (e.g., Wright, 1989; Wright et al., 1986, 1988b, 1990). Average stream-water SO_4^{2-} concentrations were reduced from 92 µeq/L in the 1985 water year to 28 µeq/L in 1992. This reduction in SO_4^{2-} was compensated primarily by decreased base cation concentrations, from 136 to 104 µeq/L ($F = 0.5$). In addition, the average H^+ concentration decreased from 87 to 61 µeq/L and the Al concentration decreased from 12 to 3 µeq/L.

Roof manipulation studies, such as those described previously, have proven valuable for investigating the environmental effects of reduced deposition of S and N and for testing of mathematical models that predict the effects of abatement strategies. However, a variety of unintended changes have also been caused by the roof construction and experimental design in some cases (Beier et al., 1998). These can confound interpretation of the resulting data. For example, reduced light penetration by 50% to the forest floor caused a decrease in moss cover at Klosterhede (Gundersen et al., 1995). Such vegetative changes may, in turn, affect nutrient cycling. Beier et al. (1998) stressed the importance of selecting roof plates that transmit maximum light and careful and regular cleaning. In addition, the frequency and intensity of water sprinkling affects both the hydrology and input of nutrients, which in turn can affect ground flora and microbial communities (Hansen et al., 1995; Gundersen et al., 1995). The sprinkling system will also change the spatial variability of water and nutrient delivery to the plot. It is important that the quantities of nutrients that are removed by filtering are calculated and reapplied under the roof, and that the water and nutrient supply is performed as close to real time as possible (Beier et al., 1998).

8.3   Climatic Interactions

It has become increasingly evident that it is often difficult to separate the effects of excess N deposition from climatic effects. For example, it has been proposed that high N loadings result in increased susceptibility of trees to drought and frost. Elevated deposition of N can affect the water uptake of trees, via

1. Increased shoot/root ratio, thereby increasing water demand.
2. Shift in root growth from mineral soil to upper organic horizon, thereby increasing susceptibility to drought.
3. Reduced fine root length and biomass.

Forests in some areas of northern Europe and the U.S. are apparently becoming increasingly N saturated. In addition, there is concern that the climate is becoming warmer. Such a warming trend may have important implications for N cycling because N mineralization and nitrification can be enhanced by warm soil temperatures and episodic drying events, especially where soil organic N pools are large. The end result of these processes can be internal acidification via the production and leaching of NO_3^-. These processes have been under investigation at several research sites.

Nitrogen deposition increases the emissions of N_2O from forest soils and also may decrease CH_4 uptake. Both increased N_2O production and decreased CH_4 consumption would increase the concentration of greenhouse gases in the atmosphere. Thus, there are important linkages between N deposition (and consequent ecosystem effects) and the release of greenhouse gases that have been implicated in potential global climate change.

The interactions between climate change and acidic deposition, as well as the direct effects of climatic manipulations on a subalpine coniferous forest ecosystem, have been investigated in the CLIMate Change EXperiment (CLIMEX) project (Jenkins et al., 1992). The major objective of CLIMEX was to quantify the impacts of atmospheric CO_2 enrichment and temperature increase on ecosystem response, especially the plant–soil–water linkages and processes. The approach involved whole catchment manipulations of temperature and CO_2 concentration at the Risdalsheia site in southern Norway, formerly part of the RAIN project.

Jenkins and co-workers measured changes in CO_2 uptake, gas exchange, and plant phenology; forest growth and nutrient status; ground vegetation; mineralization of soil organic matter; soil fauna; and biologically mediated processes, and the quality and quantity of runoff water. Process-oriented models will be developed to link aquatic and terrestrial processes.

Several of the roof experiments in Europe have examined ecosystem responses to simulated drought and subsequent rewetting of forest plots. For example, Bredemeier et al. (1998b) reported the results of simulated intensive drought periods at Solling. The above-ground parameters describing stand growth and physiology responded rapidly to the experimental treatment; height and diameter increment decreased and photosynthetic capacity was reduced. Fine roots did not show an obvious response to simulated droughts of 10 to 25 weeks. Soil water chemistry did not show the anticipated acidification pulses owing to excess nitrification in the rewetting periods after the simulated droughts. Consistent patterns of NO_3^- peaks coincident with increases in H^+ and Al_i were not observed in response to rewetting. However, peaks in NH_4^+, K^+, and DOC were frequently observed within the first few days of rewetting, suggesting mineralization and/or cell lysis processes (Bredemeier et al., 1998b).

8.4   Results and Implications

Whole system experiments generally necessitate long-term commitments of research funding to provide useful data. A symposium in Copenhagen (May 1992) on experimental manipulations of ecosystems concluded that ecosystem response time to manipulated inputs may well be on the order of 5 to 10 years. Cost-efficiency can be improved by locating many different experiments in the same general area. Such economy of scale has been demonstrated in the Experimental Lakes Area of Ontario, Canada, and at the Gårdsjön site in Sweden. Common logistics, common control catchments, and common planning and engineering all contribute to such an economy of scale.

Clear and significant ecosystem responses to the EXMAN manipulations have not been found, for the most part, during the first few years of the experiments. The manipulated sites seem to buffer changes in the inputs to the soil, resulting in slow (or small) changes in soil solution concentrations of most analytes. Because of the observed lag time in realizing significant environmental responses to experimental manipulation, it is important to initiate long-term investigations well in advance of the anticipated need for the resulting data. Such long-term studies require substantial funding commitments for long periods of time. Current federal funding mechanisms and funding cycles available in the U.S. are generally not compatible with long-term, multidisciplinary environmental studies.

An extremely important and novel aspect of the recently initiated European research on N has been the extent of coordination among projects, institutes, and investigators. Coordination among research teams from different countries throughout Europe has been a cornerstone of the large international projects such as NITREX and EXMAN. Researchers from across Europe have shared methodologies, data, and expertise in an unprecedented fashion. Manipulation studies with somewhat different objectives often shared data and reference (control) sites. Such an atmosphere of interdisciplinary, interinstitutional research cooperation has not developed to the same level in the U.S.

The influence of historical forest management on the ability of a given forest ecosystem to process N is largely unknown. Nevertheless, forest management practices, especially those that have occurred over many generations, can have important effects on soils (i.e., erosion), nutrient supplies (i.e., harvesting), organic material (i.e., litter raking), and thereby many aspects of N cycling and N effects. European forests have typically been harvested for many generations, changed in species composition or community type (e.g., conversion from heathland to forest), and managed or manipulated in a variety of ways. The interactions between these activities and atmospheric deposition are unknown.

It is not possible to separate research on ecosystem effects attributable to acidic deposition from the effects of other ecosystem stressors. Climatic fluctuations, especially precipitation input and its effect on water availability, act synergistically with a variety of indirect effects of acidic deposition. The obvious linkages between short-term climatic fluctuation and anthropogenic inputs of N and S were incorporated into the experimental approach followed by the EXMAN program. Both drought and N and S inputs were evaluated alone and in combination under a variety of conditions. The linkage with climatic change is taken further still in the CLIMEX project that entailed simultaneous whole-ecosystem manipulation of temperature, atmospheric CO_2, and acidic deposition. Thus, not only short-term climatic fluctuations, but also long-term climatic trends (hypothesized global climate change) have been under investigation as they relate to ecosystem responses to acid deposition. Unfortunately, research conducted by federal agencies in the U.S. is seldom sufficiently interdisciplinary so as to include elements of terrestrial, aquatic, and climate change research.

Although the current level of scientific understanding of N cycling in forested ecosystems is far from complete, important strides are being made at a rapid pace. The results of the broad array of manipulation and process-level studies conducted in the NITREX and EXMAN international research networks will provide critical information to continue improving our level of knowledge regarding this complex topic.

The array of experimental manipulation projects that have recently been conducted and those that are ongoing have been enormously useful in a number of important respects. A multitude of process-based components of the ecosystem acidification response that previously had been hypothesized or inferred based on empirical evidence and hydrogeochemical principles, have now been verified at the watershed scale. As a consequence, acidification theory is no longer purely theoretical. Quantitative aspects of acidification, such as proportional changes in the various ionic constituents that collectively constitute the acidification response and rates of watershed retention of acid anions, are now much better understood. And finally, experimental databases have been provided with which to test, confirm, and improve mathematical models of acidification dynamics. The author would argue that the most important advancements in acidification science of this decade have been direct results of the experimental manipulation studies. These studies have been expensive, but the gains have far outweighed the costs. The scientists who had the foresight and fund-raising capabilities to initiate this area of research including Wright, Schindler, Norton, Rasmussen, van Breemen, Hultberg, and many others have had an enormous impact on acidification science.

9

Predictive Capabilities

The treatment of mathematical modeling in this chapter, and throughout this book, is focused almost exclusively on the Model of Acidification of Groundwater in Catchments (MAGIC, Cosby et al., 1985a,b). This is not to imply that MAGIC is necessarily the best or most accurate acid–base chemistry model available. There are several reasons for this bias in treatment of modeling approaches in favor of MAGIC for the purposes of this book:

1. MAGIC is the most widely used acid–base chemistry model in the U.S. and Europe.

2. Because the model is highly generalized, it does not have extensive input data requirements and, therefore, can be applied to a large number of potential sites without incurring inordinate costs associated with data collection.

3. In part because of the second reason, MAGIC has been extensively tested against independent databases, thereby providing an excellent example of the iterative processes of model testing and refinement that all environmental models should go through.

4. The author has far more personal experience with MAGIC than with other models.

In recent years, a number of models have been developed to simulate N dynamics in forested ecosystems, and N has recently been added in various ways to MAGIC. Several of these N models are discussed at the end of this chapter.

A number of acid–base chemistry models have been developed that focus on S-driven acidification. Three primary models were used in EPA's Direct Delayed Response Project (DDRP, Church et al., 1989) to project surface water acidification response: MAGIC, the Integrated Lake Watershed Acidification Study model (ILWAS, Gherini et al., 1985), and the Trickle Down Model (Lin and Schnoor, 1986). In addition, the Internal Alkalinity Generation (IAG) model (Baker and Brezonik, 1988) was used to generate projections for seepage lakes in the NAPAP Assessment. These and other models were reviewed by Thornton et al. (1990) and Eary et al. (1989).

9.1   Model of Acidification of Groundwater in Catchments (MAGIC)

MAGIC has been the principal model used thus far by NAPAP for making projections of likely future changes in surface and soil water chemistry in response to various levels of acidic deposition. MAGIC also provided the technical foundation for the reduced-form modeling in the aquatic and soils components of NAPAP's Tracking and Analysis Framework (TAF) and has been used to estimate critical loads of S, and more recently also N, deposition to national parks and wilderness areas in many parts of the country.

9.1.1   Background and General Structure as Used for the NAPAP 1990 Integrated Assessment

MAGIC is a lumped-parameter model of intermediate complexity (Cosby et al., 1985a,b) that is calibrated to the watershed of an individual lake or stream and then used to simulate the response of that system to changes in atmospheric deposition. MAGIC includes a section in which the concentration of major ions is governed by simultaneous reactions involving S adsorption, cation weathering and exchange, Al dissolution/precipitation/speciation, and dissolution/speciation of inorganic C. A mass balance section of MAGIC calculates the flux of major ions to and from the soil in response to atmospheric inputs, chemical weathering inputs, net uptake in biomass, and losses to runoff. The model simulates soil solution chemistry and surface water chemistry to predict the annual average concentrations of the major ions. MAGIC generally represents the watershed with one or two soil-layer compartments. These soil layers can be arranged vertically or horizontally to represent the vertical or horizontal movement, respectively, of water through the soil. A vertical two-layer configuration was used for the NAPAP assessment, and the soil compartments were assumed to be really homogeneous.

The meteorological and deposition input requirements for MAGIC include the amount and ionic concentrations of precipitation and annual average air temperature. Also needed are details of the hydrological budget for each watershed. The spatial/temporal scales in the model reflect the intended use for assessment and multiple scenario evaluations. MAGIC does not use a Gran ANC in simulating watershed response. Rather, it uses a calculated alkalinity or ANC defined as follows:

$$CALK = SBC + NH_4^+ - SSA \tag{9.1}$$

$$\text{where}\quad SBC = Ca^{2+} + Mg^{2+} + Na^+ + K^+ \tag{9.2}$$

$$SSA = Cl^- + NO_3^- + SO_4^{2-} \qquad (9.3)$$

MAGIC is calibrated using an optimization procedure that selects parameter values so that the difference between the observed and predicted measurements is minimized. The calibration exercise is a three-step process. The first step is to specify the model inputs such as precipitation, deposition (both wet and dry), an estimate of historical deposition inputs and fixed parameters or parameters whose values correspond directly to (or can be computed directly from) field measurements (e.g., soil depth, bulk density, cation exchange capacity). This approach, in effect, assigns all of the uncertainty associated with sampling and intrinsic spatial variability to the "adjustable" parameters. The adjustable parameters are those that are calibrated or scaled to match observed field measurements.

The second step is the selection of optimal values for the adjustable parameters. These adjustable parameters are specified using optimization by the method of Rosenbrock (1960). Optimal values are determined by minimizing a loss function defined by the sum of squared errors between simulated and observed values of system state variables.

The final step is to assess the structural adequacy of the model in reproducing the observed behavior of the criterion variables and parameter identifiability, or the uniqueness of the set of optimized parameters. Structural adequacy is assessed by examining the mean error in simulated values of observed state variables for those variables used in the calibration procedure as well as for an additional state variable that was not used during calibration. Parameter identifiability is assessed using approximate estimation error variances for the optimized parameters (Bard, 1974).

Model calibration to a specific catchment is accomplished by specifying deposition and hydrological forcing functions, setting the values of those parameters that can be measured (fixed parameters), and determining the values of the remaining parameters that cannot be measured (adjustable parameters) through an optimization routine that adjusts those parameters to give the best agreement between observed and predicted surface water and soil chemistry (Cosby et al., 1985a,b, 1989).

Atmospheric deposition of base cations, strong acid anions, and NH_4^+ are assumed to be uniform over the catchment. Atmospheric fluxes in the program codes are calculated from concentrations of the ions in precipitation and estimated precipitation volume measured or interpolated to each catchment. These annual average concentrations and annual precipitation are used as input parameters for the model.

Atmospheric fluxes of the mass balance ions are corrected for estimated dry deposition of particulates and aerosols. Dry deposition is represented as a proportion of wet deposition, using dry deposition factors (DDF) calculated on the basis of site-specific measurements or regional average estimates.

Average annual values for soil and surface water temperature and soil P_{CO_2} (partial pressure of CO_2) are needed as inputs to the model. Mean annual soil temperatures are set equal to the mean annual air temperatures. Soil P_{CO_2} is

derived from a regression on soil temperature constructed from mean growing season soil P_{CO_2} data from 19 regions of the world (Brook et al., 1983):

$$\log_{10}(P_{CO_2}) = 0.03 * \text{TEMP} - 2.48 \tag{9.4}$$

where P_{CO_2} is in atmospheres and TEMP is the soil temperature in degrees C.
Using this expression, mean annual soil temperature of 10°C would produce
a soil P_{CO_2} of 0.0066 atm (approximately 20 times atmospheric P_{CO_2}).

Depth, bulk density, cation exchange capacity, maximum SO_4^{2-} adsorption
capacity, and the SO_4^{2-} adsorption half-saturation constant are provided from
soil characterization studies for each soil type. All soil horizons are aggregated to reflect average soil conditions.

Sulfate uptake in the lake sediments is calculated from the Baker and
Brezonik (1988) model using the values of relative lake area to the watershed
area and the discharge. Significant amounts of S can be retained in lakes
through dissimulatory reduction, with SO_4^{2-} used as an electron acceptor and
H_2S, ester sulfates, or metal sulfides as end products (Rudd et al., 1986;
Brezonik et al., 1987). Reduction rates are approximately first order for SO_4^{2-}
at concentrations typically encountered in softwater lakes. In-lake reduction
rates are apparently limited by diffusion into the sediments (Baker et al.,
1986; Kelly et al., 1987). The process appears to be rate limited, and Baker et
al. (1986) and Kelly et al. (1987) showed that this process can be represented
effectively as:

$$\% \ SO_4 \text{ retention} = \frac{K_{SO_4} * 100}{Z/\tau_w + K_{SO_4}} \tag{9.5}$$

where

$\quad K_{SO_4} \quad$ = sulfate mass transfer coefficient (m/year)
$\quad Z \quad$ = mean lake depth (m)
$\quad \tau_w \quad$ = hydraulic residence time (year) (outflow based)

The Al solubility constants in the soil layers (KAL1, KAL2) are given as log-
arithms (base 10) and are calibrated or sometimes assumed to be equal to
9.05. The assumed value represents a solid phase of $Al(OH)_3$ intermediate
between natural and synthetic gibbsite (see Cosby et al., 1985a).

It is important to test the veracity of environmental model projections,
especially in cases where policy and/or economic interests are considerable.
As Oreskes et al. (1994) pointed out, however, verification and validation of
mathematical models of natural systems are impossible, because natural systems are never closed and model results are nonunique. Model confirmation
is possible, and entails demonstration of agreement between prediction and
observation. Such confirmation is inherently partial. It is, therefore, critical
that policy-relevant models be tested in a variety of settings and under a variety of conditions (Sullivan, 1997).

The MAGIC model has been widely used throughout North America and Europe to project changes in the chemistry of drainage waters impacted by atmospheric S deposition. MAGIC projections of the effects on surface water chemistry of various S emissions scenarios formed the technical foundation for a large part of the National Acid Precipitation Assessment Program's Integrated Assessment (IA; NAPAP, 1991). Subsequently, a research effort was conducted from 1990 to 1996 to improve the performance of MAGIC and to provide testing and confirmation of the model at multiple sites. Model evaluations have included hindcast comparisons with diatom reconstructions* of pre-industrial lake-water chemistry in the Adirondack Mountains of New York, and tests of the veracity of model forecasts using the results of whole-catchment acidification experiments in Maine (Norton et al., 1992) and Norway (Gjessing, 1992) and whole catchment acid-exclusion experiments in Norway (Wright et al., 1993).

It is critical that policy-relevant environmental models such as MAGIC be confirmed under a variety of conditions. Since 1990, the MAGIC model has been tested in a variety of settings and under quite varying environmental conditions. These analyses have elucidated a number of potentially important deficiencies in model structure and method of application, and have resulted in changes to the model and its calibration procedures. The work has included in-depth evaluation of issues related to regional aggregation of soils data, background pre-industrial S deposition, natural organic acidity, N, and Al mobilization. The result has been an improved and more thoroughly tested version of MAGIC, and one that yields different forecasts than the version that formed the technical foundation for the 1990 IA.

9.1.2 Recent Modifications to the MAGIC Model

9.1.2.1 *Regional Aggregation and Background Sulfate*

MAGIC model projections of future lake-water chemistry made by NAPAP (1991) for lakes in the northeastern U.S. were based on data collections and model calibrations performed by the EPA's Direct Delayed Response Project (DDRP; Church et al., 1989; Cosby et al., 1989). The northeastern DDRP analyses were based on a probability subsample of the 1984 Eastern Lake Survey (ELS; Linthurst et al., 1986), and included 145 low-ANC (less than 400 µeq/L) lakes, larger than 4 ha in area. These lakes provide an unbiased representation of northeastern lakes included in the DDRP statistical frame.

The MAGIC model represents the horizontal dimension of the watershed as a homogeneous unit and the vertical dimension as one or two soil layers. Watershed and soils data required as model inputs are aggregated to provide

---

* Diatoms are microscopic algae, the remains of which are incorporated into lake sediments that accumulate over time. The species composition and relative abundance of diatoms at different levels in the sediment can be used to estimate the pH of lake water in the past using sophisticated mathematical relationships.

weighted-average values for each soil layer. Within the DDRP (Church et al., 1989) that formed the technical foundation for NAPAP modeling efforts in the Northeast, soil characteristics were aggregated on the basis of attributes of soil sampling classes across the entire northeastern U.S. Subsequent to the DDRP, there was concern that Adirondack soils might differ sufficiently in their chemical properties from similar soils in other areas of the Northeast that MAGIC projections for Adirondack watersheds might be biased because they were based on soil attributes that actually reflected conditions elsewhere than the Adirondacks. The DDRP soils data, therefore, were reaggregated to characterize Adirondack watershed attributes using only soil data collected from pedons in the Adirondacks (Sullivan et al., 1991).

Modeling for the DDRP and IA also assumed that the deposition of S in pre-industrial times was limited to sea salt contributions. Based on analyses presented by Husar et al. (1991), this assumption was modified such that pre-industrial deposition of S was assumed equal to 13% of 1984 values (Sullivan et al., 1991).

Recalibration of MAGIC to the Adirondack lakes database using the regionally corrected soils and background SO_4^{2-} data resulted in approximately 10 μeq/L lower estimates of 1984 ANC. A substantial downward shift was also observed in predicted pre-industrial and current lake-water pH (approximately 0.25 pH units) for lakes having pH greater than about 5.5. These differences were attributed to lower calibrated values for lake-water SO_4^{2-} concentrations and higher pCO_2 values estimated for Adirondack lakes, compared with the Northeast as a whole (Sullivan et al., 1991).

9.1.2.2   *Organic Acids*

Concern was raised subsequent to the IA regarding potential bias from the failure to include organic acids in the MAGIC model formulations used by NAPAP. MAGIC hindcasts of pre-industrial lake-water pH showed poor agreement with diatom-inferences of pre-industrial pH (Sullivan et al., 1991), and preliminary analyses suggested that these differences could be owing, at least in part, to the presence of naturally occurring organic acids in Adirondack lake waters.

Previous projections of future lake-water chemistry in Adirondack lakes using MAGIC (Church et al., 1989; Cosby et al., 1989) did not consider the acid–base chemistry of dissolved organic acids in the model formulations or their role in the response of lake chemistry to acidic deposition. It has been suggested, however, that organic acids can make significant contributions to surface water acidity (Krug and Frink, 1983). A significant fraction of organic acids in surface waters are characterized by strongly acidic pK_a values, below 4.0 (Perdue et al., 1984; Kramer and Davies, 1988). Furthermore, considerable evidence suggested that organic acids influence the response of surface waters to changes in strong acid inputs, potentially by loss of DOC (Krug and Frink, 1983; Almer et al., 1974) and most likely by changes in the protonation of organic acid anions (Wright, 1989).

There is not a method available for direct determination of organic acid concentration in the laboratory (Glaze et al., 1990). Measures of total (TOC) and dissolved organic carbon (DOC) are commonly used to represent, in relative terms, the amount of organic acidity present (Aiken et al., 1985). Some studies report TOC (unfiltered) and others report DOC (filtered); the former are slightly higher owing to the presence in most water samples of small amounts of particulate carbon. The pool of dissolved organic material in natural waters is generally comprised largely of organic acids (McKnight et al., 1985; David and Vance, 1991). Empirical methods for laboratory determination of organic acidity generally include concentration, fractionation, isolation, purification, and titration steps (e.g., Leenheer, 1981; David and Vance, 1991; David et al., 1989, 1992; Kortelainen et al., 1992). Such methods are fairly laborious and time-consuming, and are seldom used in water quality assessments and surveys. Indirect methods available for estimating organic acid anion contributions to acidity include charge balance calculations and the empirical methods of Oliver et al. (1983) that are based on measured pH and DOC, and Driscoll et al. (1994). The latter study was based on empirical data from the Adirondack Lakes Survey (ALSC). From 1984 to 1987, the ALSC surveyed 1469 lakes within the Adirondack Ecological Zone (Kretser et al., 1989; Baker et al., 1990b). This database provided an unparalleled data resource with which to investigate questions of organic acidity in lake waters in the U.S. because of the large number of lakes sampled and abundance of survey lakes having high DOC concentrations. The median DOC of the study lakes was 500 μM C and 20% of the lakes had DOC concentrations greater than 1650 μM C.

Driscoll et al. (1994) constructed a reduced data set from the ALSC database by deleting lakes that were

1. Missing variables.
2. High in salt content (greater than 1000 μeq/L).
3. High in pH (greater than 7) or ANC (greater than 400 μeq/L).
4. Outside QA/QC guidelines.

The remaining lakes were grouped into pH intervals of 0.1 pH units from pH 3.9 to 7.0, whereby each observation represented the mean of from 12 to 94 individual lake measurements of pH and related chemistry. This data reduction procedure reduced the variability in the initial data set and allowed application of nonlinear methods for fitting the various organic acid analog models to estimates of organic anion concentration from the measured anion deficits (Σ cations - Σ anions; Figure 9.1).

To evaluate the ability of model calculations to predict lake-water pH, variable pH calculations were conducted. pH was calculated based on conditions of electroneutrality, concentrations of major solutes, and important pH buffering systems (DIC, DOC, and Al). A total of four organic acid analog representations were calibrated to the ALSC reduced data set (Driscoll et al., 1994).

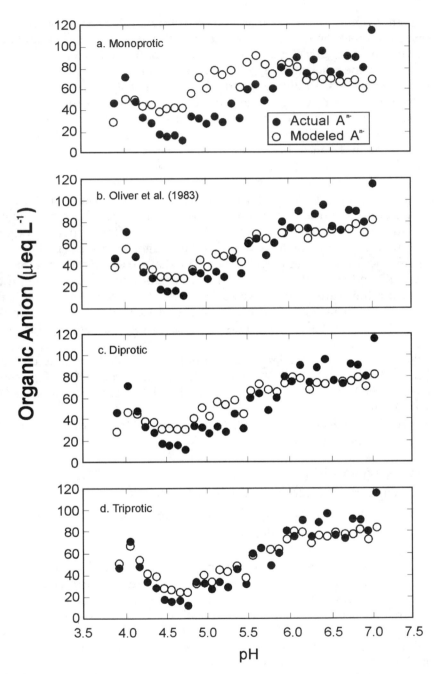

FIGURE 9.1
Comparison of calculated (from charge balance) mean organic anion concentration (A^{n-}) at
0.1 pH unit intervals with calibrated model predicted values for a. monoprotic, b. Oliver et
al. (1983), c. diprotic, and d. triprotic organic analog models. (Source: Driscoll, C.T., M.D.
Lehtinen, and T.J. Sullivan, 1994, Modeling the acid-base chemistry of organic solutes in
Adirondack, NY, lakes, *Water Resour. Res.*, Vol. 30, p. 303, Figure 2; copyright by the American
Geophysical Union. With permission.)

They included mono-, di-, and triprotic analog models and the model of Oliver et al. (1983). The model calibration involved adjustments of the H^+ dissociation constants and site density of the DOC that specifies the number of dissociation sites per mole of organic C. The object of the fitting routine was to minimize the observed differences across all lakes between the organic charge simulated by the organic acid analog model and the organic anion concentration estimated from the measured charge balance. A nonlinear least squares technique was used in the calibration, with pK_a values fit first, followed by site density. The calibration was accomplished using SAS (Driscoll et al., 1989a) for the Oliver et al. (1983) and monoprotic models, and using ALCHEMI (Schecher and Driscoll, 1994) for the diprotic and triprotic models. Additional details are provided by Driscoll et al. (1994).

The best agreement ($r^2 = 0.92$) was obtained between predicted and observed pH values using the triprotic analog representation, with fitted pK_a values of 2.62, 5.66, and 5.94, and a calibrated site density of 0.055 mol sites per mol C. The fitted values for pK_a and site density obtained by Driscoll et al. (1994) were used in the revised MAGIC applications conducted by Sullivan et al. (1996a) and described below.

In the Adirondack region of New York, 33 lakes were included in both the DDRP study and the Paleoecological Investigation of Recent Lake Acidification (PIRLA-II; Charles and Smol, 1990). This data set, therefore, provided an opportunity to evaluate the potential importance of organic acids to the modeling efforts. The hindcast comparison focused on pH reconstructions for these lakes because of the underlying importance of pH and its influence on the mobilization of potentially toxic Al and controls on the biological responses to acidification (Baker et al., 1990c).

MAGIC simulations were performed as done earlier by Cosby et al. (1989) for the DDRP (Church et al., 1989) and by NAPAP (1991), with three exceptions (Sullivan et al., 1991)

1. To remove known biases and make the MAGIC and diatom estimates as directly comparable as possible, MAGIC was recalibrated using soils data specific to the Adirondack subregion.

2. A more realistic pre-industrial S deposition, equal to 13% of 1984 values (Husar et al., 1991), was assumed.

3. The partial pressure of CO_2 in lake water was calculated from measured values of dissolved inorganic carbon (DIC) and pH.

The earlier model projections (NAPAP, 1991; Cosby et al., 1989) had been calibrated using soils and surface water data from sampling sites across the entire northeastern region of the U.S., had assumed zero pre-industrial S deposition, and had calibrated P_{CO_2} in the absence of consideration of organic acids. Changes in the first two factors improved the agreement between MAGIC and diatom estimates of historical pH, owing largely to differences in the calibrated values of strong acid anion concentrations. The last change lessened the agreement because the earlier calibration of P_{CO_2} had effectively resulted in a partial compensation for the missing organics. Additional uncertainties that might have affected the comparison between the MAGIC and diatom approaches include the failure of the process model to account for historic changes in landscape cover, disturbance, N dynamics, or changes in base cation deposition (Sullivan et al., 1991). Model scenarios using the original version of MAGIC without organic acids were designated $MAGIC_1$, and those that included the triprotic organic acid analog were designated $MAGIC_2$.

Unmodified $MAGIC_1$ hindcasts yielded pre-industrial pH values that were substantially higher than diatom-based estimates (Figure 3.3a), and the discrepancy was greatest for those lakes in the most biologically sensitive portion of the pH range (pH 5.0 to 6.0) (Baker et al., 1990c). Furthermore, $MAGIC_1$ hindcast pH estimates were greater than 6.0 for all lakes investigated, whereas diatom estimates of pre-industrial pH ranged from as low as 5.2 to above 7.0. Previous comparisons between diatom and $MAGIC_1$ (without organic acids) model estimates of historical acidification had been conducted primarily for clearwater (DOC less than 300 μM C) lakes, most of which had experienced substantial acidification (Wright et al., 1986; Jenkins et al., 1990). These comparisons generally showed somewhat better agreement for pre-industrial pH than the comparisons reported in Figure 3.3a.

The failure to consider proton binding reactions involving organic solutes in the $MAGIC_1$ hindcast simulations could contribute to the observed discrepancy between model-predicted and diatom-inferred pH values because of the influence of dissolved organic acids on the acid–base chemistry of dilute waters (Hemond, 1994). Even low concentrations of dissolved organic acids (less than 250 μM C) can appreciably affect the pH of dilute waters either in the presence or absence of strong inorganic acids (Kramer and Davies, 1988; Hemond, 1994). Although other factors might also contribute to the observed discrepancies, including, for example, uncertainties in weathering, SO_4^{2-} adsorption, base cation deposition, or hydrological routing, the pattern of effect (Figure 3.3a) suggested the importance of organic acids. Organic acids exert a disproportionately larger influence on pH at pH values below 6.5, where the greatest offset was observed.

Thus, three independent data sets (DDRP, PIRLA-II, and ALSC) and three interpretive models ($MAGIC_1$ with no organic acid representation, diatom reconstructions, and $MAGIC_2$ with Driscoll et al.'s triprotic organic acid analog) were employed to test for consistency among the results of these models for estimating pre-industrial lake-water pH (Sullivan et al., 1996a). When the organic acid model was incorporated into $MAGIC_2$ and simulated pH values were compared with diatom-inferred pH, the comparison yielded considerably closer agreement between model estimates of pre-industrial pH (Figure 3.3b) than did the simulations that did not consider the effects of organic acids (Figure 3.3a). The mean difference in $MAGIC_1$ vs. diatom estimates of pre-industrial pH was 0.6 pH units when organic acids were omitted from the modeling scenarios with the greatest discrepancy being for lakes with diatom-inferred pH less than 6.0. This mean difference was reduced to only 0.2 pH units when the triprotic organic acid model was included, and the agreement for individual low pH lakes improved by as much as a full pH unit (Figures 3.3a,b). The extent to which the incorporation of an organic acid representation into $MAGIC_1$ alters estimates of historic acidification for the population of low-ANC lakes represented by this study is illustrated in Figure 9.2. The diatom model and both versions of MAGIC resulted in cumulative frequency distributions of pre-industrial pH higher than current measured pH. The diatom model suggested the least amount of acidification, and $MAGIC_1$ without organic acids suggested the greatest acidification. $MAGIC_2$ estimates with a triprotic organic acid were intermediate, but closer to diatom estimates. Differences between the two MAGIC applications were most pronounced at the lowest end of the pH distribution, and varied by up to a full pH unit for individual lakes (Sullivan et al., 1996a).

The observed improved agreement between $MAGIC_2$ and diatom hindcasts of pre-industrial pH was attributable partly to improvement in the calibrated 1984 pH values and partly to lower estimates of ΔpH for those lakes simulated by $MAGIC_2$ to have experienced the greatest historical acidification (greater

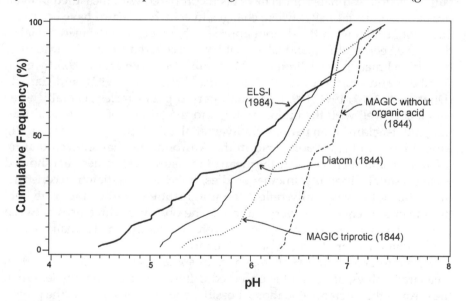

FIGURE 9.2
Cumulative frequency distributions of current measured pH from ELS and estimates of pre-industrial pH using the diatom method and the MAGIC model with and without the organic acid representation. Distributions were derived using population weighting factors developed for the DDRP. More than 40% of the lakes had measured current pH less than 6. Application of MAGIC with the triprotic organic acid suggested that one-half of these lakes also had pre-industrial pH less than 6, whereas application of MAGIC without considering organic acids suggested that all lakes had pre-industrial pH greater than 6.3. (Source: *Water Air Soil Pollut.,* Vol. 91, 1996, p. 277, Influence of organic acids on model projections of lake acidification, Sullivan, T.J., B.J. Cosby, C.T. Driscoll, D.F. Charles, and H.F. Hemond, Figure 2, copyright 1996. Reprinted with kind permission from Kluwer Academic Publishers.)

than 1 pH unit). Both effects of adding organic acids to MAGIC$_2$ are important because it is ultimately the projected endpoint pH values that are important from a policy perspective. Model forecasts are often generated to answer questions such as

- How many lakes will acidify to pH less than 5.0 if deposition is maintained at a certain level for a certain number of years?
- How much will deposition have to be reduced in order for 95% of the lakes in a region to recover to pH values above 5.5?

Even after adding organic acids to MAGIC$_2$, the model still predicted greater historical acidification of Adirondack lakes than did the diatom model (Sullivan et al., 1996a). The differences between MAGIC$_2$ and diatom-based estimates of pre-industrial pH were far more reasonable, however, when the influence of organic acids was included in the modeling effort. The remaining discrepancy may be owing to additional uncertainties in the MAGIC$_2$ model and/or a general tendency for diatom estimates to be conservative. Diatom estimates of pH have been compared with measured pH values at numerous lake sites where changes in acid–base status have occurred. Such confirmations of the diatom approach have been performed for lakes that have been acidified and lakes that have recovered from acidification or have been limed in the Adirondack Mountains (Sullivan et al., 1992), Sweden (Renberg and Hultberg, 1992), Scotland (Allott et al., 1992), and Canada (Dixit et al., 1987, 1991, 1992). Diatom-inferred pH histories generally agree reasonably well with the timing, trend, and magnitude of known acidification and deacidification periods. Sullivan et al. (1992) presented data for Big Moose Lake and Constable Pond in the Adirondacks that showed diatom inferences of mean pH close to the mean of measured pH values, that showed great seasonal variability. In other studies, however, the diatom reconstructions did not always fully reflect the magnitude of either the water pH decline or subsequent recovery, although the observed differences between predicted and measured change in pH were frequently smaller than the root mean squared error of diatom predictive models.

Diatom inferences of pH change may in some cases be slightly less than measured values, although the observed differences are generally less than the error of the inference equations. Possible explanations include the preference of many diatom taxa for benthic habitats where pH changes may be buffered by chemical and biological processes. Alternatively, such an attenuation could be an artifact of sediment mixing processes or a time-averaging artifact of sediment subsample thickness relative to the sediment accumulation rate. It is, thus, not surprising that MAGIC$_2$ model simulations that included organic acid representations estimated greater acidification than diatom-inferences for Adirondack lakes in this study. It is not possible to determine which method provides estimates closer to reality, although diatom inferences of 1984 pH agreed somewhat better with measured values than either version of MAGIC. It is reassuring that the two methods provide results that are generally in reasonable agreement.

The results of this intercomparison reported by Sullivan et al. (1996a) are important for assessment of the effects of acidic deposition in two respects. First, these results were the first to show quantitative agreement between estimates of pH of natural aquatic systems receiving acidic deposition, as derived from two independent and conceptually different approaches over a large geographic region and over a long time span. Previous model testing and evaluation studies, other than calibration exercises, had either been relatively short duration (Norton et al., 1992), site specific (Renberg and Hultberg, 1992), or had involved comparisons among two or more models that share many fundamental assumptions (Cook et al., 1992). Second, and perhaps more important, is the fact that the agreement between $MAGIC_2$ and paleolimnological model hindcast estimates of lake-water pH was dependent upon consideration of proton binding reactions involving dissolved organic acids in the process model. The latter result was obtained despite the relatively low concentrations of DOC in the study lakes ($\bar{x}$ = 313µM C). The importance of organic acids in achieving reliable model results undoubtedly increases with increasing lake-water DOC. In fact, all lakes for which estimates of ΔpH (current pH minus pre-industrial pH) decreased by more than 0.5 pH units, upon inclusion of organic acids in the model, had DOC in the range of 400 to 500 µM. Such concentrations of DOC are not considered high, but were at the upper end of measured DOC concentrations in the 33 study lakes.

Organic acids have been shown in other instances to be important contributors to, and buffers of, ecosystem acidity and, therefore, are important to include in modeling ecosystem response to acidification. For example, Lam et al. (1989) assumed a triprotic organic acid representation for observed data from Moose Pit Brook and Mersey River in Nova Scotia. The objective was to determine what specific modifications were needed to calibrate the Turkey Lakes model to colored water systems having DOC values of 800 to 3300 µM C and 400 to 1200 µM C, respectively. They assumed pK_1 = pH, for simplicity, for pH values between 4.5 and 5.5. Calibrated values for pK_2 and pK_3 were 4.8 to 5.0 and 5.0 to 5.2, respectively, for the 2 stream systems. Calibrated charge densities for DOC in both streams were about 4 µeq/mg C. They found that the assumed charge density of DOC and the assumed pK_1 value were at least as important as the SO_4^{2-} loading in influencing the pH predicted by the model. Furthermore, because the organic anions both buffer and contribute acidity to the water, the model simulations illustrated that increased or decreased SO_4^{2-} input to these two colored stream systems would not cause as large a change in pH as in clear water systems (Lam et al., 1989).

Inclusion of organic acids in the MAGIC simulations for the experimental watersheds at Lake Skjervatjern, Bear Brook, and Risdalsheia (see also Chapter 8) also had dramatic effects on model simulations of pH. In all cases, MAGIC simulated considerably higher pH values when organic acids were omitted from the model. Even at Bear Brook, where annual average DOC concentrations are very low (less than 250 μM C), incorporation of organic acids into the model reduced simulated pH by 0.1 to 0.3 pH units for the years of study. At Lake Skjervatjern and Risdalsheia, where organic acids provide substantial pH buffering, omission of the organic acid analog representation from MAGIC resulted in consistent overprediction of pH by about 0.2 to 0.5 pH units (Sullivan et al., 1994).

9.1.2.3   Aluminum

Aluminum mobilization is now widely believed to be one of the most important ecological effects of surface water acidification. Potential effects of Al mobilization from soils to surface and soil waters include alterations in nutrient cycling, pH buffering effects, toxicity to aquatic biota, and toxicity to terrestrial vegetation. MAGIC simulates Al solubility based on an assumed equilibrium with the mineral gibbsite ($Al(OH)_3$):

$$Al(OH)_3(s) + 3\ H^+ \rightleftharpoons Al^{3+} + 3\ H_2O \qquad (9.6)$$

The preceding equilibrium expression illustrates a cubic relationship between the concentrations of Al^{3+} and H^+, such that

$$[Al^{3+}]/[H^+]^3 = K_{SO} \qquad (9.7)$$

where brackets indicate activities and K_{SO} is the solubility product. For a solution in equilibrium with gibbsite, Al^{3+} changes in proportion to the change in H^+ to the third power, and a plot of pAl^{3+} vs. pH (p indicates $-\log_{10}$) will have a slope of 3 and an intercept of pK_{SO}.

The MAGIC model first calculates the total concentration of acidic cations (e.g., H^+ plus Al^{n+}) on the basis of simulated concentrations of base cations and mineral acid anions (e.g., SO_4^{2-}, NO_3^-, Cl^-) using mass balance and electroneutrality constraints. The acidic cations are then partitioned between H^+ and Al^{n+} using the gibbsite mineral equilibrium, thermodynamic equations, the partial pressure of CO_2, and the organic acid formulation. This partitioning is important because inorganic Al in solution can be highly toxic to aquatic biota, even at low concentrations (Baker and Schofield, 1982).

Model estimates of changes in the concentration of Al^{3+} in surface waters, using the MAGIC model have shown a consistent pattern of overestimating the change in Al^{3+} concentration in response to experimental treatment (Sullivan and Cosby, 1998). This overestimate of the change in Al^{3+} concentration calculated by MAGIC was owing to a combination of the cubic relationship between H^+ and Al^{3+} assumed in the gibbsite model and the model calibration procedure of selecting a gibbsite solubility product based on measured pretreatment data.

Data sets collected by the EPA in the Eastern Lake Survey-Phase II (ELS-II), National Stream Survey (NSS), and Episodic Response Project (ERP) were assessed by Sullivan and Cosby (1998) to evaluate relationships between Al_i and pH in lake and stream waters in the eastern U.S. Water samples collected within these projects had been analyzed for both total and nonlabile monomeric Al, thus allowing the labile, or inorganic, monomeric Al component (Al_i) to be determined by difference (c.f., Driscoll, 1984). Appreciable concentrations of Al_i were found in surface waters of the Adirondack Mountains in ELS-II, the Pocono/Catskill Mountains and northern Appalachian province in the National Stream Survey, and the Adirondack Mountains and Catskill Mountains in the Episodic Response Project. Speciation of the Al_i was accomplished using the chemical equilibrium model ALCHEMI (Schecher and Driscoll, 1987). With input data of pH, Al_i, total F, SO_4^{2-}, dissolved Si, and temperature, ALCHEMI estimates the concentration of the various inorganic Al species, including Al^{3+} and the Al complexes with hydroxide, fluoride, SO_4^{2-}, and silica, as well as mineral phase saturation indices. Plots of pAl_i and pAl^{3+} vs. pH were constructed to compare empirical patterns across lakes and streams with those predicted by the gibbsite formulation.

For all data sets examined, consistent relationships were evident between pAl_i and pH for the waters of interest (pH 4 to 6). The slope of this relationship was consistently near 1.0, ranging from 0.77 to 1.28. When plots of pAl^{3+} vs. pH were examined, similar results were found. The slopes of the relationships in this case were consistently near 2.0, and ranged from 1.82 to 2.34 (Table 9.1). These results illustrate that, for the surface waters in the U.S. that are of interest with respect to potential Al mobilization, a gibbsite-type equation to model Al_i concentration directly should use a power term of about 1. For predicting Al^{3+} concentration, a power term of about 2 should be used.

TABLE 9.1

Slopes of Regression Relationships Between pAl and pH for Lake and Stream Data Sets in the Eastern U.S. Analyzed by Sullivan and Cosby, 1998.

Data Sets	Reference[c]	pAl_i vs.pH			pAl^3 vs. pH		
		Slope	(s.e.)	R^2	Slope	(s.e.)	R^2
ELS-II–Adirondack lakes, spring	1	1.09	(0.20)	0.54	2.34	(0.29)	0.74
ELS-II–Adirondack lakes, fall	1	0.81	(0.09)	0.62	2.03	(0.16)	0.79
ERP–Adirondack streams	2	0.77[a]	(0.06)	0.57	N.D.	–	–
ERP–Catskill streams	2	0.84	(0.05)	0.69	N.D.	–	–
NSS-Catskill streams[b]	3	0.88	(0.13)	0.61	1.82	(0.16)	0.84
NSS–N. Appalachian streams[b]	3	1.28	(0.06)	0.85	2.26	(0.07)	0.93

[a] Regression statistics limited to streams with pH less than 5.7 because of the substantial scatter observed at higher pH.

[b] Streams having pH less than 4 were assumed to have been impacted by acid mine drainage and were deleted from the analysis.

[c] Reference 1 is Herlihy et al. (1991); Reference 2 is Wigington et al. (1993); and Reference 3 is Kaufmann et al. (1988).

None of the data we examined suggested a power term close to 3, the value previously used in model formulations.

Model simulations were also conducted by Sullivan and Cosby (1998) with the MAGIC model for two watersheds in which acidic deposition inputs have been experimentally altered: West Bear Brook, ME, and Risdalsheia, Norway (Norton et al., 1993; Wright et al., 1993). At the Bear Brook treatment catchment, ambient deposition has been augmented with additional inputs of S and N. At Risdalsheia in southernmost Norway, high ambient levels of S and N deposition have been reduced to background levels by emplacement of a transparent roof over an entire mini catchment. MAGIC projections at these sites were modified from recent applications (Cosby et al., 1995, 1996) by altering the model algorithms for predicting the Al response. The alteration was based on the results of the empirical spatial analyses described previously.

MAGIC was applied to the Bear Brook data using an exponent of 2 and an intercept, log K_{SO} equal to 4.0 that corresponded approximately to the empirical relationships derived for fall samples from Adirondack lakes (Table 9.1). The previous simulation for Bear Brook had been based on an exponent of 3 and an intercept of 10, based on calibration to the pretreatment watershed data. We judged that the log K_{SO} value derived from pretreatment data at Bear Brook was too high, based on comparison with data from other sites. At Risdalsheia, log K_{SO} equal to 2.6 was calibrated to data from the reference catchment assuming an exponent of 2.

The revised MAGIC projections of Al_i concentration at West Bear Brook agreed more closely with measured values than did the projections based on the gibbsite solubility assumption (Figure 9.3; Sullivan and Cosby, 1998). The results of comparing simulated with measured Al_i concentrations at the Risdalsheia site were not so consistent. However, the majority of the annual average measured values at Risdalsheia more closely followed the MAGIC trajectory that was constructed assuming an exponent of 2 in Eq. 9.7, rather than 3 as in the gibbsite model (Sullivan and Cosby, 1998). Neither formulation was completely satisfactory for predicting stream-water Al_i concentration at these sites. This is to be expected given the lumped-parameter nature of the model and the complexity of the Al hydrogeochemical response (Sullivan, 1994). In most cases, however, a power term of 2.0 in the model formulation for Al^{3+} provided the most reasonable projections.

9.1.2.4 Nitrogen

MAGIC, as originally formulated and applied for the studies described previously, contained an extremely simplified representation of N dynamics within catchment soils. There were no processes controlling the details of N cycling in the model. The version of the MAGIC model used for the Integrated Assessment was not appropriate for simulation of changes in atmospheric deposition of N. In light of the increasing concern about N saturation in forested ecosystems, this was a serious shortcoming in the model. A major

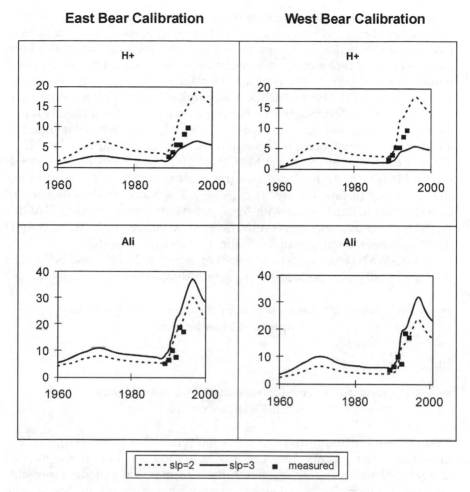

FIGURE 9.3
Observed annual average concentration (■) and MAGIC simulated values of H⁺ and Al$_i$ (in µM) where the simulations are based on gibbsite solubility with a power term of 3.0 (solid line) and a modified relationship for solubility with a power term of 2.0 (dashed line). Data are presented for the watershed manipulation experiment at West Bear Brook, ME, where sulfur and nitrogen deposition have been experimentally increased. Annual average Al$_i$ values were measured as total monomeric Al and corrected to remove organically bound Al using an empirical relationship derived from ELS-II data for the northeast U.S. ($r^2 = 0.97$, $n = 69$). The model was calibrated twice, once to East Bear Brook (left panels) and once to the manipulated stream, West Bear Brook (right panels). (Source: *Water Air Soil Pollut.*, Vol. 105, 1998, p. 654, Modeling the concentration of aluminum in surface waters, Sullivan, T.J. and B.J. Cosby, Figure 3, copyright 1998. Reprinted with kind permission from Kluwer Academic Publishers.)

uncertainty in the current modeling of ecosystem response to changing N deposition is specification of changes in the temporal dynamics and degree of N saturation. For some applications to forested watersheds, the MAGIC model has been structured to predict N saturation status from estimates of mineralization and nitrification (e.g., Ferrier et al., 1995). More recently, efforts are underway to incorporate the results of the European studies (e.g., Tietema and Beier, 1995) into the model to predict N-saturation status from forest floor C:N ratios. For many recent applications, however, it has been assumed that current measured N retention in the modeled watersheds will remain constant into the future as a percentage of N inputs (e.g., Sinja et al., 1998; Sullivan et al., 1998). Such an assumption is probably reasonable, as long as changes in N deposition in the future are modest. The difficulty is predicting the timing and magnitude of the changes in the percent N retained by a watershed that will occur if deposition changes dramatically.

A new coupled S and N model, MAGIC-WAND, was developed by extending the MAGIC model to incorporate the major ecosystem N fluxes and their changes through time (Ferrier et al., 1995). The Model of Acidification of Groundwater in Catchments With Aggregated Nitrogen Dynamics (MAGIC-WAND) represents an extension to the MAGIC model. In MAGIC-WAND the N dynamics are fully coupled to the initial S-driven model.

MAGIC-WAND considers two species of inorganic N, NO_3^- and NH_4^+. The model explicitly incorporates the major terrestrial fluxes of N, such that

$$NO_3^- \text{ leaching} = \text{deposition} + \text{nitrification} + \text{external addition}$$
$$- \text{ uptake} - \text{denitrification}$$

and

$$NH_4^+ \text{ leaching} = \text{deposition} + \text{external addition} + \text{mineralization}$$
$$- \text{ nitrification} - \text{uptake}$$

If the net result of these fluxes is positive (surplus NO_3 and/or NH_4), leaching to surface waters occurs. Nitrogen inputs to the system are in the form of inorganic N added to soil solution. Mineralization in the model represents the release of inorganic N that was formerly bound in organic matter, and the mineralization product is NH_4. Nitrogen losses from the model system are as inorganic N, and the primary output is hydrologic runoff from the soils. The runoff fluxes are calculated as the product of the simulated concentrations of NO_3 and/or NH_4 at any time step and the hydrologic discharge at that time. Provision is also made in the model for other losses of inorganic N, such as denitrification from soil or surface water. The magnitude and timing of these additional outputs of N may be specified *a priori* or they may be keyed to external inorganic N concentrations using first order reactions. The microbial-mediated transformation of NH_4 to NO_3 (nitrification) is represented in the model by a first order reaction such that the rate of loss of NH_4 (equal to the rate of production of NO_3) is given by the product of a rate constant and the concentration of NH_4 at each time step.

Plant uptake is modeled as a nonlinear process that depends upon the concentration of available NH_4 and NO_3. The equation is hyperbolic (representation of a typical Michaelis-Menten uptake function) such that

$$d(N)/dt = K_{max} * (N)/(K_{(1/2)} + (N)) \tag{9.8}$$

where (N) is the concentration of either NH_4 or NO_3, K_{max} is the maximum uptake rate (meq/m^3 per year), and $K_{1/2}$ is the half-saturation constant of the reaction (meq/m^3). The values of K_{max} and $K_{1/2}$ can be varied through time *a priori*, to represent the dynamics of ecosystem response to available N.

An important limitation of MAGIC-WAND, as described by Ferrier et al. (1995), was the necessity of specifying rates of mineralization and nitrification for the watershed being modeled. It proved difficult, time-consuming, and expensive to derive watershed-scale estimates of these process rate functions, and variability was often high. Results of experimental studies in NITREX have recently demonstrated that NO_3^- leaching can be empirically estimated based on the N concentration of various ecosystem compartments, including forest floor, soil, and foliage (see discussion in Chapter 7). The approach for modeling N in MAGIC is currently in the process of being revised to reflect these new findings (Cosby, personal communication). In the interim, recent MAGIC applications have calibrated the current watershed retention of N as a percent of total N input. These calibrated values of N retention are used to estimate N retention and leaching under future changing levels of N deposition (c.f., Sinha et al., 1998; Sullivan and Cosby, 1998).

9.1.3 Cumulative Impacts of Changes to the MAGIC Model

In order to evaluate the incremental and cumulative impact of some of the modifications to MAGIC, a suite of model simulations was conducted by Sullivan and Cosby (1995) for the Adirondack DDRP lakes. The baseline model structure was used in the DDRP and NAPAP IA studies. The changes to the model that were examined included modifying the assumption regarding background S deposition, reaggregating the soils data, recalibrating the model specifically for the Adirondack subregion, adding the organic acid model to the surface water compartment, and changing the Al^{n+}/H^+ ion relationship from cubic to quadratic. These analyses did not, however, include examination of the effects on model output of including N dynamics in the model simulations.

A suite of simulations was conducted based on the application of an assumed deposition scenario to derive a 50-year forecast using each model structure. The deposition scenario assumed constant S deposition from 1984 (the calibration year) to 1994, followed by a 30% decrease in S deposition from 1995 to 2009, with constant deposition thereafter until 2034. The modeled responses of 33 Adirondack lakes to this deposition scenario were considered. The impacts of the changes were illustrated by tabulating the percentage of lakes predicted to have pH, ANC, or Al values in excess of commonly accepted thresholds of potential biological effects.

The overall effect of the various changes to the model structure and application procedures was an increase in the percentage of lakes exceeding various biological thresholds with respect to pH, Al, and ANC subsequent to an hypothesized 30% decrease in S deposition (Table 9.2). The largest changes were observed for pH and Al; ANC projections were less affected. The modifications to the model that caused the greatest changes in projected output were the recalibration of the model to the Adirondack subregion, modification of the assumption regarding background SO_4^{2-}, and the incorporation of the organic acid model into MAGIC. The modification of the Al algorithm caused fewer lakes to be projected to exceed Al threshold values in response to the reduced deposition scenario; this change was quantitatively less important than the previous changes.

The magnitude of effect of the cumulative modifications to the model was considerable (Table 9.2). For example, 32% of the lakes had measured pH less than 5.5 in 1984, whereas only 8% were projected to still have pH less than 5.5 after the reduction in S deposition, using the original MAGIC application. In contrast, the improved version of MAGIC projected that 32% of lakes would still have pH less than 5.5 in the year 2034. Similarly, of the 30% with measured Al_i concentration greater than 50 µg/L in 1986, the original model structure projected only 4% would still have Al_i concentrations greater than 50 µg/L in 2034 compared to 30% projected to continue to have high Al_i by the improved version of MAGIC. Based on model projections using the improved version of MAGIC, little recovery of Adirondack lakes would be expected subsequent to a 30% reduction in S deposition. The number of lakes having pH less than 6.0 was actually projected to increase, and the number of lakes projected to have ANC less than 0 only decreased slightly in response to lower deposition. These estimates were independent of any possible increases in NO_3^- leaching that might occur. The lack of recovery suggested by these revised model projections was attributable partly to a decrease in the modeled base saturation of watershed soils (Sullivan and Cosby, 1995). These

TABLE 9.2

Cumulative Effects of Some of the Recent (Post-1990) Changes to the Structure and Method of Application of the MAGIC Model. MAGIC Predictions (Sullivan and Cosby, 1995) of the Percentage of Adirondack DDRP Lakes having pH, ANC, and Al Above or Below Threshold Values in the Year 2034 Subsequent to an Hypothesized 30% Decrease in S Deposition

	Percentage of Lakes having pH Below Value			Percentage of Lakes having ANC Below Value (µeq/L)			Percentage of Lakes having Al Above Value (µg/L)		
Data Type	5	5.5	6	0	25	50	50	100	200
Measured 1984 values	12	32	38	18	48	59	30	18	10
MAGIC projection of 2034 chemistry									
1990 IA version of MAGIC	0	8	20	6	34	44	4	0	0
1995 version of MAGIC[a]	8	32	44	14	40	44	30	10	4

[a] Does not include N dynamics.

results may affect expectations of recovery in response to S emission controls mandated by Title IV of the Clean Air Act Amendments of 1990.

The future response of lakes and streams to acidic deposition is also highly dependent upon the extent to which watersheds in acid-sensitive regions become N saturated. EPA scientists conducted MAGIC model simulations for 50 years into the future that effectively bounded the range of possible water chemistry responses, ranging from no watersheds reaching N saturation to all simulated watersheds reaching N saturation during the simulation period. The model projections for Adirondack lakes, for example, suggested that the percent of chronically acidic lakes in the target population in 50 years could range from 11 to 43%, depending on the number of watersheds that become N saturated (EPA, 1995a). Similarly, for mid-Appalachian streams, the modeled percent of streams acidic in 50 years ranged from 0 to 9%, depending on the extent of N saturation (EPA, 1995a).

9.1.4 MAGIC Model Testing and Confirmation Studies

MAGIC has been tested after inclusion of many of the model modifications discussed in the preceding sections. The revised model with Driscoll et al.'s (1994) organic acid model yielded reasonable agreement between model hindcast pH and diatom-inferred pH for the data set of 33 Adirondack lakes (Sullivan et al., 1996a; Figure 3.3b). Differences between diatom and MAGIC estimates of pre-industrial pH of Adirondack lakes, based on the version of MAGIC that included an organic acid representation, were well within the range of expected differences owing to annual and seasonal variability and uncertainties in the model algorithms. However, "successful" comparison of MAGIC with diatom hindcasts in one region does not constitute a sufficient verification to impart complete confidence in using MAGIC, or any process model, for predicting the response of surface water chemistry to changes in acidic inputs. Additional model confirmation in the form of comparison of model output with *measured* data, is required. This has been the focus of modeling efforts at the experimental manipulation sites at Lake Skjervatjern, Bear Brook, and Risdalsheia.

The results of these model testing efforts have been described by Sullivan et al. (1994, 1996, 1998), Cosby et al. (1995, 1996), and Sullivan and Cosby (1995) and are summarized in the following section. The experimental studies are described in Chapter 8.

9.1.4.1 Lake Skjervatjern (HUMEX)

Chemical responses of Lake Skjervatjern to the whole-catchment manipulation were simulated by Cosby et al. (1995) using the extended MAGIC model, including a representation of natural organic acidity analogous to that developed for the Adirondack lakes by Driscoll et al. (1994). The organic acid analog representation was formulated as a triprotic acid that was calibrated to empirical data collected in Norway in the 1000 Lake Survey (SFT 1987). The organic acid analog calibration procedure involved adjusting the H^+ dissociation constants and site density of the TOC that specifies the number of dissociation sites per mole of organic C. The object of the fitting routine was to minimize the observed differences across all lakes between the organic charge simulated by the analog model and the estimated organic anion concentration determined from the charge balance. A nonlinear least squares technique was employed in the calibration (Ralston and Jenrich, 1978) that was conducted using the ALCHEMI model (Schecher and Driscoll, 1994).

The MAGIC model was calibrated to the reference side of Lake Skjervatjern (Side B). After a successful calibration was achieved for the reference catchment, the inputs and parameters for the reference catchment were applied unaltered to the treatment, with three exceptions:

- Simulated applications of H_2SO_4 and NH_4NO_3 were added to the treatment catchment.

- The relative area and turnover time of the lake on the treatment side of the curtain were changed (the curtain does not divide the lake in half by volume, nor are the terrestrial drainages on either side of the lake equal in area).

- The additional water added as part of the spraying treatment was added into MAGIC.

MAGIC model projections of the response of Lake Skjervatjern to whole-catchment acid additions were close to measured values for SO_4^{2-}, NO_3^-, and NH_4^+ (Figure 9.4). Although the retention of added S within the terrestrial system was considerable, the MAGIC-simulated SO_4^{2-} concentrations at Skjervatjern in 1991 and 1992 were within 3 to 6 µeq/L of average measured concentrations (Cosby et al., 1995). The simulated values for NO_3^- and NH_4^+ in Lake Skjervatjern were also very close to measured values for both years, within about 1 µeq/L. It is important to note, however, that the observed success in modeling the N fluxes into Lake Skjervatjern lake water was not a result of N processing algorithms, which are not part of the version of MAGIC applied in this study. Rather, the simulation was run assuming that the percent retention of NH_4^+ and NO_3^- within the treated watershed would equal the calculated percent retention for each ion in the control catchment, based on estimated atmospheric inputs and lake-water concentrations. The actual percent net retention for both NH_4^+ and NO_3^- during the experimental treatment was very similar to the premanipulation percent retention of atmospheric inputs.

MAGIC simulations of base cation response were close to measured values for Na^+ and K^+ and within about 5 µeq/L for both Ca^{2+} and Mg^{2+}. The simulated sum of base cations (C_B) was within 12 µeq/L of the measured value. MAGIC predicted lower C_B than was actually observed. Lake-water ANC declined by an amount slightly greater (approximately 5 µeq/L) than was

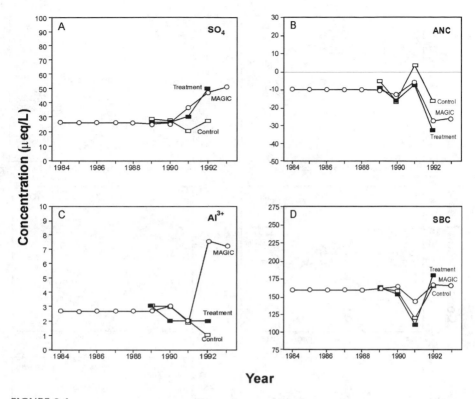

FIGURE 9.4

Volume-weighted average annual concentrations in lake water of key variables measured in Side A (treatment) and Side B (control) of Lake Skjervatjern for the period 1989 through 1992, and results of MAGIC model simulated concentrations through 1993. MAGIC simulations were based on the pretreatment chemistry of the control side, and additional inputs equal to the experimental treatments of H_2SO_4 and NH_4NO_3 applied to Side A starting at the beginning of the 1991 water year. (A) SO_4; (B) ANC defined as $C_B - C_A$; (C) Al^{3+}; (D) sum of base cations. (Reprinted from Journal of Hydrology, Vol. 170, Cosby, B.J., R.F. Wright, and E. Gjessing, An acidification model (MAGIC) with organic acids evaluated using whole-catchment manipulations in Norway, p. 117, Copyright 1995, with permission from Elsevier Science; and Sullivan et al., 1994.) *Continued*

predicted by MAGIC, whereas MAGIC predicted a more substantial pH decline than was actually observed (Cosby et al., 1995; Figure 9.4).

9.1.4.2 Risdalsheia (RAIN)

Risdalsheia provides a good parallel to Lake Skjervatjern, except high ambient S and N deposition have been experimentally decreased, whereas at Lake Skjervatjern low ambient deposition has been experimentally increased. Organic acids play a major role in the acid–base chemistry of runoff at the site, with average annual TOC values generally in the range of 800 to 1200 µM C, and in moderating pH change following reduction in acid deposition (Wright, 1989).

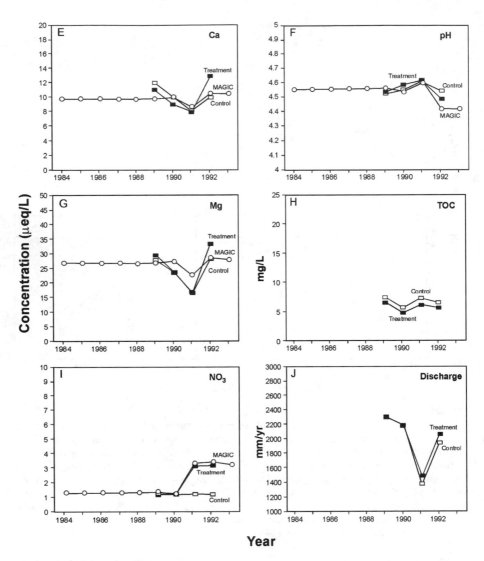

FIGURE 9.4 (Continued)
(E) Ca^{2+}; (F) Mg^{2+}; (G) pH; (H) TOC; (I) NO_3^-; (J) Discharge. (Reprinted from Journal of Hydrology, Vol. 170, Cosby, B.J., R.F. Wright, and E. Gjessing, An acidification model (MAGIC) with organic acids evaluated using whole-catchment manipulations in Norway, p. 117, Copyright 1995, with permission from Elsevier Science; and Sullivan et al., 1994.)

MAGIC was calibrated (Cosby et al., 1995) using measured inputs and out-puts for the reference catchment and was based on mean annual fluxes for the 7-year period 1986 to 1992. The organic acid analog representation we devel-oped for the Adirondack region (Driscoll et al., 1994) was recalibrated, for mod-eling at the Norwegian sites, to empirical data collected in Norway in the 1000

Lake Survey (SFT, 1987). The organic acid analog calibration procedure involved adjusting the H^+ dissociation constants and site density of the DOC that specifies the number of dissociation sites per mole of organic carbon. The object of the fitting routine was to minimize the observed differences across all lakes between the organic charge simulated by the analog model and the estimated organic anion concentration determined from the charge balance. The fitted pK_a values and site density were very close to those obtained for lakes in the Adirondack Mountains. After a successful calibration was achieved for the reference catchment, the inputs and parameters for the reference catchment were applied unaltered to the treatment catchment with three exceptions

- Deposition levels to the treated catchment were reduced to match observed.
- The N uptake dynamics of the reference catchment were modified to match those observed.
- The amount of organic acid in the treatment was adjusted using observed data on anion charge deficit.

The first change is obvious and adjusts for the experimental change in atmospheric inputs. The second change was made because there was no process-basis for N retention in MAGIC that would allow the uptake rate to vary as a function of external or internal conditions. Therefore, the changes in N retention that were observed in the treatment catchment were manually inserted into the model. The third change was made for a similar reason concerning dissolved organic material. The two catchments showed significantly different levels of TOC. Some of this difference might have resulted from the treatment; some may simply have been the result of heterogeneity in these small catchments. In either case, there is no process basis for changing TOC concentration in MAGIC (other than changes in speciation of the fixed amount of organic acid specified for the simulations). Therefore, the changes in organic C content (and anion deficit) that were observed between the catchments had to be manually inserted into the model.

No long-term historical trends in deposition were assumed for any ions except SO_4^{2-}, NO_3^-, and NH_4^+. The historical trend used for SO_4^{2-} deposition was based on the data on S emissions summarized by Bettleheim and Littler (1979) for northern Europe. The historical trends in NO_3^- and NH_4^+ deposition were assumed to parallel that of SO_4^{2-}. For the period of observation (1985 to 1992), yearly observed deposition was used in the model, preserving the year-to-year variability in this portion of the simulation. In running simulations into the future, deposition was assumed to be constant at the eight-year average (ambient deposition for the reference catchment, ROLF; reduced deposition for the experimental catchment, KIM).

The MAGIC triprotic model simulations of the responses of the treatment catchment (KIM) to reduced acidic deposition matched measured values extremely well (Cosby et al., 1995; Figure 9.5). In particular, the observed changes in SO_4^{2-}, base cation concentrations, and ANC closely paralleled the observed trends and interannual variations. Also, despite the importance of organic acids in modifying the pH of drainage waters at this site (Wright, 1989), incorporation into MAGIC of the triprotic organic acid analog, calibrated to the Norwegian 1000 Lake Survey, resulted in good agreement between modeled and measured pH (Figure 9.5).

The results of the MAGIC model testing at Risdalsheia (Cosby et al., 1995) are particularly noteworthy in two respects. First, model simulations of pH matched measured values despite using a regional model representation for organic acids to simulate pH in a system in which organic acids provide a great deal of pH buffering. Second, close agreement was demonstrated between simulated and observed chemistry, for all major variables except perhaps Al, over an 8-year period. This is a very long period of record for a model evaluation study that involved substantial change in acidic deposition inputs.

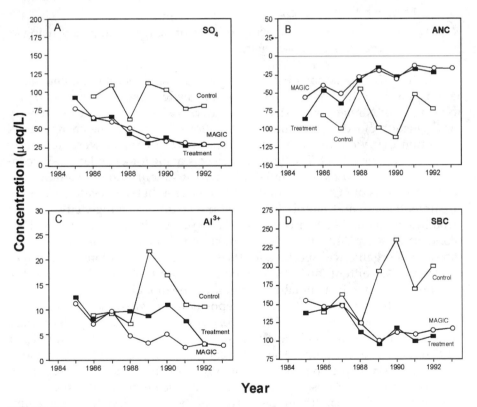

Year

FIGURE 9.5
Volume-weighted average annual concentrations in stream water of key variables in the treatment and reference catchments at Risdalsheia, and results of MAGIC model simulated concentrations. Results are provided for (A) SO_4^{2-}; (B) ANC defined as (C_B - C_A); (C) Al^{3+}; (D) C_B. (Reprinted from Journal of Hydrology, Vol. 170, Cosby, B.J., R.F. Wright, and E. Gjessing, An acidification model (MAGIC) with organic acids evaluated using whole-catchment manipulations in Norway, p. 114, Copyright 1995, with permission from Elsevier Science; and Sullivan et al., 1994.) *Continued*

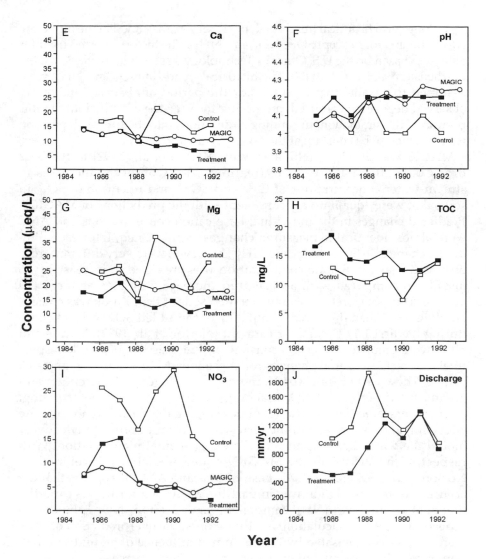

Year

FIGURE 9.5 (Continued)
(E) Ca²⁺; (F) Mg²⁺; (G) pH; (H) DOC; (I) NO₃⁻; (J) Discharge. [Reprinted from Journal of Hydrology, Vol. 170, Cosby, B.J., R.F. Wright, and E. Gjessing, An acidification model (MAGIC) with organic acids evaluated using whole-catchment manipulations in Norway, p. 114, Copyright 1995, with permission from Elsevier Science; and Sullivan et al., 1994.]

9.1.4.3 Bear Brook (WMP)

MAGIC model simulations have been conducted by Norton et al. (1992), Sullivan et al. (1994), and Cosby et al. (1996) to project the response of West Bear Brook to the watershed-scale addition of $(NH_4)_2SO_4$. This watershed manipulation provided an excellent opportunity to test the model performance under changing levels of deposition in the U.S.

No long-term historical trends in deposition were assumed in the modeling efforts for any ions except SO_4^{2-}, NO_3^- and NH_4^+. The historical trend used for SO_4^{2-} was based on the U.S. Office of Technology Assessment scenario for the northeastern U.S. (OTA, 1984). The historical trends in NO_3^- and NH_4^+ were assumed to parallel that of SO_4^{2-}. For the period of observation, yearly observed deposition was used, preserving the year-to-year variability in this portion of the simulation. In running simulations into the future, deposition was assumed to be constant at the 4-year average.

MAGIC was applied preliminarily by Norton et al. (1992) to the Bear Brook data after 1 1/2 years of treatment. They observed increases in stream-water concentrations of Ca^{2+} and SO_4^{2-}, and retention of SO_4^{2-} in soils, that were "qualitatively consistent with the predictions of MAGIC." Predicted changes in the individual base cation concentrations and ANC were almost identical to measured changes, within 2 µeq/L for each. Predicted changes in SO_4^{2-}, NO_3^-, and NH_4^+, however, were very different from observed changes. Sulfate concentration was predicted to increase by 36 µeq/L, as compared with a measured increase only 53% as large (19 µeq/L). Thus, SO_4^{2-} retention in watershed soils at Bear Brook was considerably greater than predicted; about 80% of the added SO_4^{2-} was taken up during the first 1 1/2 years of treatment (Norton et al., 1992). Nitrate and NH_4^+ were predicted to remain constant and increase by 16 µeq/L, respectively. In fact, measured changes for these ions were +10 µeq/L and -0.5 µeq/L. These early results for N, indicating stronger NH_4^+ retention and greater nitrification and NO_3^- leaching than expected, were surprising. They underscored the importance of developing the capability to simulate N cycling and export within the terrestrial system in order to effectively model the watershed response to the experimental manipulation. With respect to the success of the MAGIC model predictions developed by Norton et al. (1992) for both the base cations and ANC, it appears that the accuracy of the simulation was in part the result of a fortuitous cancellation of biases in the C_B (NH_4^+ component) and C_A (SO_4^{2-}, NO_3^- component) aspects of the ANC calculation. In other words, the overprediction of SO_4^{2-} increase was compensated by the combined influence of an underprediction of NO_3^- increase and an overprediction of NH_4^+ increase.

The annual dose of SO_4^{2-} (1800 eq ha per year) should increase the concentration of SO_4^{2-} in West Bear stream water at steady-state conditions by about 200 µeq/L above the SO_4^{2-} concentrations in East Bear (Norton et al., 1993). The measured difference between the two streams was only about 50 µeq/L through the first 2 years of manipulation, with the greatest difference during high discharge (Norton et al., 1993; Kahl et al., 1993a), increasing during hydrological episodes to 60 to 80 µeq/L. Thus, shallow flow paths and shorter contact times with watershed soils result in less S retention within the watershed, and higher concentrations of SO_4^{2-} in stream water.

Chemical responses of West Bear Brook to the whole-catchment acidification were simulated using MAGIC, coupled with Driscoll et al.'s (1994) triprotic organic acid analog model, by Cosby et al. (1996). The MAGIC model was calibrated using measured and assumed inputs and outputs from the control catchment (East Bear) and was based on mean annual fluxes for the 4-year period 1989 to 1992. Modeling efforts reported by Sullivan et al. (1994) and Cosby et al. (1996) at Bear Brook provided a continuation of the MAGIC model forecasting efforts presented by Norton et al. (1992). Comparisons of MAGIC forecasts from 1989 through 1996 and annual average measured values of key parameters in East (reference) and West (treatment) Bear Brooks from 1989 through 1992 are provided in Figure 9.6. As reported by Norton et al. (1992), MAGIC predicted a much larger increase in streamwater SO_4^{2-} concentration than was observed in the treated stream. The trajectory of MAGIC-simulated SO_4^{2-} was approaching the anticipated steady-state concentration of about 300 µeq/L projected to occur within the 8-year simulation period; i.e., about 200 µeq/L higher than pretreatment concentrations. In contrast, the trajectory of measured values was substantially lower.

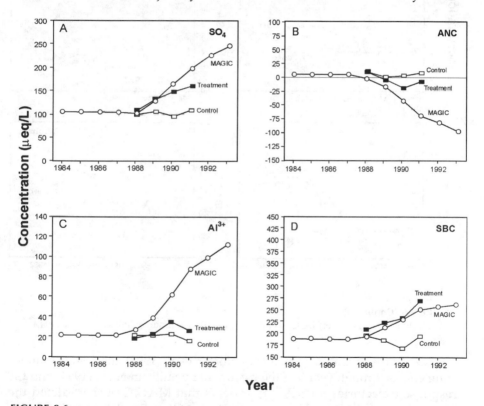

FIGURE 9.6
Volume-weighted average annual concentrations in stream water of key variables measured in East (control) and West (treatment) Bear Brooks from 1989 through 1992, and results of MAGIC model simulated concentrations through 1994. MAGIC simulations were based on the pretreatment chemistry of the control stream, and additional inputs equal to the experimental treatments of $(NH_4)_2SO_4$ applied to West Bear Brook starting at the beginning of the 1990 water year. (A) SO_4^{2-}; (B) ANC defined as $(C_B - C_A)$; (C) Al^{3+}; (D) C_B. (Source: Sullivan et al., 1994.) *Continued*

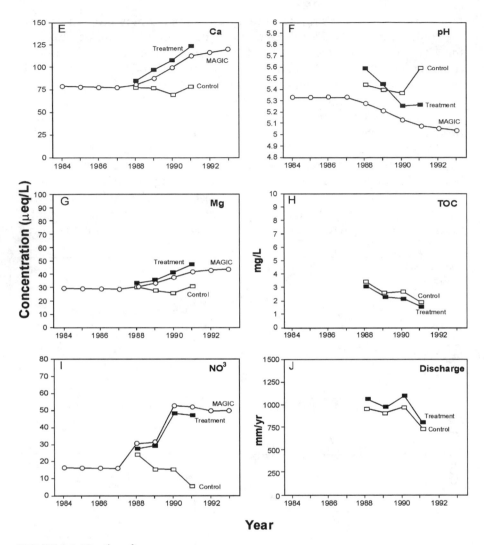

FIGURE 9.6 (Continued)
(E) Ca^{2+}; (F) Mg^{2+}; G) pH; (H) DOC; (I) NO_3^-; (J) Discharge. (Source: Sullivan et al., 1994.)

Based on the preliminary results of Norton et al. (1992), the results of modeling efforts through year 3 of the treatment by Sullivan et al., (1994) and the response trajectories for SO_4^{2-}, it appears that MAGIC overpredicted the increase in stream-water SO_4^{2-} concentrations at Bear Brook by nearly a factor of 2. This overprediction of the increase in stream-water SO_4^{2-} concentration that resulted from the manipulation experiment was owing to the high value assumed for the half saturation of S adsorption. The adsorptive behavior of the soils was effectively removed because concentrations of SO_4^{2-} never approached the one-half saturation value.

As a consequence of the overprediction of stream-water SO_4^{2-} concentration, other key variables (especially ANC and Al) were also predicted to respond to a greater degree to the experimental acidification than was actually observed. As anticipated at the onset of the project and by MAGIC, the treated stream experienced decreases in ANC and increases in Al compared to the reference stream. However, MAGIC predicted substantially larger changes in each of these parameters than was observed (Sullivan et al., 1994). The discrepancies between MAGIC predictions and measured values for ANC increased with each year of the treatment, and were attributable mainly to the errors in predicting the responses of SO_4^{2-} and, secondarily, to errors in predicting the responses of Ca^{2+} and Mg^{2+}.

MAGIC overpredicted the sum of mineral acid anions (C_A) largely because of the errors in predicted SO_4^{2-} concentrations. In addition, MAGIC underpredicted the sum of base cations (C_B), mainly Ca^{2+} and Mg^{2+}. This latter bias, however, was owing primarily to pretreatment differences between the reference and manipulated stream chemistry. This was a reflection of the extreme difficulty of finding paired catchments with identical chemistry. The combined influence of the observed biases in C_B and C_A resulted in relatively large errors in predicted ANC, because ANC is defined in MAGIC for this application as ($C_B - C_A$). The simulated decrease in ($C_B - C_A$) was too large, and in order to satisfy charge balance constraints (Σ cations = Σ anions), MAGIC calculated concentrations of acidic cations (H^+, Al^{3+}) sufficient to balance this simulated charge discrepancy. MAGIC partitioned those acidic cations between H^+ and Al^{3+}, based on the assumed solubility product (log K_{sp}) for gibbsite dissolution. Because the log K_{sp} value was assumed to be very high (10.7), MAGIC loaded most of the change in acidic cations into the Al^{3+} component, and relatively little into the H^+ component (Sullivan et al., 1994).

MAGIC projected extremely high values of Al^{3+} in response to the manipulation. Although measured Al concentrations did increase substantially in the manipulated stream, MAGIC projected increasingly larger increases each year of the experiment. By year 3 of the manipulation, the MAGIC estimate of increase in Al^{3+} was about 6 times larger than was observed, and the MAGIC estimate of Al^{3+} concentration was extremely high (approximately 90 μeq/L). The standard method for MAGIC applications was to determine the Al solubility product empirically, based on pretreatment data. For East Bear Brook, the estimated value for this model application was log $K_{sp} = 10.7$ which is much higher than is normally assumed for acidified waters. Partly as a consequence of using such a high log K_{sp} value from the control stream to predict experimental Al^{3+} concentrations in West Bear Brook, we predicted unrealistically high values of Al^{3+}.

The fundamental problem in the model application for Al is that a single value for gibbsite equilibrium cannot be used to explain variations in stream-water chemistry. This problem has been illustrated previously in several field studies. The MAGIC model forecasts presented by Sullivan et al. (1994) for Bear Brook reconfirmed the problematic nature of the commonly employed assumption of gibbsite equilibrium and a fixed K_{sp}, and illustrated the magnitude of error that can result.

There was an especially large discrepancy between simulated and observed base cation concentrations and ANC in the third year of treatment at Bear Brook (1992). The measured increases in C_B in East and West Bear Brooks were about 30 and 35 µeq/L, respectively, between 1991 and 1992, while ANC increased about 10 µeq/L. These increases in measured C_B and ANC were likely attributable in part to lower than normal precipitation during that year. Stream discharge during 1992 was only about 75% of the 3 previous years. Gross hydrological changes, such as altered contact of drainage waters with soils in response to unusually dry or wet conditions, are difficult to capture in any dynamic model. Thus, the fairly pronounced changes between 1991 and 1992 in measured values for many key variables were not represented by the MAGIC simulation.

The original calibration of MAGIC by Sullivan et al. (1994) for the Bear Brook forecast was based on 4 years of data from the reference stream, East Bear. In order to assess the degree to which discrepancies between predicted and observed stream-water chemistry at Bear Brook could be improved by correcting S dynamics and *a priori* differences between treatment and control catchments, a revised calibration was conducted. The revised final Bear Brook calibration and associated projections were reported by Cosby et al. (1996) and are described in the following section.

Recalibration of the MAGIC model for Bear Brook was accomplished using the same procedures as described for the original calibration, with two exceptions

1. The S adsorption half-saturation was set at 100 µeq/L before implementing any of the calibration steps. This value was substantially lower than that used in the original calibration (821 µeq/L), and resulted in a large increase in the effective adsorptive capacity of the soils. Selection of this lower S half-saturation value essentially substituted expert judgement in place of strictly laboratory-derived data. The 4-year volume weighted mean chemistry was then calibrated at East Bear Brook (reference stream), as had been done in the initial calibration.

2. After calibrating MAGIC to the 4-year average chemistry of the reference stream, this calibration was used as the starting point for a new calibration to the 1-year pretreatment data from West Bear Brook prior to experimental manipulation. This in essence was a fine tuning of the model calibration designed to correct for inherent biases in major ion concentrations between the two catchments.

Thus, the revised calibration and associated model forecasts (Cosby et al., 1996) corrected for the obvious large bias in effective S adsorption in watershed soils and also corrected for *a priori* differences between the treatment and reference catchments.

The resulting simulations matched measured values in West Bear Brook to a substantially greater degree (Figure 9.7) than had the forecasts based on the initial calibration that was based on the behavior of East Bear (Figure 9.6). Differences from the initial calibration were primarily related to SO_4^{2-} half-saturation, selectivity coefficients, and weathering of base cations.

Projected stream-water SO_4^{2-} concentration agreed closely with measured values in West Bear Brook for the first 3 years of manipulation (Figure 9.7). There was a suggestion, however, in the measured data of a leveling off in the annual increase in stream-water SO_4^{2-} concentration, whereas the model projected a steady increase in concentration throughout the duration of the model projection (1989 through 1994). The model simulation also showed much better agreement with measured values for the sum of base cations and ANC (Figure 9.7) than had the initial MAGIC simulation. Although the effects of the drought year (1992) on C_B and ANC were still not captured by the simulation, the overall agreement between predicted and observed C_B and ANC was much improved. Slight underestimation of pH decrease and overestimation of Al^{3+} increase were still evident in the revised projections (Figure 9.7), although the magnitudes of these biases were reduced dramatically because of the improvement in predicted SO_4^{2-} concentration and ANC.

9.1.5 Evaluation of MAGIC Projections

As described in the previous section, the accuracy of MAGIC model projections was evaluated using available paleolimnological data from the Adirondack Mountains and results of whole-catchment acidification or deacidification experiments at Bear Brook, ME, and Lake Skjervatjern and Risdalsheia, Norway. Results of these model evaluation exercises were mixed. MAGIC provided projections of lake-water chemistry in the Adirondacks and at Lake Skjervatjern and Risdalsheia that were generally consistent with independent diatom-inferred and measured data, respectively. In all cases, the agreement for pH was dependent upon inclusion of an organic acid representation in the MAGIC model, and in all cases MAGIC projected slightly greater change in pH than was inferred or measured by other means. In general, however, MAGIC performed very well in simulating the major changes that occurred at these sites. In contrast, at Bear Brook, organic acids were less important to annual average estimates of chronic chemistry, but initial MAGIC projections of stream-water acidification were substantially greater than was measured. The most significant bias was an overprediction of stream-water SO_4^{2-} concentration, attributable to underestimated S adsorption in watershed soils. Other biases were apparently related to Al solubility and hydrology.

Recalibration of MAGIC to account for the observed overprediction of SO_4^{2-} concentration and *a priori* differences between the treatment and reference catchments (Cosby et al., 1996) yielded generally good agreement.

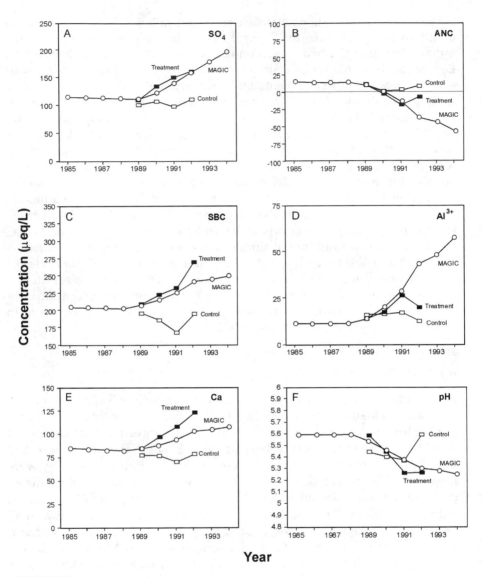

FIGURE 9.7

Results of revised MAGIC simulation for Bear Brook, after adjusting for biases in sulfur retention and *a priori* differences between treatment and control catchments. (See text for details.) (Reprinted from Science of the Total Environment, Vol. 183, Cosby, B.J., S.A. Norton, and J.S. Kahl, Using a paired-catchment manipulation experiment to evaluate a catchment-scale biogeochemical model, p. 63, Copyright 1996, with permission from Elsevier Science; and Sullivan et al., 1994.)

The modeling work conducted at Bear Brook has been especially valuable in terms of assisting in the verification of previously suspected inadequacies in model structure (see, e.g., Turner et al., 1990; Thornton et al., 1990) and application, and clarification of remaining research needs. It is unfortunate that whole-catchment manipulation experiments such as the Watershed Manipulation Project at Bear Brook were not conducted earlier in the NAPAP program. As a direct result of using Bear Brook data for the primary purpose for which the experiment was designed, that is, model testing, we were able to identify several aspects of the model in need of improvement. The Bear Brook modeling study clearly demonstrated the importance of the following problem areas

1. Lack of incorporation of N processes in the model.
2. Errors in the way Al dissolution is modeled in MAGIC (and other models).
3. Uncertainties in simulating hydrological routing and the fundamental importance of hydrology in regulating acid–base chemistry.
4. Uncertainties in S retention in watershed soils when S inputs are changed, regardless of apparent steady-state input/output conditions prior to the perturbation.
5. Differences that exist *a priori* between paired catchments selected for treatment and control basins that can introduce bias into the comparison between syn-manipulation measured variables on the treatment side vs. model projections for the reference side.

The results of the Bear Brook model evaluation exercise provided quantitative evidence of the importance of these deficiencies in a field experiment. Such model tests are seldom attempted, in part because funding agencies often fail to recognize the value of long-term whole-ecosystem manipulations and perhaps in part because few modelers have the opportunity to subject their model to such rigorous testing.

Results of modeling efforts at Bear Brook, as well as measured chemical changes at Bear Brook, and to a lesser extent Lake Skjervatjern and Risdalsheia, illustrated that a remaining major weakness of MAGIC (and other process models) relative to the needs of NAPAP was the failure to include algorithms to simulate N cycling and N retention in watershed soils and vegetation. The success of the N component of the modeling effort at Bear Brook (and Risdalsheia to some extent) was totally dependent on adjusting the N inputs to the model to match measured outputs in stream water. Nitrogen dynamics were extremely important at Bear Brook (Kahl et al., 1993a), although this had not been anticipated at the inception of the Watershed Manipulation Project. The process of evaluating and improving MAGIC is an iterative one. It now has been shown that MAGIC often yields acceptable model simulations of past and future change. It also has been shown that further improvements are needed, particularly with respect to N that is the focus of the more recent extended version of the model (c.f., Ferrier et al., 1995). The model simulations at Bear Brook also revealed uncertainties in several aspects of the model structure and/or the manner in which the model is applied to a given catchment. Results at Bear Brook verified that key remaining uncertainties related to the modeling of Al dissolution, S retention in soils, and the dependence of runoff chemistry on hydrological variations are difficult to simulate.

9.2   Nitrogen Models

A rigorous and well-confirmed process-based model of nutrient cycling in forested ecosystems is needed for answering important policy-relevant questions related to the emissions of atmospheric N, for example,

- What is the critical load of N to various terrestrial and aquatic systems?
- What is the timing of ecosystem response to increased inputs of N?
- What are the interactions between atmospheric N deposition and other anthropogenic activities, for example, forest management?
- What are the interactions between anthropogenic effects and climatic effects in regulating N cycling/saturation, especially drought, frost, and potential global warming?

The N research that has been conducted in recent years has focused on aspects of the N retention capacity of ecosystems and the ecological effects of excess N availability (e.g., Skeffington and Wilson, 1988; van Breemen and Van Dijk, 1988; Schulze, 1989; Aber et al., 1989, 1991; Stoddard, 1994; Dise and Wright, 1992). The concept of N saturation has emerged as a central focus of much of the recent research. By definition, N saturation implies the initiation of limitation on growth and biological activity by some other resource (rather than N), for example, water, light, or another nutrient (e.g., Mg^{2+}, K^+). In order to predict the timing and consequences of N saturation, therefore, forest ecosystem models are needed that simulate the availability and ecosystem processing of several potentially growth-limiting resources.

A number of models are currently available with which to simulate the response of watersheds to changing levels of N deposition. Such models were not available for the 1990 IA, but are now in various stages of model testing and application. Perhaps the most generalized is the extended version of MAGIC discussed previously. Several more detailed N models are also available, including MERLIN, NuCM, PNET_CN, and NIICCE. With the exception of a bounding study by the EPA (van Sickle and Church, 1995), regional modeling of the probable response of aquatic and terrestrial ecosystems to future changes in N deposition has not yet been conducted in the U.S., although such work is ongoing in Europe and is planned for some areas of the U.S.

Cosby et al. (1997) developed a catchment-scale mass-balance model of C and N cycling in forested ecosystems. The Model of Ecosystem Retention and Loss of Inorganic Nitrogen (MERLIN) was developed to simulate leaching losses of inorganic N and consider linked biotic and abiotic processes that affect the cycling and storage of N. Like MAGIC, the model is spatially and temporally aggregated so as to be interpretable at the catchment scale. There are four ecosystem compartments: inorganic soil, plants, labile soil organic matter, and refractory (nonlabile) soil organic matter. The time step is monthly to yearly and the model is intended to be implemented for decadal simulations or longer (i.e., greater than 50 years). Nitrogen fluxes through the ecosystem are controlled mainly by atmospheric deposition of NO_3^- and NH_4^+, hydrological discharge, plant uptake, production of litter and wood, microbial N immobilization, mineralization, nitrification, and denitrification. Nitrogen flux rates are controlled by net primary productivity, the C : N ratios of organic compartments, and the concentration of inorganic N in soil solution.

The input requirements of MERLIN are fairly comprehensive, which makes application to multiple sites or regional modeling difficult. Required inputs include the temporal sequence of C fluxes and pools, time series of hydrological discharge through the soils, historical and current deposition and fertilization, N pools in plant and soil organic compartments, soil characteristics such as those required for MAGIC, and constants specifying the N uptake and immobilization characteristics of the plant and soil organic compartments.

Cosby et al. (1997) presented a series of speculative simulation exercises to demonstrate the functioning of MERLIN. The parameter values that were chosen for the simulation were approximately those of forested systems in Europe. Simulations were presented of land use change (i.e., time-varying C dynamics) as well as steady-state C dynamics, and several levels of N deposition. Results of the simulations were found to be in general agreement with observed and hypothesized system behavior in response to changing levels of N deposition. MERLIN was further tested and evaluated by Emmett et al. (1997), who applied the model to simulate N saturation at a Sitka spruce forest at Aber, Wales.

The Nutrient Cycling Model (NuCM) was developed as part of the Electric Power Research Institute's Integrated Forest Study (Liu et al., 1991; Johnson and Lindberg, 1991). The model tracks the pools of available nutrients (including N and S) within the various soil and vegetation layers and the fluxes between ecosystem compartments. Vegetation is represented as both an overstory and an understory, each of which can be divided into canopy, bole, and roots. The model was described by Liu et al. (1991), Johnson and Lindberg (1991), and Johnson and Mitchell (1998).

Tree growth is represented in the model as a function of developmental stage of the forest. Precipitation is routed through the canopy and soil layers and the model simulates evapotranspiration, deep seepage, and lateral flow. The soil is represented by up to 10 layers, each of which can be assigned different physical and chemical characteristics. Interactions among the nutrient pools in the soil solution, soil ion-exchange complex, minerals, and soil organic matter are regulated through such processes as mineral weathering, ion exchange, decay, and nitrification. Litter decay is dependent on reactant concentrations and C : N ratios.

Model inputs are based on measurable parameters, but the input require-ments are fairly comprehensive. The model is, therefore, more applicable to intensive study sites rather than regional application.

The effects of S and N deposition in combination with tree harvest and spe-cies change at Duke Forest, NC, were simulated by Johnson et al. (1995) using NuCM. The model was used to evaluate a suite of hypotheses regarding leaching rates and changes in soil exchangeable cation pools. Johnson and Mitchell (1998) presented simulation results for a red spruce forest in the Great Smoky Mountains, NC, and a mixed hardwood stand at Coweeta Hydrologic Laboratory in the southern Appalachian Mountains of North Carolina. The effects of S adsorption characteristics on response times to changes in hypothesized S deposition were well-illustrated. At the Smokies site, simulated SO_4^{2-} leaching responded almost immediately to changes in S deposition. At the Coweeta site, in contrast, S adsorption gradually increased throughout the 30-year simulations for all scenarios except the 90% decreased deposition scenario.

The PnET family of models are generalized models that simulate the inter-actions among C, water, and nutrients in forested ecosystems. PnET-CN is an extended version of the earlier PnET models of carbon and water balances of temperate and boreal forest ecosystems (Aber and Federer, 1992; Aber et al., 1995b, 1996). PnET-CN includes several additional compartments (live biom-ass, litter, soil organic matter), N mineralization and nitrification, plant uptake of N, and leaching losses. The PnET-CN model simulates complete cycles for both C and N (Aber et al., 1997). The organic matter pool of the model is functionally equivalent to the active pool in the Century model of Parton et al. (1993). The C and N cycles interact within the model, such that increased foliar N causes increased photosynthesis, if water is not limiting, and, therefore, increased plant pools of C. Similarly, as plant C increases, the model simulates greater net primary productivity (NPP) that increases the demand for N. Nitrate leaching occurs in proportion to both the amount of NO_3^- in soil solution after satisfying plant uptake demands and also water drainage rates.

The model operates on a monthly time step and at the stand to watershed scale. The model is first applied and evaluated in comparison with data on NPP, monthly C and water balances, annual net N mineralization, nitrifica-tion, foliar N concentration, and annual and monthly N leaching losses. In this respect, it differs from models such as MAGIC in that PnET-CN is not cal-ibrated to the subject watershed. Subsequent to the evaluation (validation) step, it is then used to predict forest responses to changes in N deposition and/or land use.

Aber and Driscoll (1997) applied PnET-CN to six sites in the White Mountains of New Hampshire that had quite different land use histories and annual N leaching losses. The only model parameter that varied between sites was land use or disturbance history. The model successfully captured much of the variation in N losses. The variation was found to have two major components: differences in long-term N trends owing to land use history and interannual variability within watersheds owing to interactions between climate and biological response. Rates of change in N leaching losses were predicted by the model to be low relative to interannual variability, given the current N deposition that is only moderate (approximately 9 kg N/ha per year).

A process-oriented simulation model has been developed within the NITREX program for heat, water, N, and C cycling. The Nitrogen Isotopes and Carbon Cycling in Coniferous Ecosystems (NIICCE) model (van Dam and van Breemen, 1995) was intended as an integrative tool for interpretation of the results of the various NITREX manipulation experiments. The aim of the ^{15}N studies in NITREX was to generate narrower ranges for the parameters that describe key N transformation processes in the model. It is hoped that, with the range of experimental conditions in the NITREX project, the NIICCE model will ultimately predict ecosystem response over a large range of environmental conditions. NIICCE is planned to be linked with existing models of nutrient cycling and acidification such as RESAM and SMART (de Vries, 1990; de Vries and Kros, 1991) and MAGIC (Cosby et al., 1985a,b) and will be used to predict ecosystem response to alternative deposition scenarios (Dise and Wright, 1992).

NIICCE simulates primary production, mineralization, decomposition, root-uptake, solute transport, and isotope fractionation for a coniferous forest in a one-dimensional multi-compartment soil profile. Model inputs include fluxes of N inputs, global short-wave radiation, precipitation, and temperature.

The model was used in a provisional manner to estimate the amounts of tracer needed in experimentally enriched throughfall in order to reach levels of ^{15}N in various ecosystem pools high enough to discriminate between natural isotopic variability and increased levels after experimental addition of N (Dise and Wright, 1992). The model also has been used for preliminary simulation of N cycling at the Speuld NITREX site (van Dam and van Breemen, 1995).

The details of these and other models that simulate N dynamics can be found in the cited references. It is important to recognize that models are now available to predict the regional effects of changes in N deposition. Such models were not available at the time of preparation of the 1990 Integrated Assessment. All could benefit from further testing and require the collection of some additional field data for regional application in sensitive regions of the U.S.

10

Case Study: Adirondack Park, NY

10.1 Background and Available Data

The Adirondack Mountain region of New York is one of the most intensively studied regions of the world with respect to the effects of acidic deposition on aquatic resources. Acidic deposition effects research in the Adirondacks has played an important role in many of the areas of major scientific advancement during the last decade. For that reason, the results of research conducted in the Adirondack region are widely discussed throughout this book. These include improved understanding of N cycling and effects, results of surface water quality monitoring efforts, extensive model testing, paleolimnological inferences of both long- and short-term acidification responses, and interactions between land use and acidification processes.

This chapter is not intended to summarize current understanding of acidification sensitivities of Adirondack surface waters or the effects to date of S-driven acidification. These topics have been thoroughly discussed by Sullivan (1990), Driscoll et al. (1991), Baker et al. (1990b), and others.

Rather, the attempt here is to summarize some of the recent findings that relate to the major topics of this book. The focus of this chapter is largely on research in which the author has been involved personally. However, there has also been extensive research conducted by many other scientists in the Adirondack Mountains in recent years. Many aspects of their research are discussed and referenced in other chapters of this book.

The region is mountainous with numerous lakes, many of which have low concentrations of base cations and are, therefore, susceptible to acidification from addition of mineral acid anions such as SO_4^{2-} and NO_3^- and naturally occurring organic acid anions (Driscoll et al., 1991). Elevations range from about 30 m near Lake Champlain to 1630 m in the High Peaks area of the northeastern Adirondacks. The highlands region extends to the north, west, and south of the High Peaks and is dissected by numerous deep linear valleys formed by glacial erosion. At the center of the region is a large igneous intrusion that has undergone extensive metamorphosis. Bedrock geology includes areas of granite and granitic gneiss, anorthosite, quartz syenite, and metasediments.

Thin, acidic spodosol soils have developed on glacial sediments from the Wisconsin glaciation about 12,000 YBP. Much of the forest land is covered with northern hardwood forests, mainly yellow birch (*Betula alleghaniensis*), American beech (*Fagus grandifolia*), and sugar maple (*Acer saccharum*). Some of these hardwood forests are mixed with red spruce (*Picea rubens*), balsam fir (*Abies balsamea*), and eastern hemlock (*Tsuga canadensis*). At higher elevations, red spruce and balsam fir predominate, with extensive areas of paper birch (*Betula papyrifera*) where fire has occurred. Wetlands are common, especially at lower elevations.

Approximately 14% of the lakes represented by the Eastern Lake Survey's probability sample (Linthurst et al., 1986) were acidic (ANC less than or equal to 0). Sullivan et al. (1990b) concluded that this percentage would approximately double if lakes smaller than 4 ha had been included in the frame population, largely because a high proportion of the small lakes were organic acid systems. The Adirondack Mountains have been the focus, in part or in whole, of numerous major acid deposition research programs, including the Integrated Lake-Watershed Acidification Study (ILWAS; Chen et al., 1983), the Regionalized Integrated Lake-Watershed Acidification Study (RILWAS; Driscoll and Newton, 1985), the Adirondack Lake Survey Corporation (ALSC) survey (Kretser et al., 1989; Baker et al., 1990b), the Paleoecological Investigation of Recent Lakewater Acidification studies, PIRLA-I (Charles and Whitehead, 1986a,b), and PIRLA-II (Charles and Smol, 1990), Oak Ridge National Laboratory Watershed Assessment (Hunsaker et al., 1986a,b), the Adirondack Effects Assessment Program (Momen and Zehr, 1998), and many of the major studies conducted within the Environmental Protection Agency's (EPA) Aquatic Effects Research Program (AERP): the Eastern Lake Survey, Phase I and Phase II (Linthurst et al., 1986; Herlihy et al., 1991), Direct Delayed Response Project (Church et al., 1989), Episodic Response Project (Wigington et al., 1993), and Long Term Monitoring Program (Driscoll and van Dreason, 1993; Newell 1993).

Precipitation chemistry is generally rather uniform across the Adirondack region. Regional patterns in wet deposition of S and N are owing mainly to differences in precipitation amount (Driscoll et al., 1991). Total atmospheric deposition of S has been estimated by Sisterson et al. (1990) and Driscoll et al. (1991) to be in the range of 9 to 12 kg S/ha per year in the Adirondacks. Most available N deposition data are for low elevation sites, where deposition of N is generally less than at the moderate to high elevation sites of acidified Adirondack lakes (Friedland et al., 1991). In addition, estimates of N dry deposition are subject to considerable uncertainty (Baker, 1991). Ollinger et al. (1995) estimated total N deposition in the northeastern U.S. ranging from 3 to 4 kg N/ha per year in northern Maine to values in the range of 10 to 12 kg N/ha per year in mountainous areas of New York and southwestern Pennsylvania. Estimated total N deposition in the Adirondack region generally varied between 8 and 12 kg N/ha per year, with highest values at high elevations (Ollinger et al., 1993).

The accumulation and release of S and N in seasonal snowpacks are important factors that influence the delivery of atmospheric deposition to soils and surface waters in the Adirondacks. The SO_4^{2-} and NO_3^- pools in the snowpack generally reach maximum values in March and then decline throughout snowmelt (Rascher et al., 1987). Preferential elution of ions in the snowpack causes a pulse of SO_4^{2-} and NO_3^- to be released early in the melting process (Schaefer et al., 1990). The dominant anion in the snowpack is generally NO_3^-. This is owing to the seasonal patterns in deposition, which show highest N concentrations in precipitation during winter and highest S concentrations during summer (Driscoll et al., 1991).

Data were examined for this case study from most of the principal relevant studies that have been conducted in recent years in the Adirondack Mountain region. In addition to examining data from the ELS-I statistical survey, appropriate data were analyzed from the ALSC, ELS-II, DDRP, PIRLA-II, ALTM, and ERP. Each database provides particular kinds of information and analytical strengths, as described in the following sections (Sullivan et al., 1999).

10.1.1 ELS-I

In 1984, the EPA conducted an extensive survey of lake-water chemistry in selected areas of the eastern U.S. (Kanciruk et al., 1986). The ELS-I was based on a statistical probability design such that extrapolations with known uncertainty could be performed to estimate the number of lakes in each study region that exhibited various characteristics of acid–base chemistry. Prior to sample selection, a frame population was identified from 1 : 250,000 scale maps of the regions. The average minimum lake area was about 4 ha, corresponding to the approximate resolution of the map scale used to specify the sampling frame.

10.1.2 ALSC

The Adirondack Lakes Survey Corporation (ALSC) conducted a 6-year (1984 to 1990) survey to quantify the chemistry and fisheries of Adirondack waters (Kretser et al., 1989). Approximately 24% of the 1469 lakes surveyed had pH values of 5.0 or lower, while 26% had ANC values less than 0 µeq/L (Baker et al., 1990b). Natural inputs of organic acids are important in regulating the acidity of some lakes (Munson and Gherini, 1993; Driscoll et al., 1994), but many ALSC lakes appeared to have been acidified by acidic deposition. Lakes judged most susceptible to acidic deposition effects were drainage lakes surrounded by thin deposits of glacial till, which represent approximately 35% of all waters surveyed by the ALSC (Baker et al., 1990b).

10.1.3 ELS-II

During Phase II of the Eastern Lake Survey (ELS-II), a subset of ELS-I lakes was resampled during the spring and fall of 1986. This database provides two principal advantages over ELS-I

1. Chemical measurements were made during the spring season, when lakes are lowest in pH and ANC, and the concentrations of NO_3^- and potentially toxic inorganic monomeric Al (Al_i) are generally at their highest.

2. Aqueous Al was fractionated in the laboratory into labile and non-labile components, thereby allowing direct estimation of Al_i concentrations in lake waters. Aluminum fractionation was not performed in ELS-I.

10.1.4 DDRP

The Direct Delayed Response Project (DDRP) provided the foundation for NAPAP's surface water modeling efforts. Included in this study were 37 Adirondack lakes. They were statistically selected from the ELS-I sample, and included lakes low to moderate in ANC (less than or equal to 400 µeq/L) and larger than 4 ha in area (Church et al., 1989). MAGIC model (Cosby et al., 1985a,b) hindcasts and forecasts of lake-water chemistry have been constructed for the DDRP lakes (Church et al., 1989; NAPAP, 1991; Sullivan et al., 1992, 1996a). DDRP databases include model estimates and watershed data.

10.1.5 PIRLA

Diatom inferences of pre-industrial pH and historical acidification have been constructed for numerous lakes in the Adirondack Mountains, including the DDRP lakes. This work was conducted as part of the Paleolimnological Investigation of Recent Lakewater Acidification studies, PIRLA-I (Charles and Whitehead, 1986a,b) and PIRLA-II (Charles and Smol, 1990). Results have been discussed by Charles et al. (1990), Sullivan et al. (1990a), and Cumming et al. (1992). The PIRLA-II data provide a statistically based assessment of the magnitude of historical acidification and the spatial distribution of historically acidified lakes. In addition, the recent trends research component of PIRLA-II provided detailed information on the timing of acidification of 20 Adirondack lakes (Cumming et al., 1994).

10.1.6 ALTM

The Adirondack Long-Term Monitoring Program (ALTM) was initiated in 1982 to assess temporal changes in water chemistry in 17 Adirondack lakes

(Driscoll and Newton, 1985; Driscoll and Van Dreason, 1993). In this ongoing program, samples are collected monthly to determine long-term changes or trends in water chemistry. This database provides a basis for evaluating the seasonal variability of key chemical parameters.

10.1.7 ERP

The Episodic Response Project (ERP) was designed to investigate episodic chemical changes, and consequent biological effects, in stream water in three areas of the northeastern U.S., including the Adirondack Mountains (Wigington et al., 1993, 1996). Studied during rainfall and snowmelt hydrological events were four Adirondack streams. The ERP data provide information on changes in the chemical composition of Adirondack streamwater during periods of acute, short-term toxicity to fish. This database, therefore, provides evidence regarding the role of N as a contributor to episodic acidification and the acute toxicity to fish of short-term increases in Al^{n+} and H^+ ions.

10.2 Watershed History

There are two major aspects of watershed history that are likely to have interacted with acidification from acidic deposition in the Adirondack region: logging and forest blowdown. Each has been the focus of recent work, although the precise role of neither has been quantified. The history of Adirondack forests, and human impact on those forests, is complex. Major elements were described in the forest history compiled by McMartin (1994) and subsequently summarized by Sullivan et al. (1996b, 1999). The information presented next is taken from those publications. Some Adirondack land in the eastern and northeastern sections of the region was cleared for farming, but the amount was not significant. The major land use activity during the last century has been forestry. Early development of the lumber industry was hampered by a lack of suitable transportation. Initial logging activities focused almost exclusively on white pine that was never particularly abundant in the Adirondack forests (Ketchledge, 1965). By 1850, most of the accessible pine was gone and logging efforts had shifted to spruce. Early logging activities did not significantly alter the forest because the pines and spruce occurred mostly as scattered trees and the early logging was highly selective.

Prior to 1890, much of the future Adirondack Park had not been logged at all and the logging that had occurred had been selective. Most of the early logging was done close to waterways because the cut logs were floated downriver to the mill sites. Furthermore, there was not enough spruce in the primarily hardwood forests of the central and southwestern Adirondacks to justify logging there. Loggers would have needed roads or tracks for horses to reach the isolated pockets of spruce and these were expensive to build (McMartin, 1994).

The first railroad crossed the central Adirondacks in 1892 (Donaldson, 1921). Soon, railroad spurs became an important means of shipping lumber from mills in the interior. The pulp and paper industries appeared after 1880, and by 1920 had exhausted much of the spruce and balsam resources of the region. The combination of cut-over forests, drought, and sparks from locomotives on the railroads caused fires that seriously affected timber tracts around the turn of the century. During the fires of 1903 alone, 292,000 acres of timber and 172,000 acres of brush land burned in the Adirondacks (Middleton, 1904).

The years around the turn of the twentieth century marked unusually dramatic changes in the Adirondack forest (McMartin, 1994). The first sequence of events involved political changes that ultimately led to the creation of the Adirondack Park, and included the state's acquisition of land, establishment of the Forest Preserve, and then, finally, creation of the Park in 1892. The second important sequence of events involved changes in the forest industries. The construction of railroads within the region permitted logging over a much greater area than had previously been permitted. The use of wood in the production of pulp and paper meant that smaller logs and logs of all kinds could be used (Donaldson, 1921). Loggers returned to land from which spruce sawlogs had earlier been taken to remove virtually all of the remaining spruce as small as 5 or 6 inches in diameter. The increasing loss of forest caused growth in the preservation movement, while a growing shortage and high demand for timber further accelerated the cutting. The result was that the greatest number of trees were cut in the Adirondacks from 1890 to 1910 (Donaldson , 1921; McMartin, 1994). There was a decrease from nearly 2 million acres of virgin forest in 1885 to slightly more than 1 million acres in 1902, to just a few hundred thousand acres in 1910 (McMartin, 1994).

Many of the watersheds in the portions of the Adirondacks most impacted by recent acidification were logged around the turn of the century (Sullivan et al., 1996b). The year 1901 marked the last large log drive on the Black River in the southwestern Adirondacks, the supply of lumber nearly having been exhausted (McMartin, 1994). The steep slopes of the High Peaks region had been inaccessible to loggers in the early days of Adirondack logging. By the early 1900s, however, lumbermen were harvesting increasingly inaccessible stands, including those in the High Peaks region. High elevation sites were logged for spruce and balsam fir to support the pulp industry. The High Peaks region remained a major source of spruce and balsam fir logs through the 1920s (McMartin, 1994).

Logging has played a smaller role as an agent of change in the Adirondack watersheds that have experienced recent acidification since the early decades of the twentieth century. Within the park, forest succession is gradually restoring the natural condition (Ketchledge, 1965). However, additional changes in land cover have occurred in response to windthrow, particularly during one unusually severe storm that struck the region in 1950. This large

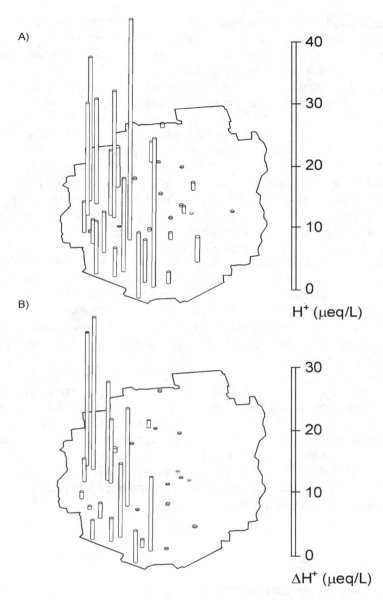

FIGURE 10.1
Map of A) measured hydrogen ion concentration, and B) diatom-inferred decreases in hydrogen ion concentration from pre-industrial times to the present for lakes in the Adirondack Park that were included in the DDRP statistical design. (Source: *Water, Air, Soil Pollut.*, Vol. 95, 1997, p. 322, Increasing role of nitrogen in the acidification of surface waters in the Adirondack Mountains, New York, Sullivan, T.J., J.M. Eilers, B.J. Cosby, and K.B. Vaché, Figure 4, Copyright 1997. Reprinted with kind permission from Kluwer Academic Publishers.)

storm, known as the Big Blow, severely damaged large tracts of forest in the Adirondacks. Most of the estimated 171,000 ha of forests that were damage occurred between the High Peaks Region and the southwestern boundary of

the park. This is the portion of the Adirondack Mountains that currently contains the majority of the acidic lakes (Baker et al., 1990a,c; Figure 10.1a) and
that has experienced the greatest acidification since pre-industrial times (Sullivan et al., 1990a, 1997). Many areas experienced in excess of 75% forest
blowdown, particularly on eastern slopes and the western shores of lakes.
This was one of the most severe storms on record for the northeastern U.S.
(Bristor, 1951).

Dobson et al. (1990) investigated the relationship between lake-water pH
and blowdown from the 1950 storm for several Adirondack data sets, including 12 lakes in the High Peaks Region, a group of 43 headwater lakes, 11 lakes
studied by the National Research Council (1986), and 23 RILWAS lakes
(Driscoll and Newton, 1985). The High Peaks Region was extensively damaged by the Big Blow, especially above 760 m elevation. Dobson et al. (1990)
found a strong spatial correlation between lake acidity and the percentage of
the watersheds that experienced blowdown.

Field reconnaissance was conducted by Dobson et al. (1990) in the High
Peaks Region, Big Moose Lake and vicinity, and at other Adirondack lakes. In
all areas, pipe networks in the soil from former tree roots were found in jux-
taposition to stumps from the 1950 blowdown and other stumps of a similar
age. Dobson et al. (1990) contended that the abundance of pipes in the soil
would short-circuit the normal infiltration processes and diminish the extent
of acid neutralization of acidic precipitation. Pipes and pipeflow appeared to
be a significant hydrologic pathway in all investigated blowdown areas.
Based on the abundance of correlative data supporting a relationship
between blowdown and low lake-water pH and the observed occurrence of
networks of pipes in the affected areas, the authors concluded that blowdown from the 1950 storm likely played an important role in recent acidification of many Adirondack lakes. It is also possible (but not demonstrated),
however, that the watersheds most susceptible to acidic deposition are also
those most susceptible to blowdown, irrespective of any cause/effect relationship (Sullivan et al., 1996b).

10.3 Lake-Water Chemistry

Regional variation in the ANC of Adirondack lakes is owing mainly to geologic factors that influence the supply of base cations to drainage waters,
rather than to inputs of SO_4^{2-} (Driscoll et al., 1991). Sulfate concentrations in
lake water are fairly uniform throughout the region, whereas base cation
(which neutralizes SO_4^{2-} acidity) concentrations are low (less than 150
µeq/L) primarily in the southwestern Adirondacks. Acidic Adirondack
lakes are generally underlain by granitic gneiss and are situated in areas of
the park that receive the greatest precipitation input (Driscoll et al., 1991).
Studies of acidity and acidification of Adirondack lakes have focused

mainly on atmospheric inputs of S and base cation release (weathering and ion exchange) from soils, and secondarily on organic anion acidity, N dynamics, and Al mobilization. Such processes constitute the central core of the various mathematical modeling efforts to predict the response of Adirondack lakes to changing levels of S and N deposition (e.g., Cosby et al., 1989; NAPAP, 1991; Sullivan et al., 1996a).

Results from the ALTM Program showed consistent decreases in SO_4^{2-} concentrations in lakes in the Adirondacks during the past two decades, with no lakes showing increasing concentrations (Driscoll and Van Dreason, 1993; NAPAP, 1998). Sulfate concentrations had been declining in Adirondack lakes since sometime in the 1970s (Sullivan, 1990). These trends of decreasing SO_4^{2-} concentration in surface waters are consistent with decreases in SO_2 emissions in the eastern U.S. and decreases in SO_4^{2-} concentration in precipitation in the Northeast (Driscoll and Van Dreason 1993). Despite this decline in atmospheric SO_4^{2-} inputs, there has been no increase in lake ANC and some of the ALTM lakes exhibited continued acidification despite the reduction in lake-water SO_4^{2-} concentrations. Nitrate concentrations had increased in nine of the ALTM lakes through about 1990. Atmospheric deposition of N in the Adirondacks has not changed appreciably since the 1970s, so the mechanism responsible for this increase in NO_3^- was unclear. However, Driscoll and van Dreason (1993) noted that elevated atmospheric deposition of N, coupled with diminished biotic demand owing to increasing stand age, could be a contributing factor.

Stoddard (1994) quantified statistically significant increases in lake-water NO_3^- over time from 1982 to 1990 in more than one-half of the ALTM lakes. The increases in lake-water NO_3^- concentration ranged from 0.4 to 1.8 µeq/L per year, with an average increase of 1 µeq/L per year. These data were interpreted to suggest that many Adirondack watersheds were becoming increasingly N saturated (Stoddard, 1994) which could cause a continued deterioration of the acid–base status of the lakes and streams in these watersheds unless in-lake and in-stream processes consume NO_3^- and generate additional ANC. However, more recent (post-1990) data for ALTM lakes show a decline in lake-water NO_3^- concentrations in recent years (Driscoll et al., 1995; Mitchell et al., 1996) that may be owing to climatic variations. It seems that an improved understanding of the factors that control NO_3^- leaching in terrestrial environments is needed before we can accurately predict the long-term effects of N deposition on lake-water chemistry.

Past changes in lake-water acid–base status have been estimated, based on analyses of diatom and chrysophyte remains in lake sediments. A large number of Adirondack lakes have been included in paleolimnological studies, especially PIRLA-I and PIRLA-II. Diatom-inferred changes in lake-water pH and/or ANC since pre-industrial times have been reported for about 70 lakes in the region (Charles et al., 1989; Sullivan et al., 1990; Cumming et al., 1992). A few lakes have also been analyzed for diatom microfossils at frequent intervals of the sediment cores, thereby allowing estimation of the timing of acidification responses in these lakes (Charles et al., 1990).

PIRLA-I studied 12 dilute, low-alkalinity Adirondack Lakes (Charles et al., 1990). Stratigraphic profiles of diatoms, chrysophytes, cladocera, and chironomids generally showed consistent patterns of change. The 8 PIRLA-I lakes that had measured pH less than 5.5 all showed diatom and chrysophyte evidence of recent acidification. The diatom-inferred onset of acidification occurred around 1900 to 1920 in four of the lakes. Diatom-inferred acidification began or accelerated around 1950 in two of those same lakes, as well as two others (Sullivan et al., 1999). Lakes with current pH greater than 6.0 and alkalinity greater than 50 μeq/L showed little or no evidence of acidification.

In addition to the diatom inferences of historical changes in lake-water acid–base chemistry provided by PIRLA-I, such inferences have also been calculated for other lakes using the sedimentary remains of scaled chrysophytes. Chrysophyte-inferred pH was estimated at frequent time intervals for PIRLA-II Adirondack lakes (Cumming et al., 1994) and can be used to establish the timing of acidification in the same manner as has been done by the diatom inferences from PIRLA-I lakes. Cumming et al. (1994) reconstructed the pH histories of 20 low-alkalinity (ANC less than 30 μeq/L) Adirondack lakes based on the species composition of chrysophytes in stratigraphic intervals from ^{210}Pb dated sediment cores. The sediment cores were sectioned at multiple intervals reflecting the period from about 1850 to about 1985. About 80% of the study lakes were inferred to have acidified since pre-industrial times. Many showed evidence of acidification beginning around the turn of the century, some of which showed evidence of some recovery since about 1970. A pattern of beginning or accelerating acidification around 1950 was also commonly observed. This was attributed by Cumming et al. (1994) to the higher levels of S deposition during that period. This pattern could also be owing, at least in part however, to the 1950 blowdown (Dobson et al., 1990; Sullivan et al., 1996b). Perhaps not coincidentally, the major logging in the areas of the Adirondacks that experienced significant acidification occurred around the turn of the century (McMartin, 1994). Thus, the onset of acidification of several study lakes for which diatom- or chrysophyte-inferred acidification chronologies are available corresponds temporally to both the onset or increase in acidic emissions and deposition and also the occurrence of major landscape disturbances associated with logging or blowdown. Recent research and assessment efforts have focused heavily on deposition aspects, with comparatively little treatment of the watershed disturbance and forest regrowth aspects other than the studies of Dobson et al. (1990), Davis et al. (1994), and Sullivan et al. (1996b, 1999).

10.4 Organic Acidity

Naturally occurring organic acids exert an important influence on the acid–base chemistry of lake waters throughout the Adirondack region. Many of the lakes that exhibit low pH and ANC contain relatively high concentrations of DOC (greater than 400 µM). The smaller lakes, in particular, often contain substantial amounts of organic acidity (Sullivan et al., 1990b). With completion of the ALSC database of over 1400 lakes, the majority of which are small, research on the role of organic acids in Adirondack lakes accelerated (e.g., Munson and Gherini, 1993; Driscoll et al., 1994).

Organic acidity of Adirondack lakes has also been the focus of recent efforts to test and improve the MAGIC model, a lumped-parameter process model of catchment-scale acidification (Cosby et al., 1985a,b). Sullivan et al. (1991, 1992) compared MAGIC model hindcasts and paleolimnological inferences of historical acidification (based on diatom-inferred pH) for a set of 33 statistically representative Adirondack lakes. This study represented the first regional and statistical model confirmation exercise conducted for a process-based acid–base chemistry model. These two assessment methods differed primarily in that MAGIC inferred greater acidification and also that acidification had occurred in all lakes in the comparison. The diatom approach inferred that acidification had been restricted to low-ANC lakes (currently less than about 50 µeq/L). The lack of organic acid representation in the MAGIC simulations was judged to be an important factor contributing to the differences in acidification implied by the two approaches. Organic acids commonly exert a large influence on surface water acid–base chemistry, particularly in dilute waters having moderate to high dissolved organic carbon (DOC) concentrations. Subsequently, MAGIC was extended by incorporating a quantitative organic acid representation, based on empirical data and geochemical considerations (Driscoll et al., 1994), and the model was tested using the paleolimnological hindcast data (Sullivan et al., 1996a). See Chapter 9 for additional discussion of this study.

Results of the hindcast comparisons between paleolimnological inferences of pre-industrial lake-water pH and MAGIC simulations, with and without the inclusion of organic acids, illustrated three important points

1. There was a systematic departure with decreasing pH between diatom and MAGIC model hindcasts of pre-industrial pH when organic acids were not included in the MAGIC simulations.

2. Inclusion of a reasonable formulation for organic acidity in the MAGIC simulations of pre-industrial pH greatly improved the agreement with diatom reconstructions, particularly for lakes having pH less than 6.0.

3. The choice among organic acid analog models (e.g., mono-, di-, triprotic, or Oliver model) did not alter the agreement with diatom reconstructions to an appreciable degree for a regional analysis.

The inclusion of organic acids in the MAGIC model simulations of pre-industrial pH did not alter the agreement between diatom and MAGIC model hindcasts for lakes with pH values greater than about 6.5. Previous hindcast comparisons (Sullivan et al., 1991) that did not include an organic acid representation had obtained good agreement for these high-pH lakes. For low-pH lakes, however, the lack of organic acid representation had resulted in an increasing level of divergence between diatom and MAGIC model hindcasts of pre-industrial pH. Thus, the lakes of greatest relevance with respect to potential biological effects of acidification, especially those having pH less than 5.5, exhibited increasingly larger discrepancies with decreasing pH between diatom and MAGIC model estimates of pre-industrial pH when organic acids were omitted from the analysis. The inclusion of an organic acid representation in the MAGIC simulations conducted in this study resulted in a great improvement in the agreement between these two modeling approaches (Sullivan et al., 1996a).

The results of these analyses of Adirondack lakes demonstrated that

1. Organic acids must be considered in modeling the response of lake waters in the Adirondack Mountains to acidic deposition.

2. Once organic acids are included in the modeling approach, reasonable agreement is obtained in hindcast comparisons with diatom-inferred pH.

The demonstrated agreement between MAGIC and paleolimnological model hindcast estimates of lake-water pH was dependent upon inclusion of organic acids in the process model despite the relatively low concentrations of DOC in the study lakes (mean value equal to 313 μM C).

10.5 Role of Nitrogen in Acidification Processes

Prior to 1990, most studies of lake-water acid–base chemistry in the Adirondack region neglected N as a potential agent of chronic acidification because acid-sensitive watersheds were believed to retain almost all atmospheric N and because lake-water NO_3^- concentrations were low relative to SO_4^{2-} in the majority of lakes sampled during the summer or fall seasons. However, recent research has suggested that the majority of Adirondack lakes have not acidified since pre-industrial times (Sullivan et al., 1990b, 1992) and, therefore, chronic NO_3 concentrations in those unacidified lakes are largely irrelevant with respect to questions of acidification causality (Sullivan et al., 1997). In addition, recent estimates of the magnitude of chronic acidification of Adirondack lakes (Sullivan et al., 1990a, 1992) are substantially lower than was widely believed prior to preparation of the NAPAP technical and assessment documents. Given these relatively recent research findings suggesting that acidification has been less widespread and of lower magnitude than previously suspected, a re-evaluation of the relative role of N has been occurring (Sullivan et al., 1997).

In a survey of N output from 65 forested plots and catchments throughout Europe, Dise and Wright (1995) found a deposition threshold of about 10 kg N/ha per year, below which significant N leaching did not occur. At deposition levels between 10 and about 25 kg N/ha per year, Dise and Wright (1995) found substantial variability in the rates of N leaching, from near zero to leaching rates that approached deposition loading rates. Estimates of total N deposition in the Adirondacks (10 to 12 kg N/ha per year; Ollinger et al., 1995) are in the range of the 10 kg N/ha per year threshold found in Europe. Thus, leaching rates in the Adirondacks may be expected to increase in the future under higher loading, at least for some lakes.

The concentration of NO_3^- in Adirondack lake waters surveyed in 1984 by the ELS and in 1985 to 1987 by the ALSC exhibited a pronounced geographic pattern, with high concentrations occurring primarily in the western and southwestern Adirondacks, and secondarily in the High Peaks region (Sullivan et al., 1997; Figures 10.2a and 10.3). During the fall season, lake-water NO_3^- concentrations in the range of 5 to 20 µeq/L were commonly found in these areas of the park by both the ELS-I (Figure 10.2a) and ELS-II (data not shown) surveys. During the spring, the geographical distribution of lake-water NO_3^- concentrations sampled by ELS-II was similar, but spring concentrations were about two-fold higher than concentrations in the fall. The observed spatial patterns in lake-water NO_3^- in both the statistically based ELS and the extensive survey conducted by ALSC corresponded closely with the distribution of lake-water acidity (Figure 10.1a) and also the distribution of diatom-inferred acidification from pre-industrial times to the present (Figure 10.1b).

The observed concentrations of NO_3^- in lake waters during the fall and summer seasons (Figures 10.2a and 10.3) were of approximately the same magnitude as the diatom-inferred increases in H^+ concentration (Figure 10.1b) and the diatom-inferred historical decline in ANC reported by Sullivan et al. (1990a). Analyses of Adirondack lakes inferred from diatom records to have acidified since pre-industrial times suggested declines in ANC averaging 18 µeq/L (quartiles 5, 23 µeq/L) for currently acidic lakes and 7 µeq/L (quartiles 0, 13 µeq/L) for lakes currently having ANC between 0 and 25 µeq/L (Sullivan et al., 1990a). Diatom-inferred increases in H^+ of about 5 to 25 µeq/L were tabulated (Figure 10.1b) from the pH-inferences reported by Cumming et al. (1992).

Both the quantitative estimates of historical acidification, inferred from diatom reconstructions of H^+ and ANC, and the spatial distribution of acidic and acidified lakes correlate closely with the magnitude and spatial distribution of lake-water NO_3^- concentrations (Sullivan et al., 1997). In contrast, lake-water SO_4^{2-} concentrations are much higher and show no spatial pattern (Figure 10.2b), although precipitation amounts and, therefore, total S loading are highest in the southwestern portion of the park. It is possible that these geographic correlations between NO_3^- concentration and acidification are coincidental. However, these patterns lend support to the hypothesis that N plays

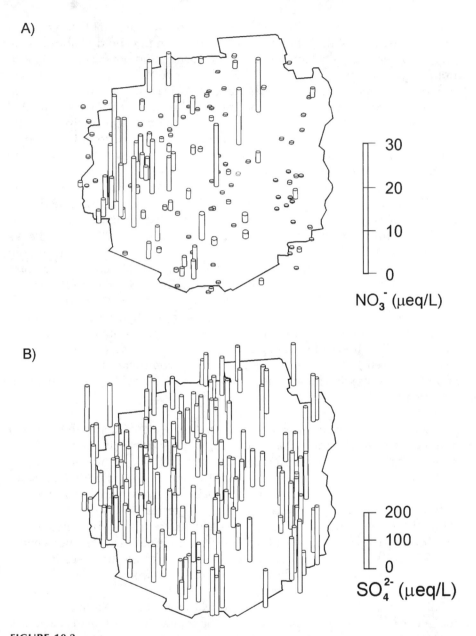

FIGURE 10.2
Map of A) NO_3^- and B) SO_4^{2-} concentrations in Adirondack lakes measured in the ELS-I statistical survey. The survey was conducted during the fall season. (Source: *Water, Air, Soil Pollut.*, Vol. 95, 1997, p. 320, Increasing role of nitrogen in the acidification of surface waters in the Adirondack Mountains, New York, Sullivan, T.J., J.M. Eilers, B.J. Cosby, and K.B. Vaché, Figure 2, Copyright 1997. Reprinted with kind permission from Kluwer Academic Publishers.)

a more significant role in chronic lake-water acidification in the Adirondacks than was believed in 1990.

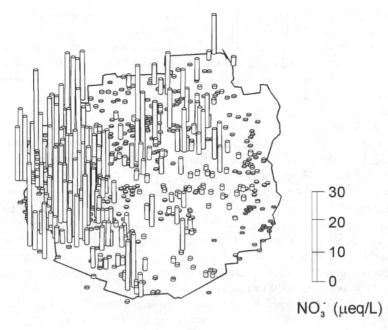

30

20

10

0

NO₃⁻ (μeq/L)

FIGURE 10.3
Map of summer NO_3^- concentrations in drainage lakes sampled by the ALSC. High concentrations of NO_3^- (greater than 5 μeq/L) were found almost exclusively in the southwestern Adirondacks and the High Peaks region. (Source: *Water, Air, Soil Pollut.*, Vol. 95, 1997, p. 321, Increasing role of nitrogen in the acidification of surface waters in the Adirondack Mountains, New York, Sullivan, T.J., J.M. Eilers, B.J. Cosby, and K.B. Vaché, Figure 3, Copyright 1997. Reprinted with kind permission from Kluwer Academic Publishers.)

Very different conclusions were drawn by Stoddard (1994), who reported median values of NO_3^- : (NO_3^- + SO_4^{2-}) ratios in surface waters of acid-sensitive regions of the U.S. sampled in EPA's National Surface Water Survey. In most regions, the median value was less than 0.2, and it was only 0.005 in the Adirondacks. Stoddard (1994) interpreted these findings as an indication that chronic acidification is more closely tied to SO_4^{2-} than to NO_3^-. Similarly, van Miegroet (1994) concluded that N is not an important contributor to chronic acidification, although she stated that the relative role of NO_3^- is expected to increase as more forested watersheds become N saturated and abatement policies for S take effect. Sullivan et al. (1997) contended that regional average NO_3^- concentrations are of little interest, however; NO_3^- concentrations in only those lakes that have acidified and the dynamics of the acidification process are more relevant.

Because SO_4^{2-} concentrations in most Adirondack lakes are so much higher (typically by approximately 100 μeq/L) than NO_3^- concentrations during the fall season, it has generally been assumed that SO_4^{2-} is the dominant cause of mineral acidity in these lakes. This is not necessarily the case, however, particularly if NO_3^- and SO_4^{2-} follow different hydrological pathways in reaching the lakes. A significant proportion of the observed lake-water NO_3^- in the fall sampling of ELS-I can be attributed to direct precipitation inputs of NO_3^- to the lake surfaces (Sullivan et al., 1997). This is consistent with the expected high retention of N in forest soils, and provides a mechanism to explain what may be the greater acidifying potential of in-lake NO_3^- as compared to in-lake SO_4^{2-} in these lakes. If most or all in-lake NO_3^- was derived from direct precipitation inputs, this NO_3^- would be expected to decrease lake-water ANC stoichiometrically. If most in-lake SO_4^{2-} was derived, in contrast, from deposition to watershed soils, this in-lake SO_4^{2-} would be expected to decrease lake-water ANC by a substantially smaller amount because base cations mobilized from watershed soils would have neutralized much of the SO_4^{2-} acidity before the SO_4^{2-} reached the lakes.

Analyses of available data for the Adirondack Mountain region suggest that increases in the concentration of NO_3^- in surface waters may be a significant contributor to recent acidification and consequent biological effects in Adirondack lakes. Although much of the evidence for a significant contribution of N leaching to chronic acidification of Adirondack lakes is circumstantial, the weight of evidence suggests that N plays an important role. Lake-water NO_3^- is commonly in the range of 5 to 25 μeq/L on a chronic basis in portions of the Adirondack Park that have experienced significant chronic lake-water acidification. Elsewhere in the park, NO_3^- concentrations in summer and fall are generally near zero. The levels of NO_3^- in the southwestern Adirondacks and the High Peaks region correspond both in space and in magnitude to inferred historical acidification (Sullivan et al., 1997). Evaluation of episodic chemistry and associated biological effects indicates that the relative importance of NO_3 as an agent of acidification increases dramatically during periods of high flow when surface waters are most toxic to fish (Wigington et al., 1993).

Such analyses do not negate the importance of SO_4^{2-} as an agent of surface water acidification in Adirondack waters. However, the role of N has not been adequately addressed in key policy-relevant modeling efforts (van Sickle and Church, 1995). Nitrogen may well be associated with chronic, as well as episodic, acidification of acid-sensitive Adirondack lakes. Past model projections of chronic surface water chemistry that did not incorporate potential future changes in lake-water NO_3^- concentrations may, therefore, be in error. Future research and modeling efforts should focus on improving predictive ability regarding both S- and N-driven acidification chronically and during periods when surface water chemistry is most toxic to aquatic biota.

10.6 Role of Landscape and Disturbance in Acidification Processes

The relationships between landscape characteristics and lake-water acid-base chemistry have been investigated through efforts to classify Adirondack lakes and their watersheds into discrete classes that reflect the dominant hydrogeochemical processes that control the susceptibility of drainage waters to acidification. Newton and Driscoll (1990) developed a watershed classification for Adirondack lakes that was based on flowpath theory, field reconnaissance, and the concentration of selected lake-water chemical parameters (Ca^{2+}, DOC, Cl^-). The resulting classification scheme included two hydrologic categories of seepage lakes (flow-through and recharge), and three major categories of drainage lakes that are determined on the basis of Ca^{2+} concentration, but presumed to reflect the presence of thin, intermediate, and thick glacial till in the watershed. The class of thick till drainage lakes (Ca^{2+} greater than 150 µeq/L) is further subdivided into watersheds that are and are not presumed to contain appreciable amounts of calcium carbonate, based on whether the Ca/Si ratio exceeds 2. Most of the Newton and Driscoll (1990) lake classes are divided into two final categories: high or low DOC, using a cutoff criterion of 5 mg/L (417 µM) of DOC.

Momen and Zehr (1998) used canonical discriminate analysis in an effort to improve the Newton and Driscoll (1990) classification system by identifying lake-water chemistry and terrestrial variables that could jointly differentiate among Adirondack lakes. Their approach is less reliant on lake-water Ca^{2+} concentration for geologic classification. This is a limitation of the Newton and Driscoll (1990) method because Ca^{2+} concentration in lakes is controlled by other factors besides watershed hydrogeology, including deposition chemistry, disturbance, and depletion of soil exchangeable Ca^{2+}. The discriminate functions of Momen and Zehr (1998) are also more useful in statistical analyses than are categorical variables (c.f., Freeman, 1987).

Interactions between acidic deposition and landscape/physical watershed characteristics were recently evaluated by Sullivan et al. (1999) for a group of 50 Adirondack lakes. Landscape characteristics were compiled and examined relative to paleolimnological inferences of historical acidification. Results of model estimates of acidification using the MAGIC model and paleolimnological analyses were compared to physical, biological, and landscape change data to evaluate if inclusion of additional processes could improve model estimates.

The goal of the study was to examine the interactions between acidic deposition and landscape change for a group of Adirondack lakes that have reasonably well-known acidification histories. Two types of landscape characteristics can impact lake acidification. The first type is a function of physical and biological attributes, such as elevation, rates of atmospheric deposition, hydraulic residence time, and vegetation. The second type is a function of landscape change processes, such as logging, fire, blowdown, and watershed disturbances. Diatom and chrysophyte-inferred historical changes in lake-water pH and the current acid–base status of Adirondack lakes were examined with respect to known or suspected forest disturbances owing to logging, fire, and blowdown, the presence of roads and lakeshore cabins, and other landscape attributes. Features of landscape characterization and landscape change were evaluated relative to

paleolimnological inferences of historical acidification and model estimates of acidification using the MAGIC model. The objectives of this study
were to determine if there were statistical relationships between landscape
characteristics/processes and lake-water acidification, assess the extent to
which such relationships may reflect causality, and, finally, to determine
whether it would be worthwhile to add a component to the MAGIC model
to model these relationships.

Sullivan et al. (1999) found that diatom-inferred acidification (pH inference
from core bottom minus pH inference from core top) was positively correlated with average long-term annual precipitation and NO_3^- and SO_4^{2-} deposition ($p \leq 0.001$). Acidification was also positively, but weakly ($p \leq 0.05$),
correlated with lake elevation and the relative presence of red spruce/yellow
birch forests in the watershed (only 34 watersheds had vegetation data).
Acidification was negatively correlated with the percent wetlands along
lakeshore and inlet stream riparian areas ($p \leq 0.001$). Weak negative correlations were found ($p \leq 0.05$) between inferred acidification and average watershed slope, hydraulic residence time, and the relative presence of
hemlock/yellow birch forests.

Acidification was positively correlated with forest blowdown during the
1950 storm, and negatively correlated with the extent of fire impact
($p \leq 0.001$). A weak negative correlation ($p \leq 0.05$) was found between
inferred acidification and watershed disturbance. Of the 21 lakes in the data
set that showed acidification of more than 0.25 pH units, only 3 had lakeshore
cabins present and were ranked relatively high in the disturbance impact
ranking. Only two have been logged in recent years (logging category 5), and
only four have been logged since 1950. In contrast, all 4 lakes that showed historical increases in pH (alkalinization) of more than 0.25 pH units had cabins
present on their lakeshores and were ranked among the 10 highest in disturbance impact. In addition, all have been logged, and three have been logged
since 1950.

Diatom-inferred acidification was also highly correlated ($p \leq 0.001$) with
three measurements of surface water chemistry: fall pH, fall NO_3^- concentration, and spring NO_3^- concentration. Thus, lakes that have acidified tend to be
those currently low in pH (e.g., Sullivan et al., 1990a) and that have relatively
high concentrations of NO_3^- (e.g., Sullivan et al., 1997). As discussed previously, the relative importance of NO_3^- (vs. SO_4^{2-}) as an agent of acidification in
these lakes is uncertain, but is clearly greater than was surmised in 1990.

Results of bivariate and multivariate analyses confirmed that lakes that
have experienced historical acidification tend to be those that receive relatively high amounts of precipitation and have short hydraulic residence
times. These variables explained 58% of the diatom-inferred acidification. A
combined model of long-term precipitation, hydraulic residence time, and
blowdown explained 71% of the historic acidification in the Adirondacks
(Sullivan et al., 1999). Lakes that have increased in pH since pre-industrial
times tend to be those subject to substantial human disturbance and those
that burned during major fires recorded after the turn of the century.

The multivariate regression model indicated that the more acidified lakes had higher precipitation, higher blowdown in their watersheds during the 1950 storm, and shorter hydraulic residence times than the less acidified lakes. Interpretation of the importance of these predictor variables is somewhat subjective, however. Higher precipitation contributes to both higher total deposition of N and S and also greater long-term depletion of base cation reserves from watershed soils owing to enhanced leaching. The significance of blowdown from the 1950 storm may reflect either location of watersheds along exposed ridges, with associated thinner soils and greater sensitivity to both blowdown and acid deposition impacts, or acidification associated with forest regrowth subsequent to the blowdown (as suggested by Dobson et al., 1990). The hydraulic residence time of lakes in the Adirondacks is influenced by water flow paths and the depth of till (Driscoll et al., 1991; Newton et al., 1987). Watersheds having thin glacial soils are expected to be hydrologically flashy and exhibit proportionally lower base flow (see Driscoll et al., 1991; Figure 6.30), whereas thick glacial soils provide larger groundwater contributions to hydrologic discharge that are higher in ANC.

Sullivan et al. (1999) also examined the relationship between the magnitude of the discrepancy between MAGIC model and diatom-inferred hindcasts of pre-industrial pH and the suite of watershed variables under study. This was done to ascertain whether the disagreement between these approaches was partly a function of one or more characteristics of the watersheds. If this turned out to be the case, then it could be inferred that certain aspects of the MAGIC model or diatom inferences might be further improved, thereby, perhaps increasing the accuracy of the model projections. The difference between diatom-inferred acidification and MAGIC-simulated acidification was not significantly correlated with any of the landscape change variables. This MAGIC-diatom offset was, however, highly correlated with the percent wetlands in the riparian zone ($p \leq 0.001$), moderately correlated with lake-water DOC ($p \leq 0.01$), and weakly correlated with the occurrence of one soil type (skerry fine sandy loam) and one vegetation type (hemlock-yellow birch) ($p \leq 0.05$). None of the other watershed characteristics or land use history data showed indication of biasing the extent of agreement between the diatom-inferences and model hindcasts. This result suggests that additional modifications to the MAGIC model (cf. Sullivan et al., 1996a,b, 1998; Sullivan and Cosby, 1995) to further take into account landscape change data are not likely to appreciably improve model performance with respect to S-driven acidification. However, the negative correlation between the MAGIC-diatom offset and the two indicators of organic acid influence (percent wetlands in riparian zone and lake-water DOC) suggest that additional work on the role of wetland vegetation and associated organic acidity in modifying the acidification response (e.g., Hemond, 1994; Sullivan et al., 1996b) may be warranted.

These results differed from results of model simulations for afforested British moorland sites. The latter have suggested that the presence and growth of forests promotes surface water acidification via increased dry and occult deposition of S, increased evapotranspiration, and increased uptake of base cations by growing trees (Neal et al., 1986; Jenkins et al., 1990; Cosby et al., 1990). Model output at the afforested Loch Chon site in Scotland suggested that the most severe acidification resulted from a combination of forest growth and acidic deposition. The forest growth component was owing to the simulated decrease in soil base saturation caused by the uptake of base cations by trees. This decreased soil base saturation increased the modeled sensitivity of the soils and drainage waters to the effects of acidic deposition (Jenkins et al., 1990).

10.7 Overall Assessment

The weight of evidence suggests that lakes and watersheds in the Adirondack Mountains have been adversely impacted by atmospheric deposition of both S and N. This evidence has been widely discussed in the scientific literature (e.g., Driscoll and Newton, 1985; Sullivan et al., 1990a, 1997; Baker et al., 1991; Driscoll et al., 1991; Driscoll and van Dreason, 1993). More recent evidence also suggests that lake-water acidification in the Adirondacks has been broadly correlated with widespread changes in landscape cover that occurred in response to massive logging operations and fires around the turn of the twentieth century and unprecedented forest blowdown during a large windstorm in 1950. Although such changes in landscape cover, particularly forest regrowth, are unlikely on their own to cause lakes to become acidic (ANC less than or equal to zero), they can cause decreases in the base saturation of soils, thereby predisposing sensitive watersheds to subsequent acidification from acidic deposition (Cosby et al., 1990). The potential importance of these landscape changes has not been widely recognized in developing and applying models of surface water acidification and recovery.

Overall, the available evidence suggests that surface water acid–base status in the Adirondack Mountains is influenced by a variety of watershed characteristics, including type and age of vegetative cover. Also, it appears that changes in these watershed characteristics interact with S and N deposition to determine the extent of lake-water acidification or alkalization that occurs. Although it may not be possible to capture all of these dynamics in a process-based modeling approach, lake-water acidification is best understood when placed in the context of historical land use and landscape change.

Results of analyses of diatom-inferred historic change in pH of Adirondack lakes suggest that historical lake-water acidification and alkalinization have been functions of precipitation, S and N deposition, disturbance history, and watershed morphology. Lakes that have acidified have generally been those that receive high levels of precipitation and atmospheric deposition of S and N, have short hydraulic residence times, and many appear to have experienced forest blowdown in the large 1950 windstorm. Such lakes are relatively undisturbed by human activities and fire. Lakes that have increased in pH since pre-industrial times tend to be those that have lakeshore cabins and have experienced watershed disturbance.

11

Case Study: Class I Areas in the Mountainous West

11.1 Background

Atmospheric emissions of S and N outside national park and wilderness area boundaries in the western U.S. threaten the ecological integrity of highly sensitive ecosystems. Aquatic and terrestrial resources, particularly those at high elevation, can be degraded by existing or future pollution. Based largely on the results of EPA's Western Lakes Survey (Landers et al., 1987), NAPAP (1991) concluded that many high-elevation western lakes were extremely sensitive to acidic deposition effects. The absence of evidence of chronic acidification was attributed to the low levels of acidic deposition received by western watersheds. It was speculated that if deposition increased substantially in the future, substantial acidification would likely occur.

Previous research has focused the greatest attention on aquatic receptors in the Sierra Nevada and portions of the Cascade and Rocky Mountains (c.f., Charles, 1991). The National Park Service (NPS) initiated a series of projects to assess air quality issues within these regions. The NPS Air Resources Division commissioned several Air Quality Regional Reviews to summarize what is already known about these systems, including the Pacific Northwest region (Eilers et al., 1994a), the Rocky Mountain region (Peterson and Sullivan, 1998), and the California region (Sullivan et al., ongoing). Analyses of documented and potential ecological effects of atmospheric pollutants have been conducted, and inventories of pollution-sensitive components of ecosystem receptors in the parks have been compiled. Although generally low levels of atmospheric pollutants are measured in these mountain ranges (Sisterson et al., 1990), increasing development adjacent to protected areas has contributed to increasing air pollution. In particular, elevated emission levels of both S and N are evident adjacent to Rocky Mountain National Park and other portions of the Colorado Front Range.

The purpose of this chapter is to provide an overview of recent research results in a selected portion of the Sierra Nevada and in several national parks located in the Rocky Mountains. The aim is not to provide regional assessments for these areas, but rather to highlight the types of research that have been conducted and discuss the research results within the context of watershed processes that control ecosystem responses to acid deposition and critical loads. Research conducted in the Cascade Mountains is not presented here; the reader is referred to Eilers et al. (1994) for the most recent summary assessment treatment for that region.

The Clean Air Act (42 U.S.C. 7470), as amended in August 1977, provides one of the most important mandates for protecting air resources in Class I areas, that is national parks over 6000 acres and national wilderness areas over 5000 acres that were in existence before August 1977. In Section 160 of the Act, Congress stated that one of the purposes of the Act was to "preserve, protect, and enhance the air quality in national parks, national wilderness areas, national monuments, national seashores, and other areas of special national or regional natural, recreational, scenic, or historic value." According to the Clean Air Act and subsequent amendments (Public Laws 95-95, 101-549), Federal land managers (FLMs) have ". . . an affirmative responsibility to protect the air quality related values (AQRVs) . . . within a Class I area."

To maintain healthy ecosystems, it is increasingly imperative that land managers be prepared to monitor and assess levels of atmospheric pollutants and ecological effects in national parks and wilderness areas throughout the West. Knowledge of emissions inventories, coupled with scientific understanding of dose–response functions and critical loads assessments, will provide land managers with a framework with which to protect sensitive resources within the Class I areas from degradation owing to atmospheric deposition of pollutants.

Air quality within Class I lands is subject to the "prevention of significant deterioration (PSD)" provisions of the Clean Air Act. The primary objective of the PSD provisions is to prevent substantial degradation of air quality and yet maintain a margin for industrial growth. An application for a PSD permit from the appropriate air regulatory agency is required before construction of a new, or modification of an existing, major air pollution source (Bunyak, 1993). The role of the FLM is to determine if there is potential for additional air pollution to cause damage to a sensitive receptor. The FLM can recommend denial of a permit by demonstrating that there will be adverse impacts in the Class I area or recommend provisions for mitigation.

The following types of questions must be answered in response to PSD permit applications:

- What are the identified sensitive receptors within AQRVs in each Class I area that could be affected by the new source?
- What are the critical doses for the identified sensitive receptors?
- Will the proposed facility result in pollutant concentrations or atmospheric deposition that will cause the identified critical dose to be exceeded?

As discussed in previous chapters, atmospheric deposition of S and/or N has the potential to damage sensitive terrestrial, and especially aquatic, eco-systems by depleting the ANC of soil and surface waters, reducing the pH, and increasing the concentration of inorganic Al in solution. Such changes in water chemistry can affect the survival of in-lake and in-stream biota. A need, therefore, has arisen to assess the levels of atmospheric deposition at which such changes occur in the Class I areas so as to ensure the protection of sensitive resources.

The NAPAP SOS/T Reports and Integrated Assessment (NAPAP, 1991) provided only a cursory treatment of aquatic effects issues in the West, largely because it was well known that atmospheric deposition of S and N were generally low compared to highly impacted areas in the East and because results from the Western Lakes Survey (Landers et al., 1987) indicated that there were virtually no acidic (ANC less than or equal to zero) lakes in the West. NAPAP (1991) recognized, however, that high-elevation areas of the West contained some of the most sensitive watersheds in the world to the potential effects of acidic deposition.

It is important to determine critical loads of S and N deposition to sensitive, high-elevation watersheds in the West. It is also important to make these determinations in a timely fashion for the following reasons

1. Nitrogen deposition has been increasing at many western locations, including the Front Range of Colorado, during recent years.

2. FLMs are faced with an ongoing, and in some locations accelerating, need to provide recommendations for approval or denial of permits for increased point source emissions of S and/or N upwind of sensitive national parks and wilderness area receptors.

3. Mounting evidence suggests that adverse impacts to aquatic resources may be occurring in some areas under current deposition levels.

Because of the proximity of well-defined population centers and industrial pollution sources in the West to individual mountain ranges, it is often important to evaluate changes in emissions in the immediate vicinity of sensitive resources as well as to assess regional emissions (Sullivan and Eilers, 1994). For example, emissions in the Rocky Mountain states have no effect on resources in the Sierra Nevada, in part because emissions from these states are generally low and in part because the prevailing wind direction is from west to east. Precipitation chemistry in the far western ranges is largely influenced by local emissions, particularly emission sources to the west (upwind) of sensitive resources. In the Rocky Mountains, deposition chemistry is influenced by a more complex collection of sources, although recent evidence suggests that local sources, that is, those sources within approximately 100 km of a given mountain range, can be as important as long-range sources. In the Mt. Zirkel Wilderness of northwestern Colorado, elevated concentrations of SO_4^{2-} and NO_3^- in the snow appear to originate largely from sources in the Yampa Valley, about 75 km to the west (Turk et al., 1992). Rocky Mountain National Park may be largely influenced by emissions from the Front Range to the southeast.

Despite the uncertainties associated with existing deposition data, it is clear that atmospheric deposition of both S and N is currently low throughout most portions of the West (NAPAP, 1998). Annual wet deposition levels of S and N are generally less than about one-fourth of the levels observed in the high-deposition portions of the northeastern U.S. (Sisterson et al., 1990).

Spring snowmelt can act to flush N into lakes and streams that was deposited in the snowpack from atmospheric deposition or N mineralized within the soil during the winter. In some alpine and subalpine western lakes, the concentration of NO_3^- remains somewhat elevated throughout the growing season. This may be related to the extent of snow cover and effects of the cold temperatures on biological uptake processes, hydrological flowpaths across exposed bedrock and talus, and/or saturation of the uptake capacity of terrestrial and aquatic biota.

A substantial component of the NO_3^- in western lake waters may have been derived from mineralization of organic N and not directly from atmospheric deposition. Much of the N released from the snowpack during the melting period is retained in underlying soils. Williams et al. (1996b) contended that measurements of subnivial (under the snowpack) microbial biomass, CO_2 flux through the snowpack, and soil N pools all suggested that subnivial N cycling during the winter and spring is sufficient to supply the NO_3^- measured in stream waters.

It is likely that microbial activity under the snowpack plays an important role in both the production of inorganic N before the snowmelt begins and the immobilization of N during the initial phases of snowmelt before vegetation becomes active. For example, Brooks et al. (1996) followed soil N dynamics throughout the snow-covered season on Niwot Ridge, CO. Sites with consistent snow cover were characterized by a 3 to 8 cm layer of thawed soil that was present for several months before snowmelt began. Nitrogen mineralization in this thawed layer resulted in soil inorganic N pools that were significantly larger than the pool of N stored in the snowpack. As snowmelt began, soil inorganic N pools decreased sharply, concurrent with a large increase in microbial biomass N. As snowmelt continued, both microbial N and soil inorganic N decreased, presumably owing to increased demand by growing vegetation (Brooks et al., 1996). The recognized importance of mineralization, the production of inorganic N from the breakdown of organic material, and subsequent conversion to NO_3^- (nitrification) as a source of stream-water NO_3^- does not imply, however, that atmospheric N deposition is not driving this flux. It is likely that mineralization and nitrification processes release N to surface waters that was derived largely from deposition and cycled through the primary production of the previous growing season.

The sensitivity to acidification of surface waters in western regions is a function of regional deposition characteristics, surface water chemistry, and watershed factors (c.f., Charles, 1991). Sullivan and Eilers (1994) attempted to integrate these three elements to provide a qualitative assessment of watershed sensitivity to acidification and a quantitative assessment of the magnitude of acidification currently experienced within the western subregions. These results were then combined to provide an assessment of the likely dose–response relationships for the subregions of interest. See Chapter 5 for further discussion of this topic.

Topographic relief is also a contributing factor to acidic deposition sensitivity in the West because the mountainous terrain contributes to major snowmelt events that may cause episodic pH and ANC depressions. These snowmelt events can last up to 2 months and result in multiple exchanges of the water volume in lakes receiving significant runoff. The short residence time of many high-elevation lakes not only contributes to elevated sensitivity to snowmelt events, but also reduces the relative importance of in-lake alkalinity generation processes.

Episodic acidification is an important issue for surface waters throughout high-elevation areas of the West. A number of factors predispose western systems to potential episodic effects (Peterson and Sullivan, 1998), including

1. The abundance of dilute to ultradilute lakes (i.e., those having extremely low concentrations of dissolved solutes), exhibiting very low concentrations of base cations, and, therefore, ANC throughout the year.

2. Large snowpack accumulations at the high elevation sites, thus causing substantial episodic acidification via the natural process of base cation dilution.

3. Short retention times for many of the high-elevation drainage lakes, thus enabling snowmelt to rapidly flush lake basins with highly dilute meltwater.

Thus, the physical characteristics (e.g., bedrock geology, lake morphometry) and climate throughout high elevation areas of the West provide justification for considering potential episodic acidification to be an important concern. In addition, the few studies that have been conducted to date confirm the general sensitivity of western lakes to episodic processes.

In most cases, episodic pH and ANC depressions during snowmelt are driven by natural processes (mainly base cation dilution) and nitrate enrichment (cf. Wigington et al., 1990, 1993; Stoddard, 1995). Where pulses of increased SO_4^{2-} are found during hydrological episodes, they are usually attributable to S storage and release in streamside wetlands. More often, lake and stream-water concentrations of SO_4^{2-} decrease or remain stable during snowmelt. This is probably attributable to the observation, based on ratios of naturally occurring isotopes, that most stream flow during episodes is derived from pre-event water. Water stored in watershed soils is forced into streams and lakes by infiltration of meltwater via the "piston effect." This is not necessarily the case for high-elevation watersheds in the West, however. Such watersheds often have large snowpack accumulations and relatively little soil cover. Selective elution of ions in snowpack, therefore, can result in relatively large pulses of both NO_3^- and SO_4^{2-} in drainage water early in the snowmelt.

Data supporting the importance of SO_4^{2-} to spring episodes in the West were presented by Reuss et al. (1995). It appears likely that S deposition will contribute to episodic acidification of sensitive western surface waters at deposition levels below those that would cause chronic acidification (Sullivan and Eilers, 1994). Episodes have been so little studied within the region, however, that it is not possible to provide quantitative estimates of episodic S standards for the western subregions of concern.

The N loading to alpine and subalpine systems may be functionally much higher than is reflected by the total annual deposition measured or estimated for the watersheds. It may, therefore, be misleading to compare total N loading estimates of 3 to 7 kg N/ha per year, for example, of some alpine systems in the Front Range with the higher loading rates found in parts of the eastern U.S. and northern Europe. There are several reasons for this. First, the actual N loading to both soils and drainage waters at high-elevation sites during summer is comprised of both the ambient summertime atmospheric loading and also the loading of the previous winter that was stored in the snowpack and released to the terrestrial and aquatic systems during the melt period, often largely occurring during May through July. For this reason, the N loading from atmospheric deposition during the summer can actually be substantially higher than the annual average atmospheric loading. Second, soil waters are often completely flushed during the early phases of snowmelt in alpine areas. Such flushing can transport to surface waters a significant fraction of the N produced in soils during winter by subnivian mineralization of the primary production of the previous summer. This N load from internal ecosystem cycling will generally be larger in areas that receive significant N deposition because the gross primary production of alpine ecosystem often tends to be N limited (Bowman et al., 1993). Thus, the functional N loading to terrestrial and aquatic runoff receptors in alpine and subalpine areas during the summer growing season is much higher than the annual average N loading for the site. This is especially true during the early phases of snowmelt, when soil waters are flushed from shallow soils and talus areas and when a large percentage of the ionic load of the snowpack is released in meltwater.

The Sierra Nevada and Rocky Mountains contain an abundance of Class I areas, the majority of which are wilderness areas administered by the US Forest Service. Fairly extensive surface water chemistry data are available for some of the Class I national parks in the Rocky Mountains and Sierra Nevada. Some of these data were synthesized by Melack and Stoddard (1991), Turk and Spahr (1991), Peterson and Sullivan (1998), and Melack et al. (1998). These data, together with additional ancillary or more recent data, are summarized in the sections that follow. Although much of the information presented here is specific to a small number of national parks, the resources in these parks are likely representative of those in surrounding terrain in most cases.

11.2 Sierra Nevada

11.2.1 Atmospheric Deposition

The Emerald Lake Basin of Sequoia National Park in the southern Sierra Nevada has been the focus of considerable research on the effects of N and S deposition on soils, forests, and surface waters. An NADP monitoring site is located at Giant Forest in Sequoia National Park at an elevation of 1902 m. Total annual precipitation at this site ranges from about 50 to 200 cm per year. Concentrations of NH_4^+, NO_3^-, and SO_4^{2-} in precipitation have not shown a trend of increase or decrease since the early 1980s. Total wet N deposition has ranged from about 1 to 4 kg/ha per year, whereas wet S deposition has ranged from less than 1 to about 2 kg/ha per year. Deposition of N and S appears to vary from year to year primarily as a function of the total quantity of precipitation. The source of these pollutants is thought to be the Central Valley, with some influence from the San Francisco Bay area (Bytnerowicz and Fenn, 1996).

Wet deposition was monitored near treeline (elevation 2800 m) at the Emerald Lake watershed during the water years 1985 through 1987 by Williams and Melack (1991b). Precipitation amounts ranged from one of the wettest years on record (1986) to one of the driest (1987). Volume-weighted pH was 4.9 for rainfall and 5.3 for snowfall. Volume-weighted mean annual concentrations of SO_4^{2-}, NO_3^-, and NH_4^+ throughout the study were all about 4 to 5 µeq/L. Average total wet deposition of N and S were 2.3 and 2.1 kg/ha per year, respectively. Low Cl^- and high NH_4^+ concentrations in rain, compared with snow, suggest that localized convective systems (as opposed to oceanic frontal systems during the winter) are the main sources of ions in rainfall. Afternoon upslope air flow, induced by heating of air along the mountain slopes, transports air masses from the San Joaquin Valley to the upper reaches of Sequoia National Park on a daily basis during summer (Williams and Melack, 1991b).

Extensive monitoring of wet deposition to high elevations of the Sierra Nevada was initiated in 1990 at nine sites (Melack et al., 1997). The upper Marble Fork of the Kaweah River, which drains Sequoia National Park, was added to the monitoring program in 1992. Snow chemistry summarized by Melack et al. (1998) for eight of the (mainly alpine and subalpine) Sierra Nevada watersheds was dilute and similar among the watersheds. Mean concentrations of NO_3^- and NH_4^+ in snow were 2.4 and 2.7 µeq/L, respectively. Mean SO_4^{2-} concentration was 2.0 µeq/L (range about 1.0 to 3.0 µeq/L). However, NO_3^- and NH_4^+ concentrations in nonwinter precipitation were 8 to 9 times greater than in the snowpack (mean values, 20.7 and 23.4 µeq/L, respectively). The SO_4^{2-} concentration in nonwinter precipitation was also high, with a mean of 15.1 µeq/L. In contrast, the mean Cl⁻ level measured in nonwinter precipitation (4.2 µeq/L) was only slightly higher than the mean Cl⁻ concentration in winter snowfall.

Mean annual NH_4^+ deposition was 0.70 kg/ha NH_4^+ -N and mean annual NO_3^- was 0.63 kg/ha NO_3^--N for the 36 water years of record. For both ions, the maximum loading rates were measured at Emerald Lake during water year 1987 (3.6 kg N/ha).

Concentrations of N measured in winter snow in the Emerald Lake watershed were among the most dilute measurements of N recorded in wet precipitation (Williams et al., 1995). Nitrogen concentrations in winter snow of about 2 µeq/L each for NH_4^+ and NO_3^- were comparable to measurements from central Alaska (Galloway et al., 1982). However, mean concentrations of N in rainwater of about 55 µeq/L for NH_4^+ and 42 µeq/L for NO_3^- were comparable to N concentrations in rainfall in areas having considerable anthropogenic sources of N, such as the Adirondack and Catskill Mountains of New York (Stoddard, 1994).

Brown and Lund (1994) studied the influence of dry deposition and foliar interactions on the chemical composition of throughfall in the Emerald Lake watershed. Summer dry deposition was a substantial component of total annual deposition and was generally in excess of summer wet deposition.

11.2.2 Surface Water Chemistry

High-elevation lakes and streams in the Sierra Nevada are among the most dilute, poorly-buffered waters in the U.S. (Landers et al., 1987; Melack and Stoddard, 1991). The catchments that supply runoff to these waters are underlain primarily by granitic bedrock and have poorly-developed soils and sparse vegetation. The hydrologic cycle is dominated by the annual accumulation and melting of a dilute, mildly acidic (pH 5.5) snowpack (Melack et al., 1997).

During the 1980s, an Integrated Watershed Study (IWS) was conducted at the Emerald Lake watershed (2800 to 3400 m elevation), the purpose of which was to investigate the possibility of acid-induced damage to the watershed and to determine the consequences of acidification on Sierran surface waters (Tonnessen, 1991). The IWS included studies of deposition, terrestrial systems, and aquatic systems. Focus shifted in the late 1980s to a larger group of watersheds. Research on the catchments of Pear, Topaz, Crystal, and Ruby Lakes was initiated in 1986 (Sickman and Melack, 1989). Spuller and Lost Lakes were added to the monitoring program in 1990 (Melack et al., 1993). Results of these monitoring studies were summarized by Melack et al. (1998).

The eight water quality monitoring sites are located in alpine and subalpine settings across a majority of the north–south extent of the Sierra Nevada.

There are four located on the western slope, all within Sequoia National Park (Emerald, Pear, and Topaz Lakes, and Marble Fork of the Kaweah River). The other four are located to the north, and along the eastern slope of the Sierra Nevada range.

The volume-weighted mean pH for lake outlet streamflow during the 36 water years of record examined by Melack et al. (1998) for 7 lakes in the Sierra Nevada was 6.05, and ranged from 5.6 to 6.7. Lost Lake had the lowest pH; Ruby and Spuller Lakes had the highest. Lost, Pear, and Emerald Lakes had volume-weighted mean ANC in the range of 15 to 30 µeq/L and were classified by Melack et al. (1998) as low in ANC. Moderate ANC waters (Topaz, Spuller, and Marble Fork) exhibited mean ANC in the range of 30 to 50 µeq/L. Crystal and Ruby Lakes had mean annual ANC greater than 50 µeq/L.

Sulfate concentrations were most consistent of the ions measured. With the exception of Ruby and Spuller Lakes, annual average SO_4^{2-} concentration ranged from 5 to 7 µeq/L. Ruby and Spuller Lakes had annual average SO_4^{2-} concentration of 8 to 10 µeq/L.

11.2.3 Seasonality and Episodic Processes

The hydrologic cycle in the Sierra Nevada is dominated by snowfall and snowmelt, with over 90% of the annual precipitation falling as snow between November and April. Through the process of preferential elution (Johannessen and Henriksen, 1978), the relatively small loads of acidic deposition in Sierran snowpacks can supply high concentrations of SO_4^{2-} and NO_3^- during snowmelt (Stoddard, 1995). In most cases, lake-water pH decreases with increasing runoff, reaching a minimum near peak snowmelt discharge. Most other solutes exhibit temporal patterns identified by Melack et al. (1998) either as dilution, or a pulse of increased concentration followed by dilution (pulse/dilution) or biological uptake (pulse/depletion). Nitrate and SO_4^{2-} often declined at peak runoff. Nitrate peaks of 5 to 15 µeq/L were common, although they were usually less than 2 µeq/L in the N-limited lakes (Crystal and Lost Lakes). Patterns of change in SO_4^{2-} concentration were similar to NO_3^- patterns but much smaller in magnitude. Except in watersheds thought to have bedrock sources of S (Spuller and Ruby Lakes), the differences between SO_4^{2-} maxima and minima were generally within 2 µeq/L.

The concentrations of base cations and ANC generally exhibited a dilution pattern and reached minima near peak runoff. Outflow ANC declined by 24 to 80% during the spring, with an average decline of 50%. Lowest ANC was generally between about 15 µeq/L (Lost and Pear Lakes) and 30 µeq/L (Ruby and Crystal Lakes). Seasonal ANC depressions were greatest during years with deep snowpacks and high snowmelt runoff.

In some catchments, NO_3^- concentration declined throughout the snowmelt period (dilution). A second pattern was observed as a NO_3^- pulse during Stage 2 of snowmelt (i.e., 25 to 50% of cumulative runoff). This was seen in the Ruby and Emerald Lakes watersheds and was described by Melack et al.

(1993) as a pulse of NO_3^- early in the melt followed by depletion caused by biological uptake (pulse/depletion).

No long-term trends in the pH or ANC of surface water were identified for the eight waters studied by Melack et al. (1998). This was despite the fact that one lake (Emerald) had 12 years of monitoring data.

Concentrations of NO_3^- in the Emerald Lake outlet increased from 2 to 3 µeq/L in the fall to 10 to 13 µeq/L during spring runoff. The observed increases in NO_3^- and SO_4^{2-} were attributed to preferential elution from the snowpack and low retention rates in the watershed. In-lake reduction of NO_3^- and SO_4^{2-} within Emerald Lake was relatively small, and most of the acid anions passed through the lake outlet. The increase in SO_4^{2-} concentrations in surface water during snowmelt in the Sierra Nevada contrasts with observations in the Adirondacks where snowmelt runoff diluted SO_4^{2-} as well as base cation concentrations (Schaefer et al., 1990).

Williams and Melack (1991a,b) documented an ionic pulse in meltwater concentrations in the Emerald Lake watershed 2- to 12-fold greater than the snowpack average. Sulfate and NO_3^- concentrations in meltwater decreased to below the initial bulk concentrations after about 30% of the snowpack had melted. The ionic pulse was variable spatially dependent on the rate of snowmelt. At a site with relatively rapid snowmelt, the pulse lasted only 2 days, whereas at a site with a slow rate of melt, the pulse lasted about 10 days. The first fraction of meltwater draining from the snowpack had concentrations of NO_3^- and NH_4^+ as high as 28 µeq/L, compared to bulk snowpack concentrations less than 5 µeq/L (Williams et al., 1995). Stream-water NO_3^- concentrations reached an annual peak during the first part of snowmelt runoff, with maximum stream-water concentrations of 18 µeq/L. During the summer growing season, stream-water NO_3^- concentrations were always near or below detection limits (0.5 µeq/L).

Melack et al. (1993) reported 2 years of intensive research at the 7 high-elevation lakes. Solute concentrations, particularly ANC and base cation concentrations, were greatest during winter, declined to minima during snowmelt, and gradually increased during summer and autumn. Sulfate concentrations varied most in lakes with lowest volumes. Nitrate concentrations increased during snowmelt in most lakes owing to inputs of stream water enriched with NO_3^-. Zooplankton species known to be intolerant of acidification were found in all seven lakes, and Melack et al. (1993) concluded that their presence is evidence that Sierra Nevada lakes are not currently showing chronic biological effects of acidic deposition.

Kattelmann and Elder (1991) developed a water balance for 2 years for the Emerald Lake watershed that provides insight into the hydrology of headwater catchments in the Sierra Nevada. Snow dominated the water balance and accounted for 95% of the precipitation. Direct short-term runoff from snowmelt accounted for more than 80% of the streamflow in both years.

Snowmelt typically dilutes lake outflow solute concentrations in the Sierra Nevada by 30 to 40%, as measured by decreases in Na^+, Cl^-, or silica (Melack et al., 1998). In contrast, SO_4^{2-} concentrations are only reduced by about 10%.

Relative to dilution, episodic SO_4^{2-} concentrations showed an increase of 128 to 150% (Melack et al., 1998). Nitrate is also relatively increased in comparison with other ions in a similar fashion.

In the Sierra Episodes Study, 10 lakes and their watersheds were selected for study. Results for two of the lakes were reported by Stoddard (1995), one of which typified the response of the majority of high elevation lakes in the study (Treasure Lake) and one whose response was most extreme (High Lake). At Treasure Lake, ANC began to decline at the onset of snowmelt and reached a minimum at peak runoff, corresponding with minimum base cation, NO_3^-, and SO_4^{2-} concentrations (Figure 11.1). At no point did Treasure Lake become acidic. High Lake watershed contained a deeper snowpack, and began melting later in the season. ANC fell to 0 and below twice during the first 10 days of snowmelt. The ANC minimum corresponded with maximum concentrations of base cations, NO_3^- and Al (Figure 11.2). The High Lake watershed produced snowmelt that is both later and more rapid than other lakes included in the Sierra Episodes Study, perhaps due to its high elevation and small watershed area (Stoddard, 1995). This caused increases in NO_3^- concentration to values greater than 40 µeq/L, exceeding concurrent increases in base cations and causing the lake to become acidic for brief periods. High Lake appears to be representative of the most extreme conditions of episodic acid-sensitivity in the Sierra Nevada.

Dilution of base cation concentrations during snowmelt is the primary factor responsible for seasonal ANC depressions in lakes in the Sierra Nevada. Melack et al. (1998) classified the lakes into two classes according to snowmelt response. In shallower lakes having short residence times (i.e., lakes that are rapidly flushed during snowmelt), NO_3^- and SO_4^{2-} accounted for less than 10% of the ANC decline (Lost, Topaz, Spuller Lakes). In lakes having larger water volumes or lower snowmelt rates, NO_3^- and SO_4^{2-} accounted for 25 to 35% of the ANC decrease (Emerald, Pear, Ruby, Crystal Lakes). In the latter group, NO_3^- and SO_4^{2-} contributed about equally during the first half of snowmelt, but SO_4^{2-} was more important during the latter half of snowmelt.

Melack et al. (1998) found that the relationship between minimum ANC during snowmelt and fall overturn ANC was linear, and the equation remained unchanged with the addition of data from earlier surveys. Application of this equation to the population of lakes in the Sierra Nevada that were included in the WLS statistical frame suggested that none of the lakes represented by the WLS are acidified by snowmelt under current levels of acidic deposition. However, the confidence limit of this empirical relationship allowed the possibility that up to 1.8% of lakes in the Sierra Nevada are acidified to ANC less than or equal to 0 during snowmelt.

Melack et al. (1998) tabulated premelt and fall index ANC and episodic ANC values for each of their study watersheds and years using both lake outflow and in-lake chemistry samples. They found that the in-lake sampling every other month captured the snowmelt ANC depression almost as well as the more intensive outflow sampling. This was because the interval of minimum ANC was so extended that, typically, any sample collected

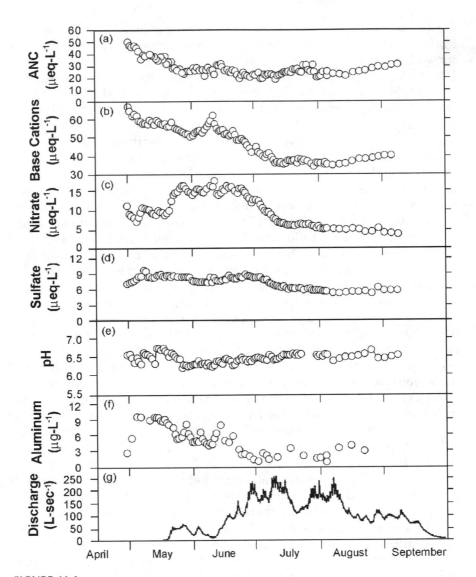

FIGURE 11.1

Time series of major ions and discharge in Treasure Lake during snowmelt in 1993. (Source: *Water, Air, Soil Pollut.*, Vol. 85, 1995, p. 356, Episodic acidification during snowmelt of high elevation lakes in the Sierra Nevada Mountains of California, Stoddard, J.L., Figure 1, Copyright 1995. Reprinted with kind permission from Kluwer Academic Publishers.)

from late May to early August captured most of the ANC depression. They developed an equation for minimum lake-water ANC as a function of fall overturn ANC, using a simplified version of the two-compartment chemical mixing model proposed by Eshleman et al. (1995):

$$\text{minimum ANC} = m\{\text{index ANC}\} - b \qquad (11.1)$$

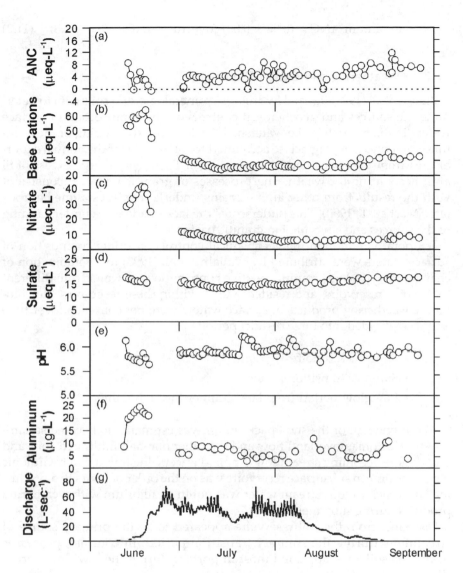

FIGURE 11.2

Time series of major ions and discharge in High Lake during snowmelt in 1993. (Source: *Water, Air, Soil Pollut.*, Vol. 85, 1995, p. 357, Episodic acidification during snowmelt of high elevation lakes in the Sierra Nevada Mountains of California, Stoddard, J.L., Figure 2, Copyright 1995. Reprinted with kind permission from Kluwer Academic Publishers.)

The parameter values of the model were estimated by Melack et al. (1998) using 7 of the 8 monitoring lake sites,* plus 15 additional lakes in the Sierra Nevada. The resulting equation ($r^2 = 0.98$, se = 8.0 µeq/L) was

---

* Data from Spuller Lake were omitted from the analysis. This lake and other similar shallow lakes with appreciable year-round groundwater inflows have unusually large snowmelt ANC depressions.

$$\text{minimum ANC} = (0.88 \pm 0.03) \,\{\text{overturn ANC}\} - (8.6 \pm 2.8) \quad (11.2)$$

11.2.4 Weathering and Cation Exchange

Williams et al. (1993) utilized hydrologic, mineralogic, and soils data to determine the sources and geochemical controls on the composition of surface waters in the Emerald Lake watershed. Preferential weathering of the anorthite component of plagioclase accounted for about 60% of the annual export of Ca^{2+} from the watershed in stream water. There was a large excess of Si after accounting for weathering processes of granitic bedrock, consistent with the results from other alpine basins underlain by crystalline bedrock (e.g., Mast et al., 1990). This underscores the lack of adequate understanding of the sources and sinks of Si in granitic basins.

Temporal variation in the geochemical controls on solute composition of surface waters were attributed by Williams et al. (1993) to a combination of changes in hydrologic routing, relative contributions of water from different hydrologic reservoirs, and residence time within these reservoirs. Stoichiometric weathering products of surface waters were not consistent over time and were divided into three distinct periods

- Snowpack runoff
- Summer transition period
- Low flow period from late summer through winter

About one-half of the snowpack runoff was estimated to become streamflow as Hortonian overland flow and the other one-half infiltrated soils and talus before becoming streamflow. Applied tracers illustrated that hydrologic residence time in subsurface reservoirs was on the order of hours to days during snowpack runoff. Stream water was not in equilibrium with weathering products during snowmelt.

Discharge from the soil reservoirs appeared to be the primary source of streamflow during the summer transition period. In contrast, processes below the soil zone appeared most important during the low-flow period (Williams et al., 1993).

Based on three independent analyses, Williams et al. (1993) suggested that cation exchange occurred during snowmelt

1. LiBr tracer experiment.
2. Increased molar ratio of Na : Ca in both soil water and stream water, consistent with the selectivity coefficients of soil cation exchangers.
3. Modeled vs. measured HCO_3^- suggested that during the period of snowpack runoff more cations were present than would have been produced by mineral weathering.

Thus, despite the short residence time of meltwater in the soil compartment during snowmelt, acid neutralization via cation exchange appears to be important. Improved understanding of cation exchange processes in both soil and talus compartments of these watersheds is needed. It will provide the foundation for further refinement of models to predict the response of these systems to future changes in acidic deposition.

11.3 Rocky Mountains

The Rocky Mountain states are sparsely populated compared to Eastern states. Wyoming is the least populated state in the nation, and by the year 2000 the population is expected to reach only 522,000 inhabitants. Montana has fewer than 1 million people and expects moderate increases in population over the next 30 years. There are few large urban areas in Wyoming and Montana, and none qualify as metropolitan areas (larger than 1 million people). Colorado is the most populated of the Rocky Mountain states and contains a metropolitan area (Denver) and several smaller urban areas, Boulder, Colorado Springs, and Fort Collins. The Front Range area of Colorado is currently experiencing a growth boom with urban and suburban development expanding at a rapid pace.

Most of the population growth in the Rocky Mountain region is occurring near urban centers. Whereas most of the Class I areas are remote from urban areas (Rocky Mountain National Park is an exception; Figure 11.3), regional transport of pollutants from urban areas to wildland areas may pose a threat to the air quality of the Class I areas.

Colorado and Utah have the highest total NO_x emission levels for the Rocky Mountain states, mainly from fossil fuel combustion by electric power utilities and on-road vehicles. Colorado and Wyoming have annual SO_2 emissions exceeding 100,000 tons per year. In these states, electric utilities are the major sources of SO_2, followed by industrial fuel combustion (including oil and gas refining) and mining operations (Peterson and Sullivan, 1998).

The Rocky Mountain region from Colorado to northern Montana encompasses a wide variety of landscapes and ecosystems. Geology, soils, aquatic systems, vegetation, and fauna are highly variable at both large and small spatial scales owing to the complex mountainous topography of this region. The Rocky Mountains are rugged glaciated mountains with many peaks up to 4,500 m in elevation. Mountainous topography is generally highly dissected with intervening valleys and plateaus. Geology is spectacularly varied with a great diversity of igneous, metamorphic, and sedimentary bedrock of various ages. Glacial till is found in many locations as a result of various glacial advances during the Pleistocene. The presence of glaciers in many high mountain valleys and cirques attests to the geomorphically dynamic landscapes of the Rockies.

FIGURE 11.3

National parks and major cities of the Rocky Mountain and northern Great Plains regions. (Source: Peterson and Sullivan, 1998.) Map produced by the National Park Service Air Resource Division.

Soils in the Rocky Mountains are diverse with respect to topography, parent material, vegetation, climate, and time of formation. Many different soil orders are found, with inceptisols, entisols, alfisols, spodosols, aridisols, and mollisols being most common. Because of the influence of gravity on steep slopes, colluvium is the most common surficial component of soils at most locations in the mountains. Alluvium is also common in river valleys. It is difficult to generalize about the nutrient status and biogeochemical cycling properties of soils in the Rockies. These factors, in conjunction with analyses of potential impacts of air pollutants, should generally be assessed on a watershed basis.

The climate of the southern and central Rockies is considered to be a semi-arid steppe regime in which there is considerable variation in precipitation with altitude. Total precipitation is moderate but greater than in the plains regions to the west and east. Foothill regions annually receive only 25 to 50 cm of precipitation, whereas higher elevations typically receive over 100 cm. In the higher mountains, a major portion of annual precipitation is snow. Climate is strongly affected by prevailing winds, resulting in generally wetter western slopes and drier eastern slopes. Average annual temperatures range from 2 to 7°C, with higher temperatures in lower valleys (Bailey, 1980). The climate of the northern Rockies is considerably colder. Annual precipitation ranges from 50 to 100 cm, with much of it falling as snow. Summers generally are dry because prevailing westerly winds during this season transport relatively dry air masses from the Pacific Northwest (Bailey, 1980).

Fairly extensive data are available regarding the acid-sensitivity of aquatic resources in four national parks within the region: Glacier, Yellowstone, Grand Teton, and Rocky Mountain. Of these, aquatic resources are most sensitive in Rocky Mountain and Grand Teton National Parks and both aquatic and terrestrial resources appear to be at greatest risk of future damage in Rocky Mountain National Park. Available data and recent findings for these parks are discussed next. Additional data are also available for the Wind River Mountains, WY, and Niwot Ridge, CO.

11.3.1 Glacier National Park

Glacier National Park (GLAC) encompasses 410,000 ha in northwestern Montana. GLAC includes rugged topography, active glaciers, clear streams and lakes, and spectacular scenic vistas. GLAC has several unique geological features. The layers of the Precambrian Belt Supergroup are extremely well-delineated in the portion of the park above treeline, and the layered sedimentary structures are unusually well-preserved on the dry eastern slopes. The Lewis Overthrust Fault is also highly visible in the park. Found in the higher elevations are 50 small alpine glaciers of relatively recent post-Pleistocene age. Current and former glacial activity in GLAC has resulted in many hanging valleys, cirques, and arretes, as well as an extensive hydrological system of lakes and streams.

The landscape of GLAC was created, in part, by an overthrust fault of ancient sedimentary substrates. Glaciers and streams have eroded the sedimentary strata in a dendritic pattern, radiating outwards from the central axis of high ridges and mountain peaks. Glacial moraines are prominent, especially in the northwestern part of the park (Martinka, 1992).

The park contains three distinct physiographic areas: the valleys of the North and Middle Forks of the Flathead River in the west, the central high mountains, and the plains in the eastern portion of the park. The central portion is dominated by two mountain ranges that run northwest to southeast and contain many small glaciers and snowfields.

Limited soils data are available. At the higher elevations, soils are sparse and found in pockets of variable depth and rock contents. At the lower elevations,

glacial ice mixed and deposited materials in a complex pattern. Volcanic erup-
tions of Mt. Mazama provided as much as 15 cm of volcanic ash to most local
soils. As a result of ash influence, many soils in the park have loamy textures.

The western slopes in the park are influenced by maritime air masses that
provide a moderate amount of precipitation. To the east, continental air
masses modify the maritime influence, and create more variable conditions,
including colder and drier winter months. Annual precipitation ranges from
about 59 cm on the periphery to 250 cm or more in the central highlands
(Martinka, 1992). Winds are generally from the west or southwest.

The average concentrations of SO_4^{2-} and NO_3^- in precipitation at the
NADP/NTN monitoring site in GLAC (at Apgar) from 1980 through 1995
were 9 and 6 µeq/L, respectively. The average annual precipitation amount
received during that time period was 77 cm. The average concentration of SO_4^{2-}
in precipitation has decreased since the early 1980s when it fluctuated between
9 and 20 µeq/L. Other parameters have remained relatively unchanged.

Wet deposition of both S and N is very low at this site. In recent years, wet
deposition of S has fluctuated between 0.8 and 1.0 kg S/ha per year. Total N
wet deposition has been between 0.9 and 1.4 kg N/ha per year (Peterson and
Sullivan, 1998).

Snowpack samples were collected at Apgar Lookout in GLAC on March 18,
1996, and April 3, 1997. Concentrations of SO_4^{2-} and NH_4^+ in the snow samples were consistent between the 2 years, 5.2 and 4.9 µeq/L, respectively for
SO_4^{2-}, and 1.9 and 2.3 µeq/L, respectively for NH_4^+. Nitrate concentrations
differed by a factor of two, however, and were 2.5 µeq/L in 1992 and 5.2
µeq/L in 1993 (Ingersoll, personal communication). Concentrations of all
three potentially acidifying ions were generally somewhat lower in the samples collected at GLAC than in comparable snow samples collected at Yellowstone and Grand Teton National Parks.

Atmospheric concentrations of S and N species have been measured at
GLAC since 1995 as part of the CASTNET dry deposition network. Dry deposition flux calculations are not yet available, but are expected to be available
in the near future. Dry deposition of both N and S is expected to be low, however, probably less than 50% of wet deposition. Thus, the total deposition of
both S and N is very low at GLAC, and shows no indication of increasing.

GLAC contains 131 named lakes, 635 unnamed lakes, approximately 2103
km of intermittent streams, and 2506 km of perennial streams. The park is
bordered on the west and south by the North and Middle Forks of the Flathead river, both of which are within the National Wild and Scenic River system. Aquatic resources within the park are outstanding.

Many of the lakes and streams within the park have characteristics generally associated with acid sensitivity, and these have been summarized by
Peterson and Sullivan (1998). They tend to be high in elevation, with little or
no soil development in their watersheds, have steep slopes and flashy
hydrology, and are hydrologically dominated by spring snowmelt. The
majority of these surface waters are, however, actually not at all sensitive to
acidification from acidic deposition owing to the preponderance of glaciers

within their watersheds and the occurrence of sedimentary bedrock (Peterson and Sullivan, 1998). Both of these features contribute buffering in the form of base cations to drainage waters in sufficient quantity to neutralize any amount of SO_4^{2-} and NO_3^- that might be reasonably expected to be deposited from acidic deposition. There are some waters in the park that receive only modest contributions of base cations, however. These do not receive glacial meltwater contributions to any significant extent and are situated in watersheds with relatively inert bedrock. These sensitive waters are relatively rare within the park (Peterson and Sullivan, 1998).

Ellis et al. (1992) monitored water quality of eight small backcountry lakes and five large frontcountry lakes in GLAC. The objective was to establish a water quality baseline for the park. Data were collected from 1984 to 1990. The backcountry lakes were located in remote alpine headwater areas across the various regions and geologies of the park. Of the lakes (Cobalt, Snyder, and Upper Dutch), three had alkalinity less than about 200 µeq/L. Cobalt Lake had the lowest alkalinity (approximately 100 µeq/L on average) and specific conductance (approximately 10 µS/cm) of the study lakes and would be expected to be sensitive to episodic effects of acidic deposition if S or N deposition to the park increased substantially. If acidic deposition in the park increased dramatically in the future, then perhaps Snyder Lake and/or Upper Dutch Lake would also be affected. The study lakes other than Cobalt, Snyder, and Upper Dutch would not be sensitive to acidification in response to any increase in deposition that would reasonably be expected to occur in the foreseeable future.

The EPA's Western Lakes Survey sampled five lakes in GLAC and ten other lakes in surrounding areas in the fall of 1986 (Landers et al., 1987). Measured values of selected important physical and chemical variables are listed in Table 11.1 for these 15 lakes and their watersheds. The lowest pH value measured in the park was 7.1, although 3 of the lakes in surrounding areas had pH between 6.5 and 7.0. One of the lakes having lowest pH (6.6) contained significant natural organic acidity (DOC equal to 10 mg/L); the others were low in pH as a consequence of their dilute chemistry. Sulfate concentrations in lake water were very low in lakes having low base cation concentrations. For example, the four lakes with total base cation concentrations less than 100 µeq/L all had SO_4^{2-} concentrations in the range of 5.7 to 10.1 µeq/L. Such concentrations of SO_4^{2-} are approximately what would be expected, based on SO_4^{2-} concentration in the precipitation (approximately 6 to 8 µeq/L), negligible dry deposition of S, 30 to 50% evapotranspiration, and minimal in-lake reduction. These lakes are highly to moderately acid-sensitive with ANC values of 21 to 84 µeq/L, although the 2 most sensitive were located outside the park boundaries. Many other lakes inside and outside the park had moderately elevated concentrations of SO_4^{2-}, in the range of 20 to 52 µeq/L. These relatively high concentrations of SO_4^{2-} are not attributable to atmospheric S deposition. They reflect geological sources of S in drainage waters, as also evidenced by the much higher concentrations (greater than 500 µeq/L) of base cations in all of the lakes that had SO_4^{2-} concentration greater than

TABLE 11.1

Results of Lake-water Chemistry Analyses by the Western Lake Survey for Selected Variables in GLAC and Adjacent Areas

Lake ID	Lake Area (ha)	Watershed Area (ha)	Elevation (m)	pH	ANC (μeq/L)	SO_4^{2-} (μeq/L)	NO_3^- (μeq/L)	Ca^{2+} (μeq/L)	C_B (μeq/L)	DOC (mg/L)
Lakes within GLAC										
4C3-004	2.8	88	1930	8.1	1210	28.6	11.6	929	1212	0.4
4C3-010	3.7	44	2298	7.3	79	5.7	4.9	49	81	0.3
4C3-011	162.0	5475	1126	8.0	543	32.1	4.5	375	613	0.5
4C3-013	4.4	62	2003	7.1	83	10.1	0.5	50	93	0.7
4C3-062	104.8	4302	1482	8.1	1142	48.7	3.0	774	1210	0.7
Lakes outside GLAC										
4C3-053	3.7	20	1979	6.6	21	8.0	0.1	29	57	1.5
4C3-016	5.4	88	1932	8.1	1388	20.1	0.0	1096	1393	1.0
4C3-017	6.2	173	1828	8.0	1209	32.3	0.7	832	1218	0.8
4C3-021	1.8	108	2104	7.9	492	27.4	1.5	768	1092	0.3
4C3-022	8.7	51	2050	7.5	360	10.0	2.1	295	386	0.8
4C3-026	167.9	2188	1229	8.3	1409	52.4	0.1	965	1435	4.5
4C3-055	2.3	7	1921	7.6	426	8.0	0.7	360	453	4.3
4C3-060	12.7	77	1228	6.6	152	3.0	0.0	81	213	10.0
4C3-031	6.4	54	2006	6.9	72	8.2	0.1	40	78	1.2
4C3-059	1.8	31	2226	7.4	292	10.5	4.3	170	319	2.7

20 μeq/L. Based on these data, it appears that GLAC and surrounding areas contain lakes that exhibit a mixture of acid sensitivities (Peterson and Sullivan, 1998). Some lakes that have low concentrations of SO_4^{2-} (less than about 10 μeq/L) that can be reasonably attributed to atmospheric deposition inputs also have low concentrations of base cations. These lakes tend to be relatively acid-sensitive, and many have ANC values below 100 μeq/L. The lowest measured ANC in the park during the WLS was 79 μeq/L. Other lakes are characterized by higher concentrations of base cations and SO_4^{2-} of geologic origin. These lakes are not acid sensitive and have ANC values greater than 500 μeq/L. Nitrate concentrations were generally below 5 μeq/L. One lake exhibited relatively high NO_3^- concentration (11.6 μeq/L) but this lake was not acid-sensitive.

11.3.2 Yellowstone National Park

Yellowstone National Park (YELL) in northern Wyoming and southern Montana contributes to two of the nation's farthest reaching drainages: the Missouri and Columbia Rivers. Surface water resources in the park include about 600 streams and 175 lakes. There are about 4400 km of free flowing rivers and streams. There are 4 large lakes (Yellowstone, Shoshone, Lewis, and Heart Lakes) that account for about 94% of the park's lake surface. Water quality varies throughout the park, mostly as a function of geologic terrain and the influence of thermal features. Natural geothermal discharges that are quite common in many portions of the park, affect the pH, ANC, temperature, salinity, SO_4^{2-} concentrations, and base cation concentrations of drainage waters. Snowmelt is an important contributor to hydrologic budgets of watersheds in the park, and water quality, therefore, tends to vary seasonally.

Precipitation volume and chemistry have been monitored at Tower Junction since 1980 by NADP/NTN. Annual precipitation amounts are generally in the range of 30 to 45 cm per year at this site. The concentrations of SO_4^{2-}, NO_3^-, and NH_4^+ in precipitation are low, with each generally below 10 µeq/L. The combined low amount of precipitation and low concentrations of acid-forming precursors result in very low levels of S and N deposition. Sulfur deposition is generally well below 1 kg S/ha per year and N deposition is seldom above this amount (Peterson and Sullivan, 1998).

Snowpack samples were collected in March or April of 1993 through 1999 at five sites in YELL: West Yellowstone, Lewis Lake Divide, Sylvan Lake, Canyon, and Twenty-One Mile. Sulfate concentrations in snow were low at all sites, with five-year (1993 to 1997) mean concentrations ranging from 4.7 µeq/L at Sylvan lake to 8.0 µeq/L at West Yellowstone. Mean concentrations of NO_3^- in the snowpack ranged from 4.2 µeq/L at Canyon to 5.0 µeq/L at Lewis Lake Divide. Ammonium concentrations were similar to NO_3^- concentrations, with mean values ranging from 4.2 µeq/L at Canyon to 7.0 µeq/L at West Yellowstone (Ingersoll, personal communication). Although higher than comparable measurements made in GLAC, these snowpack ionic concentrations were similar to those in Grand Teton National Park and are considered low.

There are no data currently available on dry or occult deposition fluxes of S or N to sensitive resources within YELL. However, it is expected that the contributions of both dry and occult deposition of S and N are low relative to the wet deposition. This is because there are no significant emission sources in close proximity to the park and because the amount of wet deposition are low. Atmospheric concentrations of S and N species have been measured in YELL since 1996 as part of CASTNET. Dry deposition flux calculations are not yet available, but are expected to be available in the near future.

Yellowstone Lake is noteworthy in a number of respects. The largest high-altitude lake in North America, it lies mostly within the Yellowstone Caldera that some of the highest measured geothermal heat fluxes in the world (Klump et al., 1988). Hydrothermal springs and hot gas fumeroles occur within the lake. Enhanced biological activity occurs around the geothermal vents that are characterized by high temperature, anoxia, and high concentrations of dissolved nutrients.

Native to the park are two subspecies of cutthroat trout (*Oncorhynchus clarki*), the Yellowstone cutthroat (*O. c. bouvieri*) and the west slope cutthroat (*O. c. lewisi*). The Yellowstone cutthroat was originally widely distributed in the intermountain region, although its range is now greatly reduced. In addition, nonnative fish, including lake trout (*Salvelinus namaycush*), have been introduced. Fishing management is now attempting to manage aquatic resources as functional components of the Greater Yellowstone Ecosystem and to preserve and restore native species and aquatic habitats.

The WLS (Landers et al., 1987) sampled six lakes in YELL and nine lakes in surrounding areas (Table 11.2). One of the lakes in the park was acidic (ANC = -24 μeq/L). This acidity was attributable to geothermal inputs, as evidenced by the extremely high concentrations of SO_4^{2-} (818 μeq/L) and sum of base cations (1330 μeq/L). One lake had an ANC of 139 μeq/L and its pH was also relatively low (6.6), mainly as a consequence of high concentration of DOC (11 mg/L). Several of the lakes in surrounding areas surveyed by the WLS were more acid-sensitive than those in the park; 4 out of 9 had ANC less than 100 μeq/L and one was below 50 μeq/L. Most of these had low SO_4^{2-} concentrations that could reasonably be attributed to atmospheric inputs (approximately 8 to 10 μeq/L) and low base cation concentrations. Thus, there are some lakes in and surrounding YELL that are acid-sensitive. Within the park, however, the vast majority of the aquatic resources are insensitive to any foreseeable increases in acidic deposition (Peterson and Sullivan, 1998).

11.3.3 Grand Teton National Park

Grand Teton National Park (GRTE) consists of 126,530 ha located in northwestern Wyoming. GRTE is surrounded by Bridger-Teton and Targhee

TABLE 11.2
Results of Lake-water Chemistry Analyses by the Western Lake Survey for Selected Variables in YELL and Adjacent Areas

Lake ID	Lake Area (ha)	Watershed Area (ha)	Elevation (m)	pH	ANC (μeq/L)	SO_4^{2-} (μeq/L)	NO_3^- (μeq/L)	Ca^{2+} (μeq/L)	C_B (μeq/L)	DOC (mg/L)
Lakes within YELL										
4D3-013	11.3	75	2006	9.4	1510	17	0.4	1092	1618	5.5
4D3-016	4.6	523	2287	5.7	1332	2909	3.5	599	6682	3.5
4D3-017	38.8	297	2514	4.8	-24	818	0.3	243	1330	6.2
4D3-019	20.8	168	2392	6.6	139	6	0.0	112	220	11.2
4D3-052	15.5	367	2198	8.3	705	30	0.3	311	980	4.8
4D3-073	3.4	119	2677	8.5	416	8	0.0	356	429	1.9
Lakes outside YELL										
4D2-050	1.5	64	2920	7.0	59	8	0.3	35	75	1.6
4D2-003	4.2	49	2793	7.5	161	7	0.7	66	184	1.3
4D3-001	76.2	7127	2037	8.6	630	33	0.4	436	737	1.3
4D3-002	3.2	80	2915	6.8	57	28	5.0	54	88	0.6
4D3-004	4.9	38	2935	7.1	79	9	0.5	45	101	1.5
4D3-006	3.0	75	2904	7.7	250	31	0.4	142	278	0.7
4D3-056	3.5	178	2935	6.9	45	11	0.3	32	66	1.8
4D3-028	2.1	31	2482	9.4	3795	109	1.6	1335	2484	16.7
4D3-024	20.7	481	3028	7.5	214	9	0.2	136	236	0.7

National Forests, and lies 10 km south of YELL. The Teton Mountains, a 67 km long range, stretch along a north–south line and reach a height of 4230 m. The Teton Mountain Range slopes steeply down to Jackson Hole, an intermountain valley about 75 km long and 10 to 20 km wide.

Late in the Tertiary age, faulting uplifted the mountains on the east side of Jackson Hole. One of the last events of diastrophism was the faulting and uplift of the Teton Mountains and the down dropping of the Jackson Hole valley floor. This large vertical displacement exposed granite, gneiss, and schist, from which Teewinot soils have developed. The sedimentary rocks have all been removed by erosion in the central portion of the range. Sedimentary rocks, and associated Starman and Tongue River soils, persist at the northern and southern ends of the range. They are mainly limestone, sandstone, and clay shale. The granite, gneiss, and schist of the high mountain areas of the park are expected to be fairly resistant to weathering and would not be expected to contribute significant amounts of base cations to drainage waters.

The dozen smaller glaciers now present in the Teton Range are relatively new, having been formed in the last few thousand years. Most surface waters within drainages fed by these glaciers likely receive sufficient contributions of base cations from glacial scouring to render them insensitive to the adverse effects of acidic deposition.

Average annual precipitation varies from about 41 cm at Jackson to about 154 cm near the summit of the Teton Mountains. Average annual snowfall varies from about 2 m at Jackson to over 7.7 m at high elevation. Snowmelt generally peaks in May and June. Surface winds display a wide range of prevailing directions and mean speeds depending on the topography and elevation of the site (Dirks and Martner, 1982). At the higher locations, the prevailing winds are consistently from the southwest.

There is no NADP deposition monitoring station in GRTE for S and N. However, there is a NADP monitoring station in YELL to the north. Both parks are exposed to the same general air masses, and both experience prevailing winds mostly from the southwest. There are no large point sources of N or S adjacent to either park that might cause major differences in local deposition. The author, therefore, relies on deposition data from YELL to evaluate deposition issues for GRTE. In addition, snowpack samples were collected in late March and April from two sites (Rendevous Mountain, Garnet Canyon) in GRTE in 1993 through 1998. Data are currently available for the period through 1997. Sulfate concentrations in snow were similar at the two sites and ranged between 3 and 9 µeq/L (mean 5 µeq/L at both sites) and NH_4^+ concentrations were somewhat lower (mean 3 and 4 µeq/L at Rendevous Mountain and Garnet Canyon, respectively). There were no apparent trends from year to year at either of the sites for any of the variables (G.P. Ingersoll, personal communnication).

There are about 90 alpine and subalpine lakes and ponds in the park. They are located above about 2700 m elevation. The majority are in remote areas that are difficult to access (Gulley and Parker, 1985). Most are less than 10 ha in area. Larger lakes are found at lower elevation. The multitude of small lakes and streams are distributed throughout the mountainous areas of the park, especially in the central and southern portions of the range.

The small size, shallow depth, and circular shape that are typical of the average lake located at high elevation in the Tetons, are indicative of glacially carved lake basins (Hutchinson, 1957). The mean maximum depth of alpine lakes in the park is 6.8 m, and the average mean depth is 3.6 m, based on bathymetric measurements of 46 alpine lakes (Gulley and Parker, 1985).

Alpine lakes in GRTE exhibit a range of characteristics that contribute to their sensitivity to potential acidic deposition impacts (e.g., Marcus et al., 1983; Peterson and Sullivan, 1998): bedrock resistant to weathering, shallow soil, steep slope, low watershed to lake surface area ratio, high lake flushing rate, high precipitation, high snow accumulation, and short growing season.

Surface water ANC values tend to be high throughout most of the low elevation areas of the park. Lakes and streams with ANC less than 400 μeq/L are generally restricted to the high mountain areas near the western border of the park (Peterson and Sullivan, 1998).

Miller and Bellini (1996) evaluated the trophic status of 17 lakes in GRTE. Phosphorus and chlorophyll *a* concentrations were measured in an effort to detect aspects of lake-water quality suggestive of eutrophication. A review of the literature did not yield earlier data on lake trophic status in the park, and the data collected in this study, therefore, will constitute the baseline for future evaluations of eutrophication in the park. Of the study lakes, six were located in the mountains. Samples were collected in July and August and analyzed for specific conductance, pH, and total phosphorus concentrations. Specific conductance was below 20 μS/cm in all of the mountain lakes, suggesting low concentrations of dissolved ions. Amphitheater and Surprise Lakes had very low specific conductance (less than 10 μS/cm) and had pH in the range 6.0 to 6.5.

Amphitheater and Surprise Lakes, the smallest of the lakes studied, are located in close proximity to each other, approximately at treeline. Neither is fed by a glacier. The absence of glacial meltwater contributions would be expected to predispose these lakes to acidic deposition effects. Key elements of acid–base chemistry were not measured by Miller and Bellini (1996), however, including ANC and the concentrations of base cations, SO_4^{2-} and NO_3^-.

Water quality was also measured by Miller and Bellini (1996) for seven moraine lakes, all of which had pH greater than 8.0 and specific conductance greater than about 17 μS/cm. The moraine lakes tend to be larger than the mountain lakes; the smallest was Bradley Lake at 28 ha and the largest, Jenny Lake at 486 ha. These lakes are insensitive to acidification from acidic deposition. Similarly, all of the four valley lakes sampled had high specific conductance (greater than 100 μS/cm) and pH greater than 8.5 and would not be sensitive to acidic deposition.

Based on the phosphorus and chlorophyll *a* measurements, mountain lakes were oligotrophic to mesotrophic, as were moraine lakes. Trophic status of the valley lakes was more variable, with some in the eutrophic range.

Sampled as part of the WLS were two lakes within GRTE and three lakes in proximity to the park. None were particularly sensitive to acidic deposition. The lowest measured ANC value was 153 μeq/L, in a lake with pH of 7.3 (Table 11.3). Gulley and Parker (1985) surveyed 70 lakes and ponds in GRTE during June, July, and August of 1982 and 1983. Sampled were 46 alpine lakes and ponds and 24 lower-elevation lakes. The majority of the lakes were relatively dilute, with specific conductance less than about 30 μS/cm. Of the lakes, 22 were highly dilute, with specific conductance less than or equal to 10 μS/cm. They were all located at high elevation; all were above 2500 m and most were above 2900 m elevation. All except one (montane) lake were located in alpine settings. Dilute lakes tended to be small; most were less than 2 ha in area. The largest was 6 ha. Watershed areas were also small in most cases (less than 100 ha). pH values were generally in the range of 7.0 to 8.0. Only 4 lakes had pH below 7.0 and 2 below 6.0. Calcium concentrations were below 30 μeq/L in 6 of the dilute lakes.

Williams and Tonnessen (in review) sampled 17 high-elevation headwater lakes throughout and adjacent to GRTE in August 1996. Although none were acidic, about one-half had ANC values near 50 μeq/L, and almost all had ANC below 200 μeq/L. About one-third had pH in the range of 5.8 to 6.0. These data do not suggest that chronic lake-water acidification has occurred but do suggest sensitivity, especially to episodic acidification. Of particular importance was the observed concentrations of NO_3^- that were relatively high in many of the lakes sampled. Of these lakes, 6 had NO_3^- concentrations in the range of 5 to 10 μeq/L and 3 had NO_3^- concentrations greater than 10 μeq/L (to a maximum of 13 μeq/L). Given that the lakes were sampled at the end of the growing season, these data suggest that the capacity for biological

TABLE 11.3

Results of Lake-water Chemistry Analyses by the Western Lake Survey for Selected Variables in GRTE and Adjacent Areas

Lake Name	Lake ID	Lake Area (ha)	Watershed Area (ha)	Elevation (m)	pH	ANC (μeq/L)	SO_4^{2-} (μeq/L)	NO_3^- (μeq/L)	Ca^{2+} (μeq/L)	C_B (μeq/L)	DOC (mg/L)
Lakes within GRTE											
Grassy Lake	4D3-021	117.9	883	2198	7.3	153	18.9	0.3	93	188	1.5
Trapper Lake	4D3-074	1.4	367	2107	7.6	435	38.7	2.1	298	491	1.1
Lakes outside GRTE											
Hidden Lake	4D3-022	7.2	129	2214	7.3	241	22.9	0.1	139	285	1.3
Loon Lake	4D3-060	9.9	160	1970	7.0	486	21.0	0.3	320	549	4.2
Upper Lake	4D3-069	47.0	3012	2022	8.8	2197	55.7	0.8	1481	2399	1.4

uptake of N has been reached at current levels of N inputs. It is likely that future increase in N deposition to these watersheds would contribute to higher lake-water NO_3^- concentrations and possible chronic acidification. Furthermore, the observed high NO_3^- concentrations suggest that additional water quality data should be collected during the snowmelt period.

Of the lakes sampled more recently in GRTE by Williams and Tonnessen (in review), those having the lowest ANC were Surprise and Amphitheater (ANC equal to 50 and 63 µeq/L, respectively). These were also the two lakes having the lowest specific conductance (5 and 6 µS/cm, respectively). Both had very low concentration of SO_4^{2-} (6 µeq/L) and no measurable NO_3^-. There were also, however, some relatively low-ANC (less than 100 µeq/L) lakes that had NO_3^- greater than 5 µeq/L (Delta, Holly, and LOTC). Amphitheatre and Surprise Lakes are currently the subjects of a modeling study to estimate the critical loads of S and N to sensitive aquatic resources in GRTE and three other western national parks (Sullivan et al., in preparation).

Available data suggest that there are a number of alpine lakes and ponds in the park that are sensitive to potential acidification from acidic deposition, as reflected in the low measured values of specific conductance and Ca^{2+} concentration, including Amphitheatre and Surprise Lakes. Measured pH values were generally circumneutral, with two lakes having pH less than 6.0. Chronic acidification probably has not occurred to any significant extent (as of the sampling dates), although sensitive lakes are located throughout most high elevation portions of the park, especially in the northcentral portion of the Grand Teton Mountain Range.

11.3.4 Rocky Mountain National Park

A great deal of work has been conducted during the past several years on the effects of acidic deposition on aquatic resources along the Front Range of Colorado. Much of this research has focused on sensitive receptors in Rocky Mountain National Park (ROMO) and nearby Niwot Ridge, located northwest of Denver. The Front Range receives modest levels of S and N deposition and contains some highly sensitive resources, a combination that makes this region an ideal candidate for monitoring and research within the West.

The discussion that follows focuses primarily on Rocky Mountain National Park, subject of part of a recent air quality review conducted for the National Park Service (Peterson and Sullivan, 1998). Conclusions for this park are also relevant for many of the surrounding areas of the Front Range, however.

The climate of ROMO is temperate montane. Snowpack accumulation is highly variable, with little accumulation on exposed ridges and south-facing slopes. Accumulation is heaviest during April and May. Depths of one to several meters are common in sheltered locations. Temperature ranges from about -5°C in winter to 25°C in summer at the lower elevations. Sub-freezing temperatures can be encountered at the higher elevations during any month.

   The hydrologic cycle of high-elevation watersheds in ROMO is character-ized by a lengthy period of snowpack accumulation during autumn, winter, and early spring, followed by a snowmelt period during late spring and early summer. The runoff period in late summer and early fall is predomi-nantly baseflow, with periods of stormflow from precipitation events (Campbell et al., 1995).

   The predominant direction of air mass movement over the Front Range is from west to east (Barry, 1973), with periodic upslope movement from the east (Kelley and Stedman, 1980). Wind rise data from Rocky Mountain National Park from 1989 through 1995 showed a distinct pattern of predom-inant air movement from the northwest, with relative frequencies of wind direction in excess of 2% of observations typically ranging from the west-southwest clockwise to the north-northeast. However, a second frequent wind direction was from the south and southeast, from the general direction of the Denver metropolitan area. This is important because air masses that move directly from the Denver area to Rocky Mountain National Park have the potential to transport high levels of both N and S compounds to the park. The easterly, upslope storm tract carries air masses across agricultural (live-stock and fertilized cropland) as well as industrial/metropolitan areas of Col-orado before reaching the vicinity of ROMO (Bowman 1992). Higher atmospheric concentration of NH_3 and NO_x gasses and HNO_3 particulates have been measured during upslope events (Parrish et al., 1986; Langford and Fehsenfeld, 1992).

   Atmospheric deposition of N and S in snow and rain along the Front Range in northern Colorado is among the highest of any area in the Rocky Mountains (Turk et al., 1992). Annual inorganic N loading in wet deposition in the Colorado Front Range is about twice that of the Pacific states and is similar to some states in the Northeast (Williams et al., 1996a). The volume-weighted mean annual concentrations of N in precipitation at Loch Vale, in the southeastern portion of the park, were 11 µeq/L NO_3^- and 7 µeq/L NH_4^+ from 1984 through 1995. The mean SO_4^{2-} concentration during that period was 12 µeq/L.

   Nitrate concentrations in the snowpack at maximum snowpack accumu-lation in the northern Colorado Front Range were among the highest mea-sured in Colorado (Turk et al., 1992). Concentrations of NO_3^- and SO_4^{2-} in the snowpack along the Continental Divide in northern Colorado were found to be twice the regional background level found throughout the Rocky Mountains in 1991 and 1992 (Turk et al., 1992). Synoptic snow survey data were collected at three sites within ROMO during March and April 1995. Concentrations of NO_3^- in snowpack were in the range of 10 to 13 µeq/L at all 3 sites. Snowpack concentrations of SO_4^{2-} and NH_4^+ were also slightly elevated above the expected background concentrations, in the range of 7 to 10 µeq/L for SO_4^{2-} and 4 to 6 µeq/L for NH_4^+ (Ingersoll, unpublished data). Such snowpack monitoring data are useful for examining spatial and temporal variation in deposition at areas other than where NADP sites are located. At many sensitive high-elevation sites, snowpack chemistry data are the only kind of deposition data available. Even though snowpack data do not provide deposition information for the entire year, they can help to identify hot spots for further research and to quantify better broad regional patterns in deposition in a cost-effective fashion.

ROMO has a high-elevation NADP/NTN site located in the Loch Vale watershed at an elevation of 3159 m and a lower elevation site at Beaver Meadows (2490 m). The Beaver Meadows site has been in operation since 1980 and the Loch Vale site since 1983. Both sites receive precipitation with elevated levels of S and N, compared with western parks in general.

Wet S deposition at Loch Vale decreased from 3.1 kg S/ha per year in 1985 to values around 2 kg S/ha per year from 1987 through 1995 (Peterson and Sullivan, 1998). The pattern of S loss in discharge (that can provide some information about inputs) was similar to the yearly water discharge pattern during the past decade, with most S losses occurring during snowmelt. Total S losses from the Loch Vale basin were considerably higher than wet inputs, and ranged from 3.3 to 4.2 kg S/ha per year (Baron et al., 1995). This information, coupled with discovery of small pyrite deposits within the basin, suggests a significant mineral source of S in the Loch Vale basin (Mast et al., 1990). Wet N deposition at Loch Vale from 1983 through 1995 has generally been in the range of 2 to 3 kg N/ha per year, with a maximum of 3.9 kg N/ha per year in 1994 (Peterson and Sullivan, 1998). There has been no trend in seasonal or annual inputs. Greater wet inputs of N occurred during years with higher precipitation, particularly those years with higher precipitation during the winter. Nitrogen deposition was statistically correlated with patterns of precipitation using a Pearson product-moment correlation ($p \leq 0.01$; Baron et al., 1995). Wet deposition at the lower elevation NADP/NTN site at Beaver Meadows is considerably lower than at Loch Vale for both S and N (Peterson and Sullivan, 1998).

Annual loading of inorganic N in wet deposition to the Colorado Front Range is about 3 kg N/ha per year at Loch Vale and near 5 kg N/ha per year at Niwot Ridge, which is quite high for the western U.S., although still relatively modest by comparison with many areas of the eastern U.S. and Europe. Annual NO_3^--N loading at Niwot Ridge, an alpine research area south of ROMO, has approximately doubled over the last decade, from 2 to 4 kg NO_3^- N/ha per year, based on NADP data. Niwot Ridge is located at 3500 m elevation about 60 km northwest of the Denver Metropolitan area and is exposed to the same general airshed as ROMO. An increase in precipitation amount during that period of time explained about one-half of the observed variation in annual wet NO_3^- deposition (Williams et al., 1996a). At the GLEES site to the north of ROMO in southeastern Wyoming, annual average NO_3^- wet deposition has also increased since measurements began in 1986 from about 1 to 2 kg NO_3^- N/ha per year.

Whereas dry deposition in the Rocky Mountains contributes less than 25% of total deposition of most chemical species (with the exception of Ca^{2+}) in winter, based on measurements from the maximum snowpack accumulation (Campbell et al., 1995), the contribution of dry deposition in summer is uncertain. A comparison of the chemical composition of lakes and wetfall suggested no significant dry deposition of SO_4^{2-} across the Rocky Mountain region in general (Turk and Spahr, 1991).

Bulk precipitation and throughfall chemistry were measured in an old growth Engelmann spruce/subalpine fir forest in the lower section of the Loch Vale watershed by Arthur and Fahey (1993). They calculated that dry deposition represented 56% of the atmospheric input of NO_3^- to the forest from May through October 1986 and 1987. Values for dry deposition as a fraction of total deposition for cations ranged from 57% for NH_4^+ to 66% for Mg^{2+}, and 83% for K^+. In contrast, no dry deposition of SO_4^{2-} was calculated.

Atmospheric concentrations of S and N species have been measured in ROMO since 1995 as part of CASTNET. Dry deposition flux calculations are not yet available but are expected to be available in the near future. The scientific consensus is that dry deposition of S at ROMO is not a major component of total deposition, and that the observed high concentrations of SO_4^{2-} in stream water at Loch Vale are owing largely to the occurrence of sulfide-bearing minerals in the watershed (Campbell et al., 1995; Baron, personal communication). Dry deposition to exposed bedrock surfaces appears to be important, however, at least during the snow-free season. Volume-weighted concentrations of NO_3^- and SO_4^{2-} in runoff from a bedrock catchment at Loch Vale were two to four times higher than in precipitation (Clow and Mast 1995). About 15% of the solute increase could be accounted for by evaporation from the rock surface. However, it is unclear to what extent runoff NO_3^- concentrations were increased by N fixation of lichens or the extent of S contribution from mineral deposits in the bedrock. Thus, the data of Clow and Mast (1995) cannot be used to quantify dry deposition fluxes to this watershed.

Cress et al. (1995) measured dry deposition of N to a snowpack in 1993 at Niwot Ridge. Changes in the concentration and mass of NO_3^- and NH_4^+ were measured in buckets filled with excavated snow and installed level with the snow surface. Exposure times ranged from 3 to 48 h. Ambient samples of HNO_3, particulate NO_3^- and particulate NH_4^+ were collected using filter packs with a Teflon prefilter (for particulates) and a nylon filter for HNO_3. Ambient concentrations were determined by dividing the measured mass of N by the volume of air sampled. A large increase was observed of all three N species after April 16. This increase in ambient N correlated in general with the seasonal change in late spring from westerly winds to upslope winds, and the arrival of convective air masses from the Denver urban corridor (Baron and Denning, 1993; Cress et al., 1995).

The concentration of N in snow samples increased by 100% between calendar days 146 and 155, an increase attributed by Cress et al. (1995) to increased dry deposition to the snowpack as a result of the extremely high ambient atmospheric concentrations of HNO_3 and particulate NO_3^- that were measured on day 153. The meteorological data showed a southeasterly flow directly from the Denver area on day 153. These data suggest that contamination of the atmosphere of alpine areas in the vicinity of Niwot Ridge (and presumably also ROMO) can result in the deposition of a significant mass of N to the snowpack in a very short period of time (Cress et al., 1995). Thus, it appears that deposition to ROMO can be strongly influenced by patterns of air movement within the region.

Total N deposition was estimated by Sievering et al. (1989) for two alpine sites at Niwot Ridge from 1987 to 1989. Comparisons were made of wet plus dry inputs for the 4-month growing season and the remaining 8-month dormant period. During the growing season, atmospheric N input was estimated to be greater than 1 mg N/m^2 per day directly from the atmosphere with a similar amount contributed from snowmelt as a result of deposition to the previous winter snowpack.

At a nearby subalpine forested site, wet plus dry deposition of N to a lodgepole pine canopy was found to vary from less than 1 mg N/m^2 per day to 2 mg N/m^2 per day (Sievering et al., 1989). These fluxes correspond to total annual N loading estimates of about 3 to 7 kg N/ha per year.

Concentrations of atmospheric particulate NO$_3^-$ (pNO$_3^-$) appear to have increased during the last decade at some sites along the Front Range. For example, Rusch and Sievering (1995) compared concentrations of pNO$_3^-$ measured during the month of July in several research efforts at the C-1 research site at Niwot Ridge (e.g., Fahey et al., 1986; Parrish et al., 1986; Marquez, 1994). The data suggested an approximate doubling of ambient pNO$_3^-$ in the last decade. Wet N deposition also doubled during that period of time at some sites.

Studies conducted by Langford and Fehsenfeld (1992) at Niwot Ridge and also 25 km to the east, near the eastern edge of the forest at Boulder, illustrated that the forest canopy will act as both a source and a sink for atmospheric NH$_3$. During periods of westerly flow (low in NH$_3$), the forest acted as a source of NH$_3$ with mean NH$_3$ emission rates of about 1.2 ng/m^2 per sec. Periods of easterly (upslope) flow induced by insulation of the mountain surfaces often occur between mid-morning and late afternoon during the summer. During these periods, the forest (especially the eastern edge) is exposed to NH$_3$-enriched air masses from the agricultural plains to the east. During upslope conditions, the forest became a net sink for NH$_3$, with a mean uptake rate of about 10 ng/m^2 per sec (20°C) near Boulder and decreasing from east to west as NH$_3$ was depleted from the air masses.

Ratios of NO$_3^-$ to SO$_4^{2-}$ in wetfall (0.8) and bulk precipitation (1.1) were high at Loch Vale compared to other mountainous sites in the region (Arthur and Fahey, 1993). This may be owing to the interception by Loch Vale and surrounding areas of southeasterly and easterly winds from the Denver area and agricultural areas east of the park that are enriched in nitrogen compounds.

As discussed previously, for a variety of reasons the N loading to alpine and subalpine systems in ROMO may be functionally much higher than is reflected by the total annual deposition measured or estimated for the watersheds. The end result is that the functional N loading to terrestrial and aquatic runoff receptors in alpine and subalpine areas during the summer growing season is much higher (perhaps more than double) than is represented by the annual average N loading for the site. This is especially true during the early phases of snowmelt, when a large percentage of the ionic load of the snowpack is released in meltwater.

Aquatic resources in Rocky Mountain National Park include a wealth of lakes and streams of exceptional quality. The natural lakes and stream valleys were formed by glaciation, evidence of which is abundant throughout the park. The majority of the surface waters are found in alpine and subalpine settings, most of which are accessible only on foot or horseback. Many high-elevation surface waters are fed by small glaciers. Because of the proximity of so many ROMO surface waters to the Continental Divide, human impacts on the water quality are minimized. With the exception of atmospheric contributions of pollutants, human impacts on most lakes and streams in the park, especially those in remote locations, are restricted to the impacts of hiking, camping, and horseback riding activities. Atmospheric deposition of air pollutants, therefore, represents one of the most important potential threats to aquatic resources in this park.

Lakes and streams in ROMO tend to be clear-water, low ionic strength, oligotrophic systems. Concentrations of virtually all dissolved constituents except oxygen (e.g., nutrients, organic material, major ions, weathering products) tend to be very low. The surface waters can be categorized as clear, cold, dilute systems. They are highly sensitive to potential degradation by human activities.

Owing to the high elevation of much of the park and barriers to fish migration, many of the lakes and streams in the park were historically fishless (Rosenlund and Stevens, 1988). Park-wide stocking of both native and non-native fish species occurred throughout this century until about 1965. By 1969, stocking of non-native fish species was abandoned. Lakes not capable of maintaining fish reproduction were allowed to revert to their fishless status, and management emphasis shifted to the restoring of native fish species. Of the 147 lakes within ROMO, about 59 had fish populations maintained by natural reproduction or stocking in 1969. By 1987, that number had dropped to 49 lakes, of which 13 were populated with pure strains of native fish in their native drainages (Rosenlund and Stevens, 1988). Several of the drainages thought to be highly sensitive, based on lake-water ANC and pH (Gibson et al., 1983), to potential adverse impacts of acidic deposition support viable fish populations.

Greenback cutthroat trout (*Salmo clarkii stomias*) constitute an important native fishery resource in the park and surrounding areas. The historical distribution of the greenback cutthroat included the Front Range of Colorado from about the Wyoming border south to the Arkansas River drainage in southern Colorado. The subspecies was known to occur only on the east side of the Continental Divide. The most critical factor in the decline of the greenback cutthroat trout has been the introduction of nonnative fish species. Greenback cutthroat have a tendency to hybridize with other species of spring-spawning salmonids and are unable to compete successfully with fall-spawning brook trout within subalpine and montane habitats. Habitat degradation owing to acidic deposition is also an important concern. Woodward et al. (1991) conducted laboratory bioassays of greenback cutthroat trout exposed to 7-day pH depressions to simulate episodic acidification. They concluded that the threshold for effects of H^+ ion concentration on greenback cutthroat trout in the absence of Al (which increases toxicity) was pH 5.0 (Woodward et al., 1991). The alevin life stage was most sensitive to mortality from low pH.

Analyses of lakes in ROMO and in adjacent wilderness areas from the Western Lake Survey (Eilers et al., 1988a) showed that the median ANC for these lakes was 80 µeq/L, with 20% of the lakes having ANC less than or equal to 41 µeq/L (Table 11.4). The minimum ANC value measured in this subpopulation was 19 µeq/L (Table 11.5). These ANC values are similar to ANC values for other sensitive areas of the West. Minimum pH and base cation concentrations were 6.48 and 47 µeq/L, respectively. Sulfate concentrations ranged from 10 to 113 µeq/L. This illustrates the importance of watershed sources of S to many lakes in the area, because concentrations would be more similar if atmospheric deposition was the primary source (e.g., Turk and Spahr, 1991). Nitrate concentrations ranged from 0 to 16 µeq/L in these Front Range lakes, with a population-weighted mean of 4 µeq/L. However, the lakes were sampled by WLS in the fall when NO_3^- concentrations are expected to be low relative to concentrations measured during other times of the year, especially spring snowmelt.

The acid–base chemistry of lake and stream waters in ROMO appears to be primarily a function of the interactions among several key parameters and associated processes: atmospheric deposition, bedrock geology, the depth and composition of surficial deposits and associated hydrologic flowpaths, and the occurrence of soils, tundra, and forest vegetation. High concentrations of base cations, ANC, and silica occur in the upper Colorado River

TABLE 11.4
Population Statistics (20th percentile, Q_1; median, M; and 80th percentile, Q_4) for ANC, C_B, SO_4^{2-}, and DOC, for Wilderness Lakes within Selected Geomorphic Units of the West Described by Eilers et al. (1988a)

Wilderness Subpopulation	Lakes Sampled	Estimated Population Size	ANC (µeq/L)			C_B (µeq/L)			SO_4^{2-} (µeq/L)			DOC (mg/L)		
			Q_1	M	Q_4	Q_1	M	Q_4	Q_1	M	Q_4	Q_1	M	Q_4
Sierra Nevada, CA	71	1787	29	53	104	42	67	115	4	7	14	0.4	0.7	1.5
Oregon Cascades	21	217	29	86	169	38	92	184	<1.5	<1.5	3	1.2	1.9	2.2
Northern Washington Cascades	52	537	47	106	193	67	134	239	8	17	42	0.3	0.6	1.2
Bitterroot Mountains, ID/MT	44	394	42	79	176	52	99	205	5	10	15	0.8	1.2	1.9
Wind River Range, WY	44	830	73	104	165	113	146	237	21	24	32	0.6	1.1	3.0
Front Range, CO	36	144	41	80	330	77	130	413	18	24	90	0.7	1.0	1.2

TABLE 11.5

Results of Lake-water Chemistry Analyses by the Western Lake Survey for Selected Variables in Rocky Mountain National Park and Adjacent Areas

Lake Name	Lake ID	Lake Area (ha)	Watershed Area (ha)	Elevation (m)	pH	ANC (µeq/L)	SO_4^{2-} (µeq/L)	NO_3^- (µeq/L)	Ca^{2+} (µeq/L)	C_B (µeq/L)	DOC (mg/L)
			Lakes within ROMO								
Spectacle Lake (NW)	4E1-012	4	62	3458	6.5	21	16	9.4	31	52	0.5
Chiquita Lake	4E1-013	1	36	3458	7.4	41	19	5.0	43	79	1.7
Arrowhead Lake	4E1-014	14	324	3397	6.5	24	18	7.8	30	58	0.8
Haynach Lakes (NW)	4E1-018	2	130	3373	7.1	103	17	0.5	80	137	1.2
No Name	4E1-019	2	36	3373	6.9	53	20	0.4	61	93	1.0
Lake Jaiyaha	4E1-022	5	290	3117	6.8	40	23	14.0	54	88	0.5
Black Lake	4E1-025	3	546	3239	6.6	31	19	13.5	48	73	0.7
Lake Powell	4E1-026	5	122	3519	6.5	19	20	15.5	42	68	0.5
Lake Nanita	4E1-027	14	275	3288	6.8	74	21	1.7	80	115	0.8
Lake Verna	4E1-028	13	1251	3105	6.8	88	24	2.9	94	142	1.4
Fifth Lake	4E1-029	2	150	3312	7.0	78	37	5.6	109	145	0.8
Box Lake	4E1-030	2	26	3275	7.0	56	18	4.1	62	98	1.0
Keplinger Lake	4E1-032	3	54	3564	6.7	33	17	11.4	41	69	0.5
Sandbeach Lake	4E1-033	5	65	3136	7.1	68	23	1.2	58	122	5.3
No Name	4E1-035	1	54	3312	7.8	80	33	0.3	94	130	0.7
Finch Lake	4E1-038	2	26	3023	7.4	171	36	1.1	102	258	4.4
Bluebird Lake	4E1-051	9	290	3348	7.1	80	26	3.4	85	132	1.5
Mirror Lake	4E1-053	7	300	3361	7.2	80	24	0.6	64	116	0.9
Fourth Lake	4E1-054	3	401	3166	6.9	77	33	3.0	92	135	1.2
Timber Lake	4E1-057	3	124	3373	6.8	52	47	1.2	52	119	1.6
Snowdrift Lake	4E1-058	3	44	3373	6.7	41	14	0.7	36	66	0.7
Mills Lake	4E1-060	6	1189	3031	6.7	42	23	9.6	57	94	1.4
Emmaline Lake	4E1-009	1	73	3361	7.9	118	24	6.3	90	155	1.2
Island Lake	4E1-036	7	158	3483	7.3	103	25	0.1	89	131	0.9
Coney Lake	4E1-039	3	500	3227	7.2	127	90	13.4	193	243	1.0

Continued

Aquatic Effects of Acidic Deposition

TABLE 11.5 (Continued)

Lakes in the Colorado Front Range Outside ROMO

Lake Name	Lake ID	Lake Area (ha)	Watershed Area (ha)	Elevation (m)	pH	ANC (μeq/L)	SO_4^{2-} (μeq/L)	NO_3^- (μeq/L)	Ca^{2+} (μeq/L)	C_B (μeq/L)	DOC (mg/L)
Blue Lake	4E1-040	9	259	3446	6.9	25	21	5.8	41	57	0.7
Crater Lake	4E1-041	10	275	3141	8.5	68	14	0.3	64	78	0.8
Green Lakes (NE)	4E1-043	4	47	3434	7.7	288	146	6.3	353	439	1.0
Caribou Lake	4E1-045	2	57	3400	7.5	307	34	4.7	218	390	0.9
No Name	4E1-046	4	49	3610	7.0	27	10	8.7	30	47	0.3
Upper Diamond Lake	4E1-047	2	49	3580	7.6	110	13	8.2	79	133	0.5
Jasper Lake	4E1-048	7	254	3298	7.6	148	33	1.4	133	193	1.3
King Lake	4E1-049	4	18	3486	6.9	56	21	1.4	52	88	0.6
Woodland Lake	4E1-050	2	171	3346	7.5	177	39	2.4	157	242	1.2
No Name	4E1-055	9	212	3422	6.8	33	19	7.6	43	65	0.6
Bob Lake	4E1-056	3	31	3532	7.3	60	29	7.7	64	97	0.7
Red Deer Lake	4E1-059	6	85	3163	7.4	79	30	0.7	80	118	1.8
Stapp Lakes (Large, North)	4E2-015	2	21	2861	7.2	111	38	2.2	83	195	11.3
Pumphouse Lake	4E2-021	2	28	3457	7.4	154	20	0.1	103	197	1.1
Crater Lakes (NW)	4E2-022	3	54	3361	7.4	129	17	5.8	100	168	0.4
Clayton Lake	4E2-023	2	166	3349	7.4	169	30	6.3	132	230	1.8
Murray Lake	4E2-024	4	142	3690	7.6	302	37	1.0	198	354	1.2
Duck Lake	4E2-025	18	570	3392	7.3	274	55	0.8	207	368	2.1
Chicago Lakes (Large, North)	4E2-026	9	342	3483	7.2	226	73	9.9	241	335	1.8
No Name	4E2-051	1	10	3510	9.3	692	31	0.3	553	808	9.9
Forest Lakes (Large, North)	4E2-055	2	98	3300	7.7	192	21	0.5	142	233	0.9
Panhandle Reservoir	4E3-005	20	4903	2592	7.8	377	30	0.0	240	420	3.2
Dowdy Lake	4E3-006	43	228	2481	8.0	952	67	1.5	696	1183	6.9
Parika Lake	4E3-009	1	70	3471	8.2	889	113	2.6	682	1050	1.2
No Name	4E3-016	2	21	2848	6.9	363	37	0.1	245	451	3.6
North Catamount Reservoir	4E3-035	85	1585	2848	7.6	418	139	0.4	414	633	1.5

basin, an area underlain by highly weatherable ash flow tuff and andesite. In contrast, the ANC and base cation concentrations are an order of magnitude lower in Glacier Creek, a watershed underlain by Silver Plume granite (Gibson et al., 1983). Using ANC as a criterion of acidification sensitivity, the Glacier Gorge basin and two subbasins (Ypsilon Creek and Roaring River) of the Fall River basin are the most sensitive areas of ROMO to potential acid deposition impacts (Gibson et al., 1983).

The NO_3^- concentrations in both precipitation and surface waters of ROMO are relatively high, likely the result, in large part, of upslope transport of NO_x from the Denver metropolitan area (Kelley and Stedman, 1980). Highest NO_3^- concentrations are above timberline, where biological activity and, therefore, NO_3^- uptake by terrestrial and aquatic biota is lowest (Gibson et al., 1983).

A great deal of research has been conducted on the interactions between atmospheric pollutants and water quality at the Loch Vale watershed. Biogeochemical and hydrological processes have been studied intensively at this site since 1983 (e.g., Baron, 1992; Denning et al., 1991; Campbell et al., 1995). A general description of the watershed is as follows. Loch Vale watershed is a 7 km^2 basin situated along the Continental Divide in the southeastern portion of ROMO. Of the surface area, 55% is exposed bedrock. Talus comprises 26% where large boulders are interspersed with tundra underlain by thin, minimally developed Entisol soils (Walthall, 1985). Alpine tundra covers 11% of the watershed, and the remainder is glaciers and lakes (2%), well-developed subalpine forest soils (5%), and alluvial and bog soils located in saturated areas and adjacent to streams (1%) (Walthall, 1985; Baron, 1992).

Loch Vale is located 80 km northwest of Denver, and ranges in elevation from 3000 to 4000 m. The 5-year average (1984 to 1988) precipitation amount is 113 cm per year (Baron, 1992). A spruce–fir forest and small subalpine meadows dominate the landscape at the lower elevations on Cryoboralf soils (Arthur and Fahey, 1992; Walthall, 1985).

The Loch is generally representative of moderately acid-sensitive lakes in the Front Range, but clearly is not representative of the most sensitive lakes. It has an ANC that is near the median value for wilderness lakes in the Front Range (Eilers et al., 1989b).

A number of factors predispose watersheds in ROMO such as Loch Vale to potential adverse effects of N deposition. These include

- Steep watershed gradient.
- Short hydrologic residence time of lake waters.
- Large input of N to lakes and streams during the early phases of snowmelt.
- High percentage of watershed covered by exposed bedrock and talus.
- Small percentage of watershed covered by forest.
- Phosphorus limitation of aquatic ecosystem primary production in some surface waters.

Thus, it is not surprising that the Loch Vale watershed leaches relatively high amounts of NO_3^- under only moderate levels of N deposition. In order to understand the response of this watershed (and other similar watersheds in the park) to atmospheric N deposition, it is important to consider a variety of hydrologic and biogeochemical processes that occur in different parts of the basin.

Campbell et al. (1995) studied the water chemistry of the two major tributaries to The Loch, Andrews Creek and Icy Brook. The catchments for the two streams are entirely alpine, consisting of rock outcrops, talus slopes, and some tundra. Only 5 to 15% of the catchments are covered by well-developed soil. Total storage of soil water was estimated to be less than 5% of the total outflow at The Loch (Baron and Denning, 1992). Volume-weighted mean annual concentrations of NO_3^- in the streams were 21 and 23 µeq/L. Total N export was approximately equivalent to atmospheric inputs, assuming about 25% evapotranspiration. Nitrate concentrations in individual samples ranged from 12 µeq/L in late summer to 39 µeq/L during snowmelt.

The Loch Vale watershed can, for all practical purposes, be considered N saturated (e.g., Aber et al., 1989; Stoddard, 1994). It is not clear to what extent the terrestrial and aquatic systems are receiving N inputs in excess of the assimilative capacities of biota, however. The apparent N saturation may be entirely hydrologically mediated. In other words, hydrologic flowpaths, extent of exposed bedrock and talus, and brief soil water residence times may limit the opportunity for biological uptake to the extent that the ecosystems may be N limited but still be unable to utilize atmospheric inputs of N (Campbell et al., 1995). Nevertheless, the implications of this apparent N saturation are important with respect to the estimation of critical loads of N deposition (Williams et al., 1996a). For example, critical loads for N deposition have been estimated to be 10 kg N/ha per year for northern Europe, based on empirical results that showed no N leaching to surface waters below this level (Dise and Wright, 1995), which is about 2 1/2 times the N deposition received by the Loch Vale watershed. Clearly, leaching of NO_3^- to surface waters occurs at much lower levels of N deposition at Loch Vale and other areas of the Front Range.

Emmett et al. (1998) found NO_3^- leaching at sites included in the European NITREX project (Wright and van Breemen, 1995) increased when the C : N ratio in the forest floor decreased below 24. This is well below the C : N ratio of Loch Vale soils (33; Baron, personal communication). Thus, it is likely that watershed soils at Loch Vale and similar watersheds continue to take up most incoming N under current N deposition and will continue to do so under moderately increased deposition. An explanation other than N saturation of alpine soils is needed to explain the high rates of NO_3^- leaching observed (c.f. Williams et al., 1996a,b). One possible explanation is the preponderance of exposed bedrock and talus in such watersheds.

Whereas the median lake-water NO_3^- concentration measured in the Western Lake Survey (Landers et al., 1987) was less than 1 µeq/L, the annual mean NO_3^- concentration at the outlet to The Loch ranges from less than 1 µeq/L to about 31 µeq/L during the peak snowmelt period (Baron, 1992). High NO_3^- concentrations have been observed in all lakes and streams in the Loch Vale watershed. Lake-water NO_3^- concentrations of nearby lakes are similar to those in The Loch. Baron (1992) reported median NO_3^- concentration from 1983 to 1988 for The Loch, Sky Pond, and Glass Lake of 16, 15, and 13 µeq/L, respectively. Surface water NO_3^- concentrations are also high in similar terrain outside the park (c.f., Toetz and Windell, 1993; Caine and Thurman, 1990).

During the last 10 years, the annual minimum concentrations of NO_3^- in surface waters during the growing season have increased from below detection limits to about 10 µeq/L in high-elevation catchments at Niwot Ridge and in the Glacier Lakes Ecosystem Experiment Site (GLEES) in southeastern Wyoming (Williams et al., 1996a,b). Wet NO_3^- deposition to adjacent NADP collectors has more or less doubled during that time period at both sites.

Williams et al. (1996b) sampled 53 ephemeral streams during snowmelt runoff in the Green Lakes Valley in 1994 and an additional 76 sites from the central Colorado Rocky Mountains to the Wyoming border in 1995. Nitrate concentrations in stream water during snowmelt ranged to 44 µeq/L in the Green Lakes Valley and during the growing season ranged to 23 µeq/L in the regional sampling conducted in 1995. Landscape type had a significant effect on NO_3^- concentrations ($p < 0.01$) in drainage waters throughout the Colorado Rocky Mountains. Tundra areas had significantly lower NO_3^- concentrations than talus and bedrock areas, suggesting that tundra ecosystems are still N limited, and that nitrification combined with limited plant uptake account for the high concentrations of NO_3^- observed in waters draining talus and bedrock areas (Williams et al., 1996b).

In response to an hypothesized overall pattern of global warming, it has been suggested that high-elevation environments may be expected to experience cooler temperatures and increased precipitation. Such a trend has been observed during the past 45 years at Niwot Ridge (Brooks et al., 1995a). Deeper snowpack accumulation and longer period of snowpack cover would be expected to result in warmer soil temperatures and higher rates of subnivian mineralization. This hypothesis was tested by Brooks et al. (1995a) who constructed a 2.6×60 m snow fence at the Niwot Ridge Long-Term Ecological Research site. The fence resulted in a snowpack that was significantly deeper than reference areas and also deeper than at the same area during the previous winter. The average period of continuous snow cover in the main snow drift increased by 115 days compared to reference sites. The deeper and earlier snowpack insulated soils from the extreme ambient air temperatures, resulting in a 9°C increase in minimum soil surface temperature and a 12°C increase in minimum soil temperature at 15 cm depth. Warmer soils contributed to greater microbial activity, measured as CO_2 flux through the snowpack, which continued through much of the winter. CO_2 production was 55% greater than production before construction of the snow fence (Brooks et al., 1995a). Such effects of snowpack are important with respect to N mineralization in alpine and subalpine environments. Soil heterotrophic respiration under seasonal snowpack has been shown to mineralize 20 to 50% of yearly above-ground primary production at alpine and subalpine sites in Wyoming (Sommerfeld et al., 1993). Brooks et al. (1995a) concluded that the timing of snowpack development is the most important factor controlling microbial activity in alpine soils during winter.

Inputs to the soil inorganic N pool at Niwot Ridge owing to mineralization and nitrification under deep snowpack (17 to 20 kg N/ha) were an order of magnitude higher than inputs directly from snowmelt (less than or equal to 1.5 kg N/ha; Brooks et al., 1995b). Nitrogen mineralization seemed to be a function of the severity of the freeze and the length of time the soils were insulated by snowpack. Mineralization was often higher under deeper, earlier accumulating snowpacks. Under shallower, late accumulating snowpacks, N mineralization was lower and also more variable (5 to 15 kg N/ha; Brooks et al., 1995b). The severity with which the soils freeze may also be an important determining factor of the amount of N mineralization. Brooks et al. (1995b) found the highest mineralization inputs under a shallow snowpack that experienced a severe freeze, followed by an extended period of snow cover. Such a result may be attributable to the release of labile C and N compounds from cell membranes that were ruptured by the freeze/thaw process followed by an extended period of mineralization under snowpack (Schimel et al., 1995).

In alpine and subalpine terrestrial ecosystems, the limiting factor for primary production is often N supply, which is largely determined by the ability of soil microbes to fix atmospheric N_2 and to mineralize organic N. Most terrestrial ecosystems are considered N limited (Friedland et al., 1991; Bowman et al., 1993). Inputs of anthropogenic atmospheric N to alpine plant communities also have the potential to alter plant community structure and increase sensitivity to water stress, frost, and herbivory (Bowman et al., 1993).

A substantial component of the NO_3^- in alpine and subalpine surface waters may have been derived from mineralization of organic N and not directly from atmospheric deposition. Much of the N released from the snowpack during the melting period is retained in underlying soils. Williams et al. (1996b) contended that measurements of subnivial microbial biomass, CO_2 flux through the snowpack and soil N pools all suggest that subnivial N cycling during the winter and spring is sufficient to supply the NO_3^- measured in stream waters.

It is clear that talus and exposed bedrock surfaces are not inert, although few studies have been conducted to examine changes in runoff chemistry attributable to chemical and biological reactions in these important watershed compartments. Baron and Campbell (1997) inferred that rock outcrops and talus in the Loch Vale watershed were biologically active and strongly influenced stream chemistry. Mass balance calculations suggested that about 10% of atmospheric N inputs may be immobilized or stored in the bedrock compartment of the watershed. Clayton et al. (1998) showed increased ANC of snowmelt and rain after 15 to 50 m transport over rock, lichens, and thin pockets of saprolite and soil in the Wind River Mountains, WY. Similar findings were reported by Eilers and Vaché (1998) in Goat Rocks Wilderness, WA.

Owing to topographic heterogeneity in elevation, slope, aspect, and vegetative cover, the onset and progression of snowmelt are not uniform throughout the Loch Vale watershed. South- and west-facing slopes begin melting earliest, and lower-elevation portions of the watershed melt before higher-elevation portions. Because some areas begin melting earlier than others, the pulse of solutes from snowmelt is smeared at the watershed outlet (Denning et al., 1991; Baron et al., 1995).

Stream discharge in the Loch Vale watershed typically increases by an order of magnitude during snowmelt. In contrast, concentrations of dissolved weathering products are generally only diluted by a factor of two or three. This suggests that perhaps enough snowmelt is routed through subsurface flowpaths or in contact with surface bedrock to react with soil/rock minerals to mitigate the dilution effects of direct snowmelt runoff to surface waters. Alternatively, snowmelt may displace relatively highly concentrated soil waters that have resided within the catchment throughout winter. These concentrated solutions then mix with meltwater to yield stream-water chemistry that is intermediate between the two. Thus, it is difficult to ascertain, on the basis of the concentration of dissolved weathering products alone, the relative importance of direct snowmelt runoff to the chemistry of stream water. Sueker (1995) measured Na^+ and Cl^- concentrations in three headwater catchments in ROMO (Spruce Creek, Fern Creek, Boulder Brook), all located near Loch Vale on the east side of the Continental Divide and underlain by granite and biotite schist. Using Na^+ as a tracer in a 2-component mixing equation, average subsurface water contribution to streamflow was estimated to range from about 33 to 80% throughout the melt period in Spruce Creek and Fern Creek. These creeks drain deeply incised valleys (1300 and 800 ha) with only about 20 to 38% of the valleys covered by talus and colluvium, mostly in shallow deposits. Subsurface sources were estimated to provide over one-half of the streamflow on the rising limb of the snowmelt-driven hydrograph until early June; subsequently, surface runoff water was estimated to dominate streamflow on the falling limb of the hydrograph.

Uncertainties in these estimates are undoubtedly quite high, owing to probable violation of some key assumptions of the model (e.g., subsurface concentrations constant over time and well characterized). Nevertheless, it is clear that subsurface contributions to streamflow are important in the early stages of snowmelt in hydrologically flashy alpine basins with minimal surficial deposits. Subsurface flows appear to dominate streamflow in less deeply incised basins that have considerable deposits of talus and colluvium. Pore spaces in deep surficial deposits provide a reservoir for the mixing of infiltrating snowmelt with resident subsurface water, with subsequent slow release of these mixed waters to stream channels (Sueker, 1995). As a consequence, Na^+ concentrations were generally two to three times higher in Boulder Brook (which has deep surficial deposits) than in either Spruce or Fern Creeks, which flow through deeply incised valleys. Unfortunately, the chemical composition of the major source waters (soils, talus fields, snow-pack) changes at the same time that their mixing ratio in streams changes, confounding use of end-member mixing models to describe the controls on ionic contributions to stream waters (Campbell et al., 1995).

Preferential elution of ionic species causes meltwater concentrations of NO_3^-, SO_4^{2-}, and NH_4^+ to be highest in early snowmelt and more dilute as the snowmelt progresses (Johannessen and Henriksen, 1978). Measured snow-melt water at Loch Vale followed this general pattern. Campbell et al. (1995) found that the peak snowmelt concentration of N was 78% greater than the average snowpack concentration. This suggested that maximum meltwater contributions to stream water would have a NO_3^- concentration of about 28 µeq/L. Measured stream-water concentrations of NO_3^- were higher, but not dramatically higher, during May and early June in Andrews Creek and to a lesser extent in Icy Brook (Figure 11.4). Nitrate concentrations increased in both streams with increased discharge in May, but the increase occurred first in Andrews Creek. Concentrations decreased steadily throughout the summer. In late August, NO_3^- concentrations increased rapidly to premelt concentrations after most of the snow had melted. This autumn increase may have been owing to a combination of an increased proportion of stream water as baseflow subsequent to the end of the snowmelt and also lessened biological demand for N in both the terrestrial and aquatic ecosystems as temperatures decreased during autumn.

The impacts of atmospheric deposition on high-elevation aquatic systems are strongly controlled by the flowpaths of water through the catchments. Hydrology is an important controlling factor for deposition impacts in virtually all environments (Turner et al., 1990), but hydrology is of overriding importance in alpine and subalpine ecosystems. The depth and make-up of soils, talus, and colluvium, and the slope of the watershed collectively determine the residence time of subsurface water within the watershed, extent to which snowmelt and rainfall runoff interact with soils and geologic materials, and, consequently, the extent of NO_3^- uptake by biota vs. NO_3^- leaching and acid neutralization within the watershed.

Chemical hydrograph separation techniques (e.g., Hooper and Shoemaker, 1986; Hinton et al., 1994) have been used to trace the movement of water through alpine and subalpine basins (Caine, 1989; Mast et al., 1995; Sueker, 1995). New water (snowmelt) often contributes more than one-half of the streamflow after seasonal peak flows have been achieved, but old water (stored from the previous year) typically dominates the hydrograph early in the snowmelt process.

Mast et al. (1995) evaluated the mechanisms that control streamflow at Loch Vale using the concentrations of [18]O and dissolved Si as chemical tracers. They concluded that streamflow is generated approximately as follows. Streamflow at the beginning of the melt period has a large component of "old water" that was displaced into the stream by the piston effect as meltwater infiltrated the soil and talus areas. After the pre-event soil waters have been flushed into streams, streamflow is mainly generated

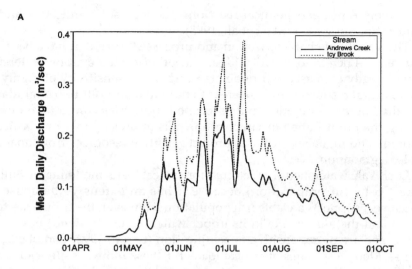

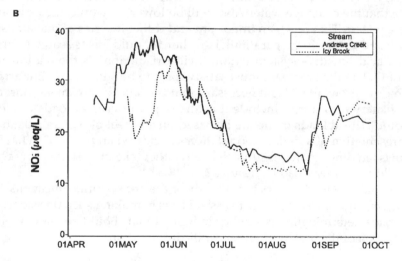

FIGURE 11.4

Daily discharge (A) and nitrate concentration (B) in Icy Brook and Andrews Creek within the Loch Vale watershed in April to September 1992. (Source: Campbell et al., 1995; IAHS Publication 0144-7815, No. 228, p. 249, Figure 3. Reprinted with permission from the International Association of Hydrological Sciences.)

by snowmelt discharge from subsurface reservoirs, with increasing amounts of surface flow during the main melting event in June. During the initial stages of snowmelt, therefore, stream chemistry reflects the months of weathering and decomposition products that accumulated in soils and talus areas under the snowpack and that were pushed into streams by the piston effect. As the melt continues, the contribution of pre-event water declines and the composition of stream water is increasingly controlled by relatively rapid geochemical reactions between soils and talus and the infiltrating snowmelt (Mast et al., 1995).

The variety of characterization and process studies that have been conducted at Loch Vale, Niwot Ridge, and other high-elevation watersheds have greatly increased our understanding of the sensitivity of highly valued natural resources in these areas to potential air pollution degradation. As the human population and associated industries continue to develop along the Front Range urban corridor, this knowledge will be critical as federal land managers strive to protect sensitive resources from unacceptable degradation.

Loch Vale watershed was selected for critical loads modeling by Sullivan and Cosby (in preparation) because it has an extensive database and because it supports a viable fish population. However, the tributaries to the Loch and the alpine lake in its upper watershed (Sky Pond) have lower ANC and are more sensitive to acidification (Baron, 1992; Campbell et al., 1995). Model estimates of critical loads for these more sensitive subwatersheds would be somewhat lower than would the model estimates of critical loads for The Loch. As you move further up in the watershed, it is often the case that the drainage water ANC will be lower, as would the estimates of critical load. This raises an important and difficult series of questions (Sullivan and Cosby, in preparation). How hard should FLMs search to find the most acid-sensitive system within their jurisdiction? Is there a lower size limit that a lake or stream must satisfy for it to be sufficiently important as to merit "protection"? Is it necessary that the lake or stream support a viable fishery in order to include it as the object of critical loads modeling? Should critical loads modeling be based on a subset of watersheds that are representative of a defined population of acid-sensitive lakes? What if one lake is ten times more sensitive than any other lake in the Class I area? How should the critical load be decided?

It should be readily apparent that these are not scientific questions. There is no scientific basis available to assist FLMs in making such decisions. Critical loads determination is a policy judgment that should be made with consideration of the best available scientific information.

12

Conclusions and Future Research Needs

Lakes and streams throughout many areas of the U.S. have experienced decreases in SO_4^{2-} concentration over the past two decades that can be attributed to decreased atmospheric emissions and deposition of S. However, the acidity levels of surface waters have not improved for the most part in response to this decline in S deposition. These documented responses to S emissions reductions would not be available without the existence of long-term monitoring networks for lakes and streams. Continued monitoring will be required to document any future recovery or acidification that might occur as S (and perhaps N) emissions and deposition continue to decline.

Significant advances have been made since 1990 in the scientific understanding of the aquatic effects of acid deposition. Models to predict the response of lakes and streams to changing levels of deposition have been extensively tested and improved. Current models predict that lakes and streams are less responsive to changes in S deposition than the models used as the basis for the NAPAP 1990 Integrated Assessment. The improved models predict that lakes in the Adirondack Mountains, for example, have been less acidified than previously thought and, therefore, will require larger reductions in S deposition to recover from existing acidity-related problems. Earlier NAPAP modeling efforts were deficient with respect to the treatment of N deposition. It is now known that N deposition can have important impacts on both aquatic and terrestrial ecosystems. Regional modeling of the effects of changing levels of N deposition in several potentially N-sensitive areas of the U.S. will be important to adequately evaluate the impacts of Title IV of the Clean Air Act Amendments of 1990 on both aquatic and terrestrial biota.

A great many conclusions can be drawn from the aquatic effects research conducted during the last decade. For example, important items discussed in this book and by NAPAP (1998) include the following

- Lakes and streams throughout many areas of the U.S. have experienced substantial decreases in SO_4^{2-} concentration during the last one to two decades in response to reduced levels of atmospheric S deposition. However, surface water ANC and pH have generally *not* recovered.

- Biological effects on fish during acidic episodes are largely attributable to increased concentrations of dissolved Al in stream water. This Al is transported from soils to streams by short-term increases in acidity that are often largely caused by N.

- Current understanding suggests that S is quantitatively less important, but N more important, as causes of both chronic and especially episodic acidification, than was believed in 1990.

- It is now known that naturally occurring organic acids are more important components of surface water acid–base chemistry and more important influences on the sensitivity of model projections of acidification than was widely believed in 1990.

- Sensitive watersheds in the northeastern U.S. have responded to atmospheric S inputs primarily by enhanced base cation (especially Ca^{2+} and Mg^{2+}) leaching. Although this means that drainage waters have been less acidified than was earlier thought, such a response has raised concerns about long-term depletion of base cation reserves from forest soils that could impact future forest productivity and the rate of watershed recovery in response to decreased levels of S deposition.

- Significant progress has been made in the scientific understanding of N cycling in forested ecosystems and the effects of N deposition on NO_3^- leaching and consequent acidification of surface waters. The ability to effectively model N dynamics on a regional basis is on the immediate horizon.

- Human land use activities, especially soils disturbance and forest management, have important effects on the responses of terrestrial and aquatic ecosystems to atmospheric inputs of S and especially N. Acid deposition is only one of many factors that affect acidity status.

- The principal model used by NAPAP in the 1990 Integrated Assessment to forecast aquatic effects from S deposition has been extensively tested and improved since 1990. It now predicts less acidification and recovery of surface waters in response to changing levels of S deposition than previously. Model forecasts for Adirondack lakes have been substantially revised.

- Earlier modeling efforts were deficient with respect to treating changes in N deposition. This is now known to have been an important oversight that should be rectified.

- The improved MAGIC model predicts that sensitive lakes and watersheds in the Adirondack Mountains are less responsive (in terms of change in ANC, pH, and inorganic aluminum) than was predicted by the version of MAGIC used for the 1990 IA. Although this implies that lakes have been less impacted than previously believed, it also implies that larger reductions in acidic deposition may be required in order for those lakes that have been affected to recover.

- Significant progress has been made regarding the establishment of critical loads of S and N for the protection of water quality in sensitive areas of the West, Northeast, and Southeast.

There has been an enormous increase during the past several years in the amount of research being conducted on the effects of atmospheric deposition of N on aquatic and, especially, terrestrial ecosystems. This research has been initiated mostly in Europe; little comparable work is ongoing in the U.S. The experimental approach has shifted heavily into the area of whole-ecosystem experimental manipulations that have been and are being conducted across gradients of atmospheric deposition and other environmental factors throughout northern Europe. Individual investigators are in many cases working at a variety of sites, thus enhancing the comparability of the resulting databases. Manipulations are focused primarily on coniferous forest ecosystems, and involve

1. Increasing ambient deposition of S and/or N.
2. Excluding ambient deposition via construction of roofs over entire forested plots or catchments.
3. Manipulating climatic factors, especially water availability.

The research is highly interdisciplinary, and not segregated into categories (e.g., aquatic, terrestrial, watershed, climate change) as is generally done in the U.S. Furthermore, experiments are designed to continue for long periods of time (i.e., 5 to 10 years). Study sites originally developed for S acidification research have been suitable for N research, thus providing rapid transitions into N ecosystem studies. Manipulation studies conducted thus far have clearly demonstrated that a long-term commitment is an essential component of whole-ecosystem research. With a few notable exceptions, research of this type and scope is generally lacking in the U.S.

The whole-ecosystem studies in Europe have been augmented by a number of detailed, process-level studies at the various manipulation sites. Key aspects include stable isotope (^{15}N) tracer studies to quantify the partitioning of N into various ecosystem pools (i.e., soil, litter, trees, ground vegetation) and to measure changes in the quantities of stored N in these pools. Other studies focus on quantifying the rates of important processes, including denitrification, mineralization, and leaching.

Results of both the broad-scale and detailed studies have been used to build, test, and validate mathematical models that simulate N processing, nutrient cycling, and water regulation in coniferous forest ecosystems under varying depositional and climatic regimes. Ultimately, these models will be used to predict N saturation, estimate the critical loads of N for European forests, and to specify emission controls needed to protect European forests from the detrimental effects of excess N deposition.

Elucidation of the major biogeochemical processes that alter the chemistry of precipitation water as it flows through watershed soils and into streams and lakes has been, and continues to be, a complex task. Hundreds of scientists in Europe, Canada, and the U.S. have focused their research efforts on this topic for the last two to three decades. The ultimate goal has been to develop an accurate, albeit generalized, understanding of these biogeochemical processes and through such understanding to develop watershed-scale predictive capabilities. Model predictions, in turn, constitute the scientific foundation for natural resource public policy that will hopefully provide for the protection or restoration of acid-sensitive aquatic resources throughout the country and elsewhere. It is unrealistic to assume that air pollution will be reduced to zero in an industrialized society. It is reasonable, however (and some would argue essential) that our society strive to reduce or limit air pollution to ecologically sustainable levels. It is the model forecasts that provide us with estimates, watershed by watershed, region by region, of how high those allowable levels should be. Because scientific understanding is and will always be incomplete, predictive models entail uncertainty. This uncertainty is reduced and confidence in the model output built through an iterative process of model testing and refinement. Good sources of databases for model testing include the results of long-term monitoring of surface water chemistry in regions that have undergone substantial changes in atmospheric inputs of S and N and also the results of plot-scale and watershed-scale experimental manipulation and paleolimnological assessments of historical acidification. Thus, the study of acidic deposition effects has involved multiple approaches, including process studies, model development and testing, monitoring, paleolimnology, and experimental manipulation (Figure 12.1). Each approach is critical in its own right, and all are required to provide the technical foundation for informed public policy decisions regarding air pollution controls and pollution mitigation.

This multipronged approach to the study of the aquatic effects of acidic deposition is an excellent prototype for addressing other environmental issues, including climate change. Because of the complexities of environmental systems, multiple approaches will always be required. The results of any one avenue of research will always be limited in its contribution to scientific understanding and informed decision-making.

In 1990, most of the major research programs in the U.S. on the effects of acidic deposition on aquatic and terrestrial ecosystems came to an end. The NAPAP Integrated Assessment (NAPAP 1991) represented the culmination of 10 years of intensive research at a cost of $500 million. Although the scientific research community had learned a great deal during those 10 years, major uncertainties remained. Our ability to accurately predict future ecosystem responses to changing levels of acidic deposition was rather limited and we lacked good information regarding the level of uncertainty inherent in our model projections. Although model calibration uncertainty was painstakingly quantified, we had little basis for evaluation of the uncertainty associated with fundamental biogeochemical process understanding and model structure.

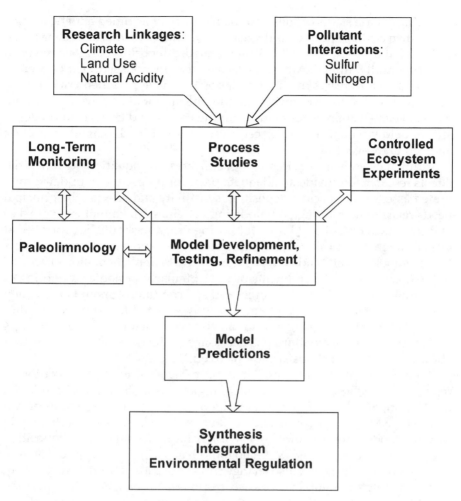

FIGURE 12.1
Major avenues of research employed in the study of the aquatic effects of acidic deposition.

Significant additional advancements have been made during the 1990s. Although research funding levels plummeted in 1990, the research that continued during the 1990s tended to be more focused and was able to capitalize on the earlier knowledge gains of the previous decade. Major advancements have been made in a number of research areas. These have included developing an understanding of the role of N in acidification processes and the interactions between acidic deposition and other features of the watershed and external forcing functions. The importance of organic acidity, Al dynamics, land use, and climate have been elucidated and reinforced. Significant gains have been made in the characterization of resource sensitivities in the mountainous portions of the West and the role of episodic acidification processes. We now recognize the importance of the depletion of base cation reserves in soils by acidic deposition. Base cation depletion can have significant impacts on the acid–base status of soils and the recovery of acidified aquatic ecosystems when acidic deposition is decreased. Quantitative information has been collected at a variety of locations where acidic deposition levels have been experimentally altered. Measured ecosystem responses to experimental increases and decreases in S and N deposition have provided critical databases with which to test, improve, and confirm our predictive models. We have developed approaches for quantifying critical and target loads of deposition, below which sensitive resources will not be damaged or will be allowed to recover from previous damage.

Additional work is required in several areas of aquatic and terrestrial effects research to provide an adequate basis for preparing an updated integrated assessment of acidic deposition sensitivity and effects. The principal needs relate to monitoring and modeling activities. The monitoring of lakes and streams in EPA's Long Term Monitoring Program should be continued at least through 2010 to allow quantification of the responses of surface waters to changing levels of S and N deposition owing to implementation of Title IV of the Clean Air Act Amendments of 1990. Similarly, periodic monitoring of terrestrial resources, including forest and soil conditions, should be initiated at periodic intervals for a suite of carefully selected forest research sites. Because of the often long lag time involved in ecosystem response to changes in deposition, future monitoring is necessary to elucidate responses attributable to emission reductions already realized.

Modeling needs fall into three general categories: re-implementation of regional S modeling, implementation of regional N modeling, and continued model testing and verification, especially of N models. Testing and improvement of the MAGIC model since 1990 have resulted in substantial changes in MAGIC model forecasts of recovery for Adirondack lakes as S deposition declines in response to Title IV. Analogous changes in model application procedures and structure will also change model forecasts for other regions but by an unknown amount. In each case, responses to changes in N deposition should be included. This effort will not require new data collection if it is assumed that the percent of incoming N that is retained by each watershed remains constant. However, in order to take advantage of the full range of N dynamics available in the current version of several models, additional field data need to be collected, for example, regarding the N and C pools in the forest floor, soils, and foliage of the model watersheds. Such efforts are currently underway in some areas.

Modeling for the 1990 Integrated Assessment did not consider the effects of changes in N deposition. It is now known that this was an important oversight and models to simulate the terrestrial and aquatic effects of N deposition are now available. One or more such models should be implemented in areas of the U.S. thought to be highly sensitive to potential N effects. Candidate models include the revised MAGIC model, NuCM, and pNET-CN. Candidate regions include the Adirondack Mountains, mid-Appalachian Mountains, southern Appalachian Mountains, portions of the West, and northern New England.

Testing of the N models should continue. Available data sets, with which to further test the models, include ecosystem manipulation data sets from Maine and Europe and long-term monitoring data sets from New York, New Hampshire, Norway, and Germany.

The development of the science of watershed-scale acidification provides an excellent example of how the scientific community can and should approach a complex environmental issue. By exploring a range of approaches, from simple long-term monitoring to the formulation, testing, and application of process-based models, we can develop an acceptable level of understanding regarding how natural systems respond to an environmental stressor. Key features in the development of the science include application of hydrogeochemical principles, experimental approaches, historical reconstructions, and empirical case study investigations. Only by the combination of approaches can we develop a level of understanding that is adequate as the scientific basis for public policy. There is much at stake in the public policy decision process. If we underprotect natural resources through such policy, there can be serious ecological consequences to some of our most prized natural resources. If we attempt to force overprotection, there can be serious economic consequences to entire industries, jobs can be lost, and the environmental research community can suffer irreparable loss of credibility. The task is complex and the stakes are high. Hopefully, we have done a good job so far.

Definitions

Acid anion—Negatively charged ion that does not react with hydrogen ion in the pH range of most natural waters.

Acid–base chemistry—The reaction of acids (proton donors) with bases (proton acceptors). In the context of this book, this means the reactions of natural and anthropogenic acids and bases, the result of which is described in terms of **pH** and **acid neutralizing capacity** of the system.

Acid cation—Hydrogen ion or metal ion that can hydrolyze water to produce hydrogen ions, for example, ionic forms of aluminum, manganese, and iron.

Acid mine drainage—Runoff with high concentrations of metals and sulfate and high levels of acidity resulting from the oxidation of sulfide minerals that have been exposed to air and water by mining activities.

Acid neutralizing capacity (ANC)—The equivalent capacity of a solution to neutralize strong acids. The components of ANC include weak bases (carbonate species, dissociated organic acids, aluminohydroxides, borates, and silicates) and strong bases (primarily, OH^-). In the National Surface Water Survey, as well as in most other recent studies of acid–base chemistry of surface waters, ANC was measured by the Gran titration procedure.

Acidic deposition—Transfer of acids and acidifying compounds from the atmosphere to terrestrial and aquatic environments via rain, snow, sleet, hail, cloud droplets, particles, and gas exchange.

Acidic episode—An **episode** in a water body in which **acidification** of surface water to an **acid neutralizing capacity** less than or equal to zero occurs.

Acidic lake or stream —A lake or stream in which the **acid neutralizing capacity** is less than or equal to zero.

Acidification—The decrease of **acid neutralizing capacity** in water or **base saturation** in soil caused by natural or anthropogenic processes. Paleolimnologists use this term to specify diatom-inferred decrease in pH, because ANC is not often reconstructed using diatoms.

Acidified—Pertaining to a natural water that has experienced a decrease in **acid neutralizing capacity** or a soil that has experienced a reduction in **base saturation**.

Acidophilic—Describing organisms that thrive in an acidic environment.

Alkalinity—The equivalent sum of $HCO_3^- + CO^{2-} + OH^-$ minus H^+, that is, buffering conferred by the bicarbonate system; the terms ANC and alkalinity are sometimes used interchangeably. ANC includes alkalinity plus additional buffering from dissociated organic acids and other compounds.

Analyte—A chemical species that is measured in a water sample.

Anion—A negatively charged ion.

Anion–cation balance—A method of assessing whether all ions have been accounted for and measured accurately; in an electrically neutral solution, such as water, the total charge of positive ions (**cations**) equals the total charge of negative ions (**anions**).

Anion deficit—The concentration of measured **cations** minus measured **anions**; usually the result of unmeasured organic anions or analytical uncertainty and often used as a surrogate for organic anion concentration.

Anion exchange/adsorption—A reversible process occurring in soil by which anions are adsorbed and released.

Anion reduction—The process by which NO_3^- is reduced to N_2O or N_2 (nitrate reduction or **denitrification**) and SO_4^{2-} is reduced to S^{2-} (**sulfate reduction**). Nitrate reduction and sulfate reduction can be mediated by plants during growth; the reduced N and S are assimilated into the growing plant. Denitrification and dissimilatory sulfate reduction are mediated by bacteria in anoxic zones in soils, sediment, or the water column. In the absence of oxygen, bacteria use the energy derived from these two reduction reactions in the decomposition of organic matter.

Anthropogenic—Of, relating to, derived from, or caused by humans or related to human activities or actions.

Assemblage—A group of taxa recorded in a sample (e.g., the group of taxa preserved in a sedimentary section).

Background sulfate—Estimated pre-industrial (pre-1850) concentration of sulfate in surface waters or precipitation.

Base cation—An alkali or alkaline earth metal cation (Ca^{2+}, Mg^{2+}, K^+, Na^+).

Base cation buffering—The capacity of a watershed soil or a sediment to supply base cations (Ca^{2+}, Mg^{2+}, K^+, Na^+) to receiving surface waters in exchange for acid cations (H^+, Al^{3+}); may occur through cation exchange in soils or weathering of soil or bedrock minerals.

Base cation supply—The rate at which base cations can be supplied to buffer incoming acid cations; this rate is determined by the relative rate of mineral weathering, the availability of base cations on exchange sites, and the rate of mobile anion leaching.

Base saturation—The proportion of total soil **cation exchange capacity** that is occupied by exchangeable **base cations**, that is, by Ca^{2+}, Mg^{2+}, K^+, and Na^+.

Bedrock—Solid rock exposed at the surface of the earth or overlain by saprolites or unconsolidated material.

Benthic—Referring to bottom zones or bottom-dwelling organisms in water bodies.

Bias—A systematic difference (error) between a measured (or predicted) value and its true value.

Bioassay—Measurement of the response of an organism or group of organisms upon exposure to *in situ* environmental conditions or simulated environmental conditions in the laboratory; also referred to as toxicity test.

Biological effects—Changes in biological (organismal, populational, community-level) structure and/or function in response to some causal agent; also referred to as biological response.

Biological significance—The quality of being important in maintaining the structure and/or function of biological populations or communities.

Calculated conductance —The sum of the products of individual ionic species and their known equivalent conductance values, measured under specified conditions. Calculated conductance often is compared with measured conductance as a quality assurance procedure.

Calibration—Process of checking, adjusting, or standardizing operating characteristics of instruments or coefficients in a mathematical model with empirical data of known quality. The process of evaluating the scale readings of an instrument with a known standard in terms of the physical quantity to be measured.

Carbonaceous particles—A collective term referring to carbonaceous spherules resulting from coal and oil combustion, as well as soot particles from wood burning. Carbonaceous particles are well preserved in lake sediments, where they can be used to infer past coal, oil, or wood-burning activities.

Catchment—See **watershed**.

Cation—A positively charged ion.

Cation exchange—The interchange between a cation in solution and another cation on the surface of any surface-active material such as clay or organic matter.

Cation exchange capacity—The sum total of exchangeable cations that a soil can adsorb.

Cation leaching—Movement of cations out of the soil, in conjunction with mobile anions in soil solution.

Cation retention—The physical, biological, and geochemical processes by which cations in watersheds are held, retained, or prevented from reaching receiving surface waters.

Chronic acidification—See **long-term acidification**.

Chrysophyte—Members of the classes Chrysophyceae and Synurophyceae that are covered by siliceous scales. Like the diatom valves, chrysophyte scales are taxonomically diagnostic and well-preserved in lake sediments. These algae have flagella and live in the open water (euplankton) of a lake.

Circumneutral—Close to neutrality with respect to **pH** (neutral pH is equal to 7); in natural waters, pH 6 to 8.

Close-interval sectioned sediment core—Refers to PIRLA-II cores whose recent sediments were sectioned at very close intervals (0.25 cm) so as to establish a fine temporal resolution.

Conceptual model—Simplified or symbolic representation of prototype or system behavior and responses.

Conductance—See **specific conductance**.

Confidence limits—A statistical expression, based on a specified probability, that estimates the upper and/or lower value (limit) or the interval expected to contain the true population mean.

Decomposition—The microbially mediated reaction that converts solid or dissolved organic matter into its constituents (also called decay or mineralization).

Denitrification—Biologically mediated conversion of nitrate to gaseous forms of nitrogen (N_2, NO, N_2O); denitrification occurs during decomposition of organic matter.

Diatom—Alga in the class Bacillariophyceae that are characterized by cell walls composed of two siliceous halves, known as valves (two valves equal a frustule). These siliceous valves are taxonomically diagnostic and well-preserved in lake sediments, so past diatom assemblages can be interpreted from their fossil remains.

Dissolved inorganic carbon—The sum of dissolved (measured after filtration) carbonic acid, bicarbonate, and carbonate in a water sample.

Dissolved organic carbon—Organic carbon that is dissolved or unfilterable in a water sample (0.45 μm pore size).

Drainage basin—See **watershed**.

Drainage lake—A lake that has a permanent surface water inlet and outlet.

Dry deposition—Transfer of substances from the atmosphere to terrestrial and aquatic environments via gravitational settling of large particles and turbulent transfer of trace gases and small particles.

Dynamic model—A mathematical model in which time is included as an independent variable.

Empirical model—Representation of a real system by a mathematical description based on experimental or observational data.

Episodes—A subset of hydrological phenomena known as events. Episodes, driven by rainfall or snowmelt, occur when **acidification** takes place during a **hydrologic event**. Changes in other chemical **parameters**, such as aluminum and calcium, are frequently associated with episodes.

Episodic acidification—The short-term decrease of **acid neutralizing capacity** from a lake or stream. This process has a time scale of hours to weeks and is usually associated with **hydrological events**.

Equivalence point—The point at which, during a titration, the concentration of proton donors equals the concentration of proton acceptors.

Equivalent—Unit of ionic concentration, a mole of charge; the quantity of a substance that either gains or loses one mole of protons or electrons.

Eutrophication—A process of accelerated aquatic primary production in response to nutrient enrichment that ultimately can result in oxygen depletion and changes in biological community structure and function.

Evapotranspiration—The process by which water is returned to the air through direct evaporation or transpiration by vegetation.

Forecast—To estimate the probability of some future event or condition as a result of rational study and analysis of available data.

Frame—A structural representation of a population providing a sampling capability.

Gran analysis—A mathematical procedure used to determine the equivalence points of a titration curve for **acid neutralizing capacity**.

Ground water—Water in a saturated zone within soil or rock.

Groundwater flow-through lake—A **seepage lake** that receives a substantial amount of groundwater input. Although there is no clear distinction between this type of lake and a **groundwater recharge lake**, groundwater flow-through lakes have been operationally defined as having silica concentrations greater than or equal to 1.0 mg L^{-1} or, in Florida, potassium concentrations greater than or equal to 15 μeq L^{-1}.

Groundwater recharge lake—A **seepage lake** that receives little or no groundwater input, but discharges water to the groundwater system. This type of lake is also known as a mounded or perched seepage lake. Operationally, groundwater recharge lakes have been defined as having silica concentrations less than 1.0 mg L^{-1} or potassium concentrations less than 15 μeq L^{-1}.

Hindcast—To estimate the probability of some past event or condition as a result of rational study and analysis of available data.

Hydraulic residence time—A measure of the average amount of time water is retained in a lake basin. It can be defined on the basis of inflow/lake volume, represented as "RT," or on the basis of outflow (outflow/lake volume) and represented as τ_W. The two definitions yield similar values for fast-flushing lakes, but diverge substantially for long-residence time seepage lakes.

Hydrologic(al) event—Pertaining to increased water flow or discharge resulting from rainfall or snowmelt.

Hydrologic(al) flow paths—Surface and subsurface routes by which water travels from where it is deposited by precipitation to where it drains from a **watershed**.

Hydrology—The science that treats the waters of the earth—their occurrence, circulation, and distribution; their chemical and physical properties; and their reaction with their environment, including their relationship to living things.

Index sample—As defined in the NSWS, a sample or group of samples taken from a certain place at each sampling unit (lake or stream reach) at a particular time of the year. For the Eastern and Western Lake Surveys, the index sample was a single sample collected from the center of each lake at a depth of 1.5 m during the fall turnover period. For the National Stream Survey, the index sample was the average of 2 or 3 samples collected during the spring baseflow period within a stream reach.

Inorganic aluminum—The sum of free aluminum ions (Al^{3+}) and dissolved aluminum bound to inorganic ligands; operationally defined by **labile monomeric aluminum**.

Labile monomeric aluminum—Operationally defined as aluminum that can be retained on a cation exchange column and measured by one of the two extraction procedures used to measure **monomeric aluminum**. Labile monomeric aluminum is assumed to represent inorganic monomeric aluminum (Al$_i$).

Liming—The addition of any base materials to neutralize surface water or sediment or to increase **acid neutralizing capacity**.

Littoral zone—The shallow, near-shore region of a body of water; often defined as the band from the shoreline to the outer edge of the occurrence of rooted vegetation.

Long-term acidification—The decrease of **acid neutralizing capacity** in a lake or stream over a period of hundreds to thousands of years, generally in response to gradual leaching of ionic constituents.

Macrophytes—Macroscopic forms of aquatic vegetation.

Macropore flow—Flow of water through large pores or voids in soil or rock (macropores), in response to the force of gravity.

Mineral acids—Inorganic acids, e.g., H_2SO_4, HNO_3, HCl, H_2CO_3. See **strong acids** and **weak acids**.

Mineralization—Process of converting organic nitrogen in the soil into ammonium which is then available for biological uptake.

Mineral weathering—Dissolution of rocks and minerals by chemical and physical processes.

Mitigation—Generally described as amelioration of adverse impacts caused by acidic deposition at the source (e.g., emissions reductions) or the receptor (e.g., lake liming).

Mobile anions—Anions that flow in solutions through watershed soils, wetlands, streams, or lakes without being adsorbed or retained through physical, biological, or geochemical processes.

Model—An abstraction or representation of a **prototype** or system, generally on a smaller scale.

Monomeric aluminum—Aluminum that occurs as a free ion (Al^{3+}), simple inorganic complexes (e.g., $Al(OH)_n^{3-n}$, AlF_n^{3-n}), or simple organic complexes, but not in polymeric forms; operationally, extractable aluminum measured by the pyrocatechol violet method or the methyl-isobutyl ketone method (also referred to as the oxine method) is assumed to represent **total monomeric aluminum**. Monomeric aluminum can be divided into labile and nonlabile components using cation exchange columns.

Monte Carlo method—Technique of stochastic sampling or selection of random numbers to generate synthetic data.

Natural acids—Acids produced within terrestrial or aquatic systems through natural, biological, and geochemical processes; that is, not a result of acidic deposition or deposition of acid precursors.

Nitrification—Oxidation of ammonium to nitrite or nitrate by microorganisms. A by-product of this reaction is H^+.

Nitrogen fixation—Biological conversion of elemental nitrogen (N_2) to organic N.

Nitrogen saturation—Condition whereby nitrogen inputs to an alpine or forested ecosystem exceed plant uptake requirements.

Nonlabile monomeric aluminum—Operationally defined as aluminum that passes through a cation exchange column and is measured by one of the two extraction procedures used to measure **monomeric aluminum**; assumed to represent organic monomeric aluminum (Al_o).

Nutrient cycling—The movement or transfer of chemicals required for biological maintenance or growth among components of the ecosystem by physical, chemical, or biological processes.

Organic acids—Heterogeneous group of acids generally possessing a carboxyl (-COOH) group or phenolic (C-OH) group; includes fulvic and humic acids.

Organic aluminum—Aluminum bound to organic matter, operationally defined as that fraction of aluminum determined by colorimetry after sample is passed through a strong cation exchange column.

Paleolimnology—The branch of limnology that deals with describing and interpreting lake histories by studying the information contained in lake sedimentary profiles. This information includes morphological and biogeochemical fossils of past lake biota, geochemistry, and physical attributes of the sediments. These sediment profiles are usually dated using radioisotopes suited for the time-scale of interest (for example, ^{210}Pb).

Parameter—A characteristic factor that remains at a constant value during the analysis, or a quantity that describes a statistical population attribute.

Pelagic zone—Referring to open-water areas not directly influenced by the shore or bottom.

Perched seepage lakes—See **groundwater recharge lake**.

Periphyton—Plants that live attached to or closely associated with surfaces (e.g., on the bottom sediments or macrophytes).

pH—The negative logarithm of the hydrogen ion activity. The **pH** scale is generally presented from 1 (most acidic) to 14 (most alkaline); a difference of one **pH** unit indicates a ten-fold change in hydrogen ion activity.

Physiography—The study of the genesis and evolution of land forms; a description of the elevation, slope, and aspect of a study area.

Piston effect—Hydrological process whereby snowmelt or storm discharge forces water that had been previously stored within the watershed out of soils and into streams.

Plankton—Plant or animal species that spend part or all of their lives in open water.

Pool—In ecological systems, the supply of an element or compound, such as exchangeable or weatherable cations or adsorbed sulfate, in a defined component of the ecosystem.

Population—For the purpose of this book, the total number of lakes or streams within a given geographical region or the total number of lakes or streams with a given set of defined chemical, physical, or biological characteristics; or an assemblage of organisms of the same species inhabiting a given ecosystem.

Precision—A measure of the capacity of a method to provide reproducible measurements of a particular **analyte** (often represented by variance).

Probability sample—A sample in which each unit has a known probability of being selected.

Project—To estimate future possibilities based on rational study and current conditions or trends.

Quality assurance—A system of activities for which the purpose is to provide assurance that a product (e.g., database) meets a defined standard of quality with a stated level of confidence.

Quality control—Steps taken during sample collection and analysis to ensure that data quality meets the minimum standards established in a **quality assurance** plan.

Reduction/oxidation (redox) reactions—Reactions in which substances gain or lose electrons, that is, in which substances are converted from an oxidized to a reduced oxidation state and vice versa.

Regionalization—Describing or estimating a characteristic of interest on a regional basis.

Retention time—The estimated mean time (usually expressed in years) that water resides in a lake prior to leaving the system. See **hydraulic residence time**.

Salt effect—The process by which hydrogen ions are displaced from the soil exchange complex by base cations (from neutral salts). The result is a short-term increase in the acidity of associated water; also referred to as sea-salt effect.

Saturated flow—Flow of water through the voids in rock or soil at a pressure greater than atmospheric, that is, under a head of pressure.

Scenario—One possible deposition sequence following implementation of a control or **mitigation** strategy and the subsequent effects associated with this deposition sequence.

Secchi disk depth—A measure of the transparency of water.

Seepage lake—A lake with no permanent surface water inlets or outlets. Seepage lakes are sometimes divided into two categories: **groundwater recharge lakes** and **groundwater flow-through lakes**.

Short-term acidification—See **episode**.

Simulation—Description of a **prototype** or system response to different conditions or inputs using a **model** rather than actually observing the response to the conditions or inputs.

Simulation model—**Mathematical model** that is used with actual or synthetic input data, or both, to produce long-term time series or predictions.

Species richness—The number of species occurring in a given aquatic ecosystem, generally estimated by the number of species caught using a standard sampling regime.

Specific conductance—The conductivity between 2 plates with an area of 1 cm^2 across a distance of 1 cm at 25°C.

Steady state—The condition that occurs when the sources and sinks of a property (e.g., mass, volume, concentration) of a system are in balance (e.g., inputs equal outputs; production equals consumption).

Steady-state model—A model in which the variables under investigation are assumed to reach equilibrium and are independent of time.

Stratified design—A statistical design in which the population is divided into strata, and a sample selected from each stratum.

Stream order—A method of categorizing streams based on their position in the drainage network. First-order streams are permanent streams with no permanent tributaries. Higher-order streams are formed by the confluence of two or more streams of the next lower stream order.

Strong acid anion sum (SAA or C_A)—Refers to the equivalent sum of SO_4^{2-}, NO_3^-, Cl^-, and F^-. The term specifically excludes organic acid anions.

Strong acids—Acids with a high tendency to donate protons or to completely dissociate in natural waters, for example, H_2SO_4, HNO_3, HCl^-, and some **organic acids**. See **acid anions**.

Strong bases—Bases with a high tendency to accept protons or to completely dissociate in natural waters, for example, NaOH.

Subpopulation—Any defined subset of the target **population**.

Sulfate adsorption—The process by which sulfate is chemically exchanged (e.g., for OH^-) or adsorbed onto positively charged sites on the soil matrix; under some conditions this process is reversible, and the sulfate may be desorbed.

Sulfate reduction—The conversion of sulfate to sulfide during the decomposition of organic matter under anaerobic conditions (dissimilatory sulfate reduction) and the formation of organic compounds containing reduced sulfur compounds (assimilatory sulfate reduction).

Sulfate retention—The physical, biological, and geochemical processes by which sulfate in **watersheds** is held, retained, or prevented from reaching receiving surface waters.

Sum of base cations (SBC or C_B)—Refers to the equivalent sum of Ca^{2+}, Mg^{2+}, Na^+, and K^+. The term specifically excludes cationic Al^{n+} and Mn^{2+}.

Surficial geology—Characteristics of the earth's surface, especially consisting of unconsolidated residual, colluvial, alluvial, or glacial deposits lying on the bedrock.

Synoptic—Relating to or displaying conditions as they exist at a point in time over a broad area.

Target population—A subset of a **population** explicitly defined by a given set of exclusion criteria to which inferences are to be drawn from the sample attributes.

Total monomeric aluminum—Operationally defined simple unpolymerized form of aluminum present in inorganic or organic complexes.

Turnover—The interval of time in which the density stratification of a lake is disrupted by seasonal temperature variation, resulting in entire water mass becoming mixed.

Validation—Comparison of **model** results with a set of **prototype** data not used for **verification**. Comparison includes the following: using a data set very similar to the verification data to determine the validity of the model under conditions for which it was designed; using a data set quite different from the verification data to determine the validity of the model under conditions for which it was not designed but could possibly be used; and using post-construction prototype data to determine the validity of the **predictions** based on model results.

Valve—One-half of a siliceous cell wall of a diatom. Diatom valves are identified and counted in lake sediments in paleolimnological studies. Two valves together are referred to as a frustule.

Variable—A quantity that may assume any one of a set of values during analysis.

Verification—Check of the behavior of an adjusted model against a set of **prototype** conditions.

WACALIB—This is a FORTRAN program by J.M. Line and H.J.B. Birks that implements regression calibration based on weighted averaging.

Watershed—The geographic area from which surface water drains into a particular lake or point along a stream.

Water Year—Hydrologic year that runs from October 1 through September 30. For example, water year 1999 began on October 1, 1998.

Weak acids—Acids with a low proton-donating tendency that tend to dissociate only partially in natural waters, for example, H_2CO_3, H_4SiO_4, and most **organic acids**. See **acid anions**.

Weak bases—Bases with a low proton-accepting tendency that tend to dissociate only partially in natural waters, for example, HCO_3^-, $Al(OH)_4^-$.

Weighted Averaging—This is a statistical method that can be used for analyzing environmental gradients with biological response variables (e.g., diatoms and chrysophytes). The optimum of each species along a gradient (e.g., pH) is estimated as the average of all pH values for lakes in which the taxon occurs, weighted by the taxon's relative abundance (WA regression). Reconstructions calculate a predicted environmental value for an algal assemblage based on the optima of the species and their abundances (WA calibration). This method is an approximation of the more formal procedure of maximum likelihood regression and calibration. Weighted averaging is computationally much easier and often performs better than maximum likelihood (Birks et al., 1990; Kingston and Birks, 1990). The theory has been developed and elaborated mainly by ter Braak (1986, 1988).

Wet deposition—Transfer of substances from the atmosphere to terrestrial and aquatic environments via precipitation, for example, rain, snow, sleet, hail, and cloud droplets. Droplet deposition is sometimes referred to as occult deposition.

References

J. Aber, W. McDowell, K. Nadelhoffer, A. Magill, G. Berntson, M. Kamakea, S. Mc-
Nulty, W. Currie, L. Rustad, and I. Fernandez, 1998, Nitrogen saturation in
temperate forest ecosystems, *Bioscience*, 48, 921.

J.D. Aber and C.T. Driscoll, 1997, Effects of land use, climate variation and N deposition on N cycling and C storage in northern hardwood forests, *Glob. Biogeochem.
Cycl.*, 11, 639.

J.D. Aber and C.A. Federer, 1992, A generalized, lumped-parameter model of photosynthesis, evapotranspiration and net primary production in temperate and
boreal forest ecosystems, *Oecologia*, 92, 463.

J.D. Aber, A. Magill, S.G. McNulty, R.D. Boone, K.J. Nadelhoffer, M. Downs, and R.
Hallett, 1995a, Forest biogeochemistry and primary production altered by nitrogen saturation, *Water Air Soil Pollut.*, 85, 1665.

J.D. Aber, J.M. Melillo, K.J. Nadelhoffer, J. Pastor, and R.D. Boone, 1991, Factors
controlling nitrogen cycling and nitrogen saturation in northern temperate forest
ecosystems, *Ecol. Appl.*, 1(3), 303.

J.D. Aber, K.J. Nadelhoffer, P. Steudler, and J.M. Melillo, 1989, Nitrogen saturation in
northern forest ecosystems: excess nitrogen from fossil fuel combustion may
stress the biosphere, *Bioscience*, 39, 378.

J.D. Aber, S.V. Ollinger, and C.T. Driscoll, 1997, Modeling nitrogen saturation in forest
ecosystems in response to land use and atmospheric deposition, *Ecol. Model.*,
101, 61.

J.D. Aber, S.V. Ollinger, and C.A. Federer, 1995b, Predicting the effects of climate
change on water yield and forest production in the northeastern U.S., *Clim. Res.*,
5, 207.

J.D. Aber, P.B. Reich, and M.L. Goulden, 1996, Extrapolating leaf CO_2 exchange to
the canopy: a generalized model of forest photosynthesis validated by eddy
correlation, *Oecologia*, 106, 257.

R.R. Aiken, D.M. McKnight, R.L. Wershaw, and P. MacCarthy, Eds., 1985, *Humic
Substances in Soil, Sediment, and Water*, John Wiley & Sons, New York.

T.E.H. Allot, R. Harriman, and R.W. Battarbee, 1992, Reversibility of lake acidification
at the Round Loch of Glenhead, Galloway, Scotland, *Environ. Pollut.*, 77, 219.

B. Almer, W. Dickson, C. Ekström, and E. Hørnstrøm, 1978, Sulfur pollution and the
aquatic ecosystem, in *Sulfur in the Environment, Part II, Ecological Impacts*, J.O.
Nriagu, Ed., John Wiley & Sons, New York, 273.

B. Almer, W. Dickson, C. Ekström, E. Hørnstrøm, and U. Miller, 1974, Effects of
acidification on Swedish lakes, *Ambio*, 3, 30.

A.P. Altshuller and R.A. Linthurst, Eds., 1984, *The Acidic Deposition Phenomenon and
Its Effects: Critical Assessment Review Papers*, Vol. 2, EPA-600/8-83-016-BF, Environmental Protection Agency, Washington, DC.

D.S. Anderson, R.B. Davis, and F. Berge, 1986, Relationships between diatom assemblages in lake surface sediments and limnological characteristics in southern Norway, in *Diatoms and Lake Acidity*, J.P. Smol, R.W. Battarbee, R.B. Davis, and J. Meriläinen, Eds., Dr. W. Junk, Dordrecht, The Netherlands, 97.

M.P. Anderson and C.J. Bowser, 1986, The role of groundwater in delaying lake acidification, *Water Resour. Res.*, 22, 1101.

D.V. Arrington and R.C. Lindquist, 1987, Thickly mantled karst of the Interlachen, Florida area, in *Karst hydrogeology: Engineering and Environmental Applications*, B.F. Beck and W.L. Wilson, Eds., A.A. Balkema, Rotterdam, 31.

M.A. Arthur and T.J. Fahey, 1992, Biomass and nutrients in an Engelmann spruce–subalpine fir forest in north central Colorado: pools, annual production, and internal cycling, *Can. J. For. Res.*, 22, 315.

M.A. Arthur and T.J. Fahey, 1993, Throughfall chemistry in an Engelmann spruce–subalpine fir forest in north central Colorado, *Can. J. For. Res.*, 23, 738.

C.E. Asbury, F.A. Vertucci, M.D. Mattson, and G.E. Likens, 1989, Acidification of Adirondack lakes, *Environ. Sci. Technol.*, 23, 362.

W.R. Aucott, 1988. Areal variation in recharge to and discharge from the Floridan aquifer system in Florida, Water Resources Investigations Rep. 88-4057, U.S. Geological Survey.

R.G. Bailey, 1980, Description of the ecoregions of the United States, USDA Forest Service Misc. Publ. 1391, Washington, DC.

J.P. Baker, 1982, Effects on fish of metals associated with acidification, in *Acid Rain/Fisheries*, T.A. Haines and R.E. Johnson, Eds., Proc. Int. Symp. Acidic Precipitation and Fisheries Impacts in Northeastern North America, American Fishery Society, Bethesda, MD, 165.

J.P. Baker and C.L. Schofield, 1982, Aluminum toxicity to fish in acidic waters, *Water Air Soil Pollut.*, 18, 289.

J.P. Baker, D.P. Bernard, S.W. Christensen, and M.J. Sale, 1990c, Biological effects of changes in surface water acid–base chemistry, Report SOS/T 13, National Acid Precipitation Assessment Program, Washington, DC.

J.P. Baker, S.A. Gherini, S.W. Christensen, C.T. Driscoll, J. Gallagher, R.K. Munson, R.M. Newton, K.H. Reckhow, and C.L. Schofield, 1990b, Adirondack Lakes survey: an interpretive analysis of fish communities and water chemistry, 1984–1987, Adirondack Lakes Survey Corporation, Ray Brook, NY.

J.P. Baker, J. Van Sickle, C.J. Gagen, D.R. DeWalle, W.E. Sharpe, R.F. Carline, B.P. Baldigo, P.S. Murdoch, D.W. Bath, W.A. Kretser, H.A. Simonin, and P.J. Wigington, Jr., 1996, Episodic acidification of small streams in the northeastern United States: effects on fish populations, *Ecol. Appl.*, 6, 422.

L.A. Baker, 1984, Mineral and nutrient cycles and their effect on the proton balance of a softwater, acidic lake, Ph.D. dissertation, University of Florida, Gainesville, FL.

L.A. Baker, 1991, Ion enrichment analysis for the Regional Case Studies Project, in *Acidic Deposition and Aquatic Ecosystems: Regional Case Studies*, D.F. Charles, Ed., Springer-Verlag, New York.

L.A. Baker and P.L. Brezonik, 1988a, Dynamic model of in-lake alkalinity generation, *Water Resour. Res.*, 24, 65.

L.A. Baker, P.L. Brezonik, E.S. Edgerton, and R.W. Ogburn, III, 1985, Sediment acid neutralization in softwater lakes, *Water Air Soil Pollut.*, 25, 215.

L.A. Baker, P.L. Brezonik, and C.D. Pollman, 1986, Model of internal alkalinity generation in softwater lakes: sulfate component, *Water Air Soil Pollut.*, 31, 89.

L.A. Baker, A.T. Herlihy, P.R. Kaufmann, and J.M. Eilers, 1991a, Acidic lakes and streams in the United States: the role of acidic deposition, *Science*, 252, 1151.

L.A. Baker, P.R. Kaufmann, A.T. Herlihy, and J.M. Eilers, 1990a, Current status of surface water acid–base chemistry, State of the Science SOS/T 9, National Acid Precipitation Program.

L.A. Baker, C.D. Pollman, and J.M. Eilers, 1988b, Alkalinity regulation in softwater Florida lakes, *Water Resour. Res.*, 24, 1069.

Y. Bard, 1974, *Nonlinear Parameter Estimation*, Academic Press, New York.

M. Barinaga, 1990, Where have all the froggies gone?, *Science*, 247, 1033.

L.A. Barmuta, S.D. Cooper, S.K. Hamilton, K.W. Kratz, and J.M. Melack, 1990, Responses of zooplankton and zoobenthos to experimental acidification in a high elevation lake (Sierra Nevada, California, U.S.A.), *Freshwater Biol.*, 23, 571.

R.B. Barnes, 1975, The determination of specific forms of aluminum in natural water, *Chem. Geol.*, 15, 177.

J. Baron, personal communication, U.S. Geological Survey, Colorado State University, Ft. Collins, CO.

J. Baron, 1992, *Biogeochemistry of a Subalpine Ecosystem: Loch Vale Watershed*, Ecological Studies 90, Springer-Verlag, New York.

J. Baron and O.P. Bricker, 1987, Hydrologic and chemical flux in Loch Vale watershed, Rocky Mountain National Park, in *Chemical Quality of Water and the Hydrologic Cycle*, R.C. Averett and D.M. McKnight, Eds., Lewis Publishers, Chelsea, MI, 141.

J. Baron and A.S. Denning, 1992, Hydrologic budget estimates, in *Biogeochemistry of a Subalpine Ecosystem*, Ecological Studies 90, J. Baron, Ed., Springer-Verlag, New York, 108.

J. Baron and A.S. Denning, 1993, The influence of mountain meteorology on precipitation chemistry of low and high elevations of the Colorado Front Range, USA, *Atmos. Environ.*, 27A, 2337.

J. Baron, E.J. Allstott, and B.K. Newkirk, 1995, Analysis of long term sulfate and nitrate budgets in a Rocky Mountain basin, in *Biogeochemistry of Seasonally Snow Covered Catchments*, K.A. Tonnessen, M.W. Williams, and M. Tranter, Eds., Proc. Boulder Symposium, July 1995, IAHS Pub. No. 228, 255.

J. Baron, A.S. Denning, and P. McLaughlin, 1992, Deposition, in *Biogeochemistry of a Subalpine Ecosystem*, J. Baron, Ed., Springer-Verlag, New York.

J. Baron, S.A. Norton, D.R. Beeson, and R. Herrmann, 1986, Sediment diatom and metal stratigraphy from Rocky Mountain lakes with special reference to atmospheric deposition, *Can. J. Fish. Aquat. Sci.*, 43, 1350.

J.S. Baron and D.H. Campbell, 1997, Nitrogen fluxes in a high elevation Colorado Rocky Mountain basin, *Hydrol. Process.*, 11, 783.

J.S. Baron, D.S. Ojima, E.A. Holland, and W.J. Parton, 1994, Nitrogen consumption in high elevation Rocky Mountain tundra and forest and implications for aquatic systems, *Biogeochemistry*, 27, 61.

R.G. Barry, 1973, A climatological transect of the east slope of the Front Range, Colorado, *Arctic Alpine Res.*, 5, 89.

L.E. Battoe and E.F. Lowe, 1992, Acidification of Lake Annie, Highlands Co., FL, *Water Air Soil Pollut.*, 65, 69.

S.E. Bayley, D.W. Schindler, B.R. Parker, M.P. Stainton, and K.G. Beaty, 1992, Effect of forest fire and drought on acidity of a base-poor boreal forest stream: similarities between climate warming and acidic precipitation, *Biogeochemistry*, 17, 191.

R.J. Beamish and H.H. Harvey, 1972, Acidification of the La Cloche Mountain lakes, Ontario, and resulting fish mortalities, *J. Fish. Res. Bd. Can.*, 29, 1131.

C. Beier and L. Rasmussen, Eds., 1993, EXMAN-experimental manipulation of forest ecosystems in Europe, project period 1988–1991, Air Pollut. Res. Rep. 7, Commission of European Communities, Brussels.

C. Beier, P. Gundersen, K. Hansen, and L. Rasmussen, 1995, Experimental manipulation of water and nutrient input to a Norway spruce plantation in Klosterhede, Denmark. 2. Effects on tree growth, vitality and nutrition, *Plant Soil*, 168/169, 613.

C. Beier, P. Gundersen, and L. Rasmussen, 1998, European experience of manipulation of forest ecosystems by roof cover: possibilities and limitations, in *Experimental Reversal of Acid Rain Effects*, H. Hultberg and R. Skeffington, Eds., The Gårdsjön Roof Project, John Wiley & Sons, Chichester.

J. Bettleheim and A. Littler, 1979, Historical trends of sulphur oxide emissions in Europe since 1865, Report CEGB PL-GS/E/1/79 Central Electricity Generating Board, London.

H.J.B. Birks, F. Berge, J.F. Boyle, and B.F. Cumming, 1990b, A paleoecological test of the land-use hypothesis for recent lake acidification in south-west Norway using hill-top lakes, *Phil. Trans. R. Soc. Lond. B.*, 327, 369.

H.J.B. Birks, J.M. Line, S. Juggins, A.C. Stevenson, and C.J.F. ter Braak, 1990a, Diatoms and pH reconstruction, *Phil. Trans. R. Soc. Lond. B.*, 327, 263.

C. Blanchard, J. Michaels, A. Bradman, and J. Harte, 1987, Episodic acidification of a low-alkalinity pond in Colorado, ERG Publication 88-1, Energy and Resources Group, Univ. Calif., Berkeley.

B.T. Bormann, F.H. Bormann, W.B. Bowden, R.S. Pierce, S.P. Hamburg, D. Wang, M.C. Snyder, C.Y. Li, and R.C. Ingersoll, 1993, Rapid N_2 fixation in pines, alder, and locust: evidence from the sandbox ecosystem study, *Ecology*, 74, 583.

W. Bowman, T. Theodore, J. Schardt, and R. Conant, 1993, Constraints of nutrient availability on primary production in two alpine tundra communities, *Ecology*, 74, 2085.

W.D. Bowman, 1992, Inputs and storage of nitrogen in winter snowpack in an alpine ecosystem, *Arctic Alpine Res.*, 24, 211.

A.W. Boxman, D. van Dam, H.F.G. van Dijk, R.F. Hogervorst, and C.J. Koopmans, 1995, Ecosystem responses to reduced nitrogen and sulphur inputs into two coniferous forest stands in the Netherlands, *For. Ecol. Manage*, 71, 7.

A.W. Boxman, P.J.M. van der Ven, and J.G.M. Roelofs, 1998, Ecosystem recovery after a decrease in nitrogen input to a Scots pine stand at Ysselsteyn, the Netherlands, *For. Ecol. Manage.*, 101, 155.

A.W. Boxman, H.F.G. van Dijk, and J.G.M. Roelofs, 1994, Soil and vegetation responses to decreased atmospheric nitrogen and sulphur inputs into a Scots pine stand in the Netherlands, *For. Ecol. Manage.*, 68, 39.

D.F. Bradford, M.S. Gordon, D.F. Johnson, R.D. Andrews, and W.B. Jennings, 1994a, Acidic deposition as an unlikely cause for amphibian populations decline in the Sierra Nevada, California, *Biol. Conserv.*, 69, 155.

D.F. Bradford, D.M. Graber, and F. Tabatabai, 1994b, Population declines of the native frog, *Rana muscosa*, in Sequoia and Kings Canyon National Parks, California, *Southwestern Naturalist*, 39, 323.

D.F. Bradford, C. Swanson, and M.S. Gordon, 1992, Effects of low pH and aluminum on two species of declining amphibians in the Sierra Nevada, California, *J. Herpetol*, 26, 369.

D.F. Bradford, F. Tabatabai, and D.M. Graber, 1993, Isolation of remaining populations of the native frog, *Rana muscosa*, in Sequoia and Kings Canyon National Parks, California, *Conserv. Biol.*, 7(4), 882.

D.F. Brakke, A. Henriksen, and S.A. Norton, 1989, Estimated background concentrations of sulfate in dilute lakes, *Water Resour. Bull.*, 25, 247.

T.E. Brandrud, 1995, The effects of experimental nitrogen addition on the ectomycorrhizal fungus flora in an oligotrophic spruce forest at Gårdsjön, Sweden, *For. Ecol. Manage.*, 71, 111.

M. Bredemeier, K. Blanck, A. Dohrenbusch, N. Lamersdorf, A.C. Meyer, D. Murach, A. Parth, and Y.-J. Xu., 1998b; The Solling roof project—site characteristics, experiments and results, *For. Ecol. Manage.*, 101, 281.

M. Bredemeier, K. Blanck, N. Lamersdorf, and G.A. Wiedey, 1995, Response of soil water chemistry to experimental 'clean rain' in the NITREX roof experiment at Solling, Germany, *For. Ecol. Manage.*, 71, 31.

M. Bredemeir, K. Blanck, Y.J. Xu, A. Tietema, A.W. Boxman, B.A. Emmett, F. Moldan, P. Gundersen, P. Schleppi, and R.F. Wright, 1998a, Input–output budgets at the NITREX sites, *For. Ecol. Manage.*, 101, 57.

M. Brenner and M.W. Binford, 1988, Relationships between concentrations of sedimentary variables and trophic state in Florida lakes, *Can. J. Fish. Aquat. Sci.*, 45, 294.

M.T. Brett, 1989, Zooplankton communities and acidification processes (a review), *Water Air Soil Pollut.*, 44, 387.

P.L. Brezonik, L.A. Baker, J.R. Eaton, T.M. Frost, P. Garrison, T.K. Kratz, J.J. Magnuson, W.J. Rose, B.K. Shephard, W.A. Swenson, C.J. Watras, and K.E. Webster, 1986, Experimental acidification of Little Rock Lake, Wisconsin, *Water Air Soil Pollut.*, 31, 115.

P.L. Brezonik, L.A. Baker, and T.E. Perry, 1987, Mechanisms of alkalinity generation in acid-sensitive softwater lakes, in *Sources and Fates of Aquatic Pollutants*, R. Hites and S.J. Eisenreich Eds., Adv. Chem. Ser. 216, American Chemical Society, Washington, DC, 229.

P.L. Brezonik, J.G. Eaton, T.M. Frost, P.J. Garrison, T.K. Kratz, C.E. Mach, J.H. McCormick, J.A. Perry, W.A. Rose, C.J. Sampson, B.C.L. Shelley, W.A. Swenson, and K.E. Webster, 1993, Experimental acidification of Little Rock Lake, Wisconsin: chemical and biological changes over the pH range 6.1 to 4.7, *Can. J. Fish. Aquat. Sci.*, 50, 1101.

O.P. Bricker and K.C. Rice, 1989, Acidic deposition to streams; a geology-based method predicts their sensitivity, *Environ. Sci. Technol.*, 23, 379.

C.L. Bristor, 1951, The great storm of November, 1950, *Weatherwise*, February 1951, 10.

G.A. Brook, M.E. Polkoff, and E.O. Box, 1983, A world model of carbon dioxide, *Earth Surf. Process. Landforms*, 8, 79.

P.D. Brooks, M.W. Williams and S.K. Schmidt, 1995b, Snowpack controls on soil nitrogen dynamics in the Colorado alpine, in *Biogeochemistry of Seasonally Snow Covered Catchments*, K.A. Tonnessen, M.W. Williams, and M. Tranter, Eds., Proc. Boulder Symposium July 1995, IAHS Pub. No. 228, 283.

P.D. Brooks, M.W. Williams, and S.K. Schmidt, 1996, Microbial activity under alpine snowpacks, Niwot Ridge, Colorado, *Biogeochemistry*, 32, 93.

P.D. Brooks, M.W. Williams, D.A. Walker, and S.K. Schmidt, 1995a, The Niwot Ridge snow fence experiment: biogeochemical responses to changes in the seasonal snowpack, in *Biogeochemistry of Seasonally Snow Covered Catchments*, K.A. Tonnessen, M.W. Williams and M. Tranter, Eds., Proc. Boulder Symposium, July 1995, IAHS Pub. No. 228, 293.

A. Brown and L. Lund, 1994, Factors controlling throughfall characteristics at a high elevation Sierra Nevada site, California, *J. of Environ. Qual.*, 23, 844.

D.J.A. Brown, 1983, Effect of calcium and aluminum concentrations on the survival of brown trout (*Salmo trutta*) at low pH, *Bull. Environ. Contam. Toxicol.*, 30, 582.

D.J.A. Brown and K.J. Sadler, 1981, The chemistry and fishery status of acid lakes in Norway and their relationship to European sulfur emissions, *J. Appl. Ecol.*, 18, 433.

A. Bulger, personal communication, University of Virginia, Charlottesville, VA.

A.J. Bulger, C.A. Dolloff, B.J. Cosby, K.M. Eshleman, J.R. Webb, and J.N. Galloway, 1995, The Shenandoah National Park: Fish in Sensitive Habitats (SNP:FISH) Project: an integrated assessment of fish community responses to stream acidification, *Water Air Soil Pollut.*, 85, 309.

K.R. Bull, 1991, The critical loads/levels approach to gaseous pollutant emission control, *Environ. Pollut.*, 69, 105.

K.R. Bull, 1992, An introduction to critical loads, *Environ. Pollut.*, 77, 173.

J. Bunyak, 1993, Permit application guidance for new air pollution sources, Natural Resources Report 93-09, National Park Service Rep. B-79-2 Department of the Interior.

A.B. Bytnerowicz and M.E. Fenn, Nitrogen deposition in California forests: a review, *Environ. Pollut.*, 1996, 92, 127.

N. Caine, 1989, Hydrograph separation in a small alpine basin based on inorganic solute concentrations, *J. Hydrol.*, 1, 89.

N. Caine and E. Thurman, 1990, Temporal and spatial variations in the solute content of an alpine stream, *Geomorphology*, 4, 55.

D.H. Campbell, D.W. Clow, G.P. Ingersoll, M.A. Mast, N.E. Spahr, and J.T. Turk, 1995, Nitrogen deposition and release in alpine watersheds, Loch Vale, Colorado, USA, in *Biogeochemistry of Seasonally Snow Covered Catchments*, Proc. Boulder Symposium, K.A. Tonnessen, M.W. Williams, and M. Tranter, Eds., July 1995. IAHS Pub. No. 228, 243.

P.G.C. Campbell, R. Bougic, A. Tessier, and J.P. Villeneuve, 1984, Aluminum speciation in surface waters on the Canadian Pre-Cambrian Shield, *Verh. Int. Verein. Limnol.*, 22, 371.

C. Carey, 1993, Hypothesis concerning the causes of the disappearance of boreal toads from the mountains of Colorado, *Conserv. Biol.*, 7, 355.

V.W. Carlisle, R.E. Caldwell, F. Sodek, L.C. Hammond, F.G. Calhoun, M.A. Granger, and H.L. Breland, 1978, Characterization data for selected Florida soils, Soil Sci. Res. Rep. No. 78-1, University of Florida, Gainesville, FL.

V.W. Carlisle, C.T. Hallmark, F. Sodek, R.E. Caldwell, L.C. Hammond, and V.E. Berkheiser, 1981, Characterization data for selected Florida soils, Soil Sci. Res. Rep. No. 81-1, University of Florida, Gainesville, FL.

D.F. Charles, 1985, Relationships between surface sediment diatom assemblages and lakewater characteristics in Adirondack lakes, *Ecology*, 66, 994.

D.F. Charles, Ed., 1991, *Acidic Deposition and Aquatic Ecosystems: Regional Case Studies*, Springer-Verlag, New York.

D.F. Charles and S.A. Norton, 1986, Paleolimnological evidence for trends in atmospheric deposition of acids and metals, in *Acid deposition: Long-term Trends*, Committee on Monitoring and Assessment of Trends in Acid Deposition, National Academy Press, Washington, DC, 335.

D.F. Charles and J.P. Smol, 1988, New methods for using diatoms and chrysophytes to infer past pH of low-alkalinity lakes, *Limnol. Oceanogr.*, 33, 1451.

D.F. Charles and J.P. Smol, 1990, The PIRLA II project: regional assessment of lake acidification trends, *Verh. Int. Verein. Limnol.*, 24, 474.

D.F. Charles and D.R. Whitehead, 1986a, Paleoecological Investigation of Recent Lake Acidification (PIRLA); Methods and Project Description, Electric Power Research Institute, Palo Alto, CA.

D.F. Charles and D.R. Whitehead, 1986b, The PIRLA project: Paleoecological investigations of recent lake acidification, *Hydrobiologia*, 143, 13.

D.F. Charles, R.W. Battarbee, I. Renberg, H. van Dam, and J.P. Smol, 1989, Paleoecological analysis of lake acidification trends in North America and Europe using diatoms and chrysophytes, in *Acid Precipitation*, Springer-Verlag, New York, 207.

D.F. Charles, M.W. Binford, E.T. Furlong, R.A. Hites, M.J. Mitchell, S.A. Norton, F. Oldfield, M.J. Paterson, J.P. Smol, A.J. Uutala, J.R. White, D.R. Whitehead, and R.J. Wise, 1990, Paleoecological investigation of recent lake acidification in the Adirondack Mountains, N.Y., *J. Paleolimnol.*, 3, 195.

C.W. Chen, S.A. Gherini, R.J.M. Hudson, and J.D. Dean, 1983, *The Integrated Lake-Watershed Acidification Study*, Vol. 1, *Model Principles and Application Procedures*, EPRI EA-3221, Project 1109-5, Electric Power Research Institute, Palo Alto, CA.

C.W. Chen, S.A. Gherini, N.E. Peters, P.S. Murdoch, R.M. Newton, and R.A. Goldstein, 1984, Hydrologic analyses of acidic and alkaline lakes, *Water Resour. Res.*, 20, 1875.

H. Chen, 1996, Object Watershed Link Simulation (OWLS), Ph.D. dissertation, Oregon State University, Corvallis, OR.

H. Chen and R. Beschta, in press, Dynamic hydrologic simulation of the Bear Brook Watershed in Maine (BBWM), *Environ. Manage. Assess.*

J. Chorover, P.M. Vitousek, D. Everson, A. Esperenza, and D. Turner, 1994, Solution chemistry profiles of mixed-conifer forests before and after fire, *Biogeochemistry*, 26, 115.

N. Christophersen, H.M. Seip, and R.F. Wright, 1982, A model for streamwater chemistry at Berkenes, Norway, *Water Resour. Res.*, 18, 977.

M.R. Church, 1984, Predictive modeling of the effects of acidic deposition on surface waters, in *The Acidic Deposition Phenomenon and its Effects: Critical Assessment Review Papers*, Vol. II, *Effects Science*, EPA-600/8-83-016BF, EPA Office of Research and Development, 4-113.

M.R. Church, 1999, The Bear Brook Watershed Manipulation Project: watershed science in a policy perspective, *Environ. Monitor. Assess.*, 55, 1.

M.R. Church and R.S. Turner, 1986, Factors affecting the long-term response of surface waters to acidic deposition: state of the science, EPA/600/3-86/025, NTIS PB 86 178 188-AS, Environmental Protection Agency, Corvallis, OR.

M.R. Church and J. Van Sickle, 1999, Potential relative future effects of sulfur and nitrogen deposition on lake chemistry in the Adirondack Mountains, United States, *Water Resour. Res.*, 35, 2199.

M.R. Church, C.L. Schofield, J.N. Galloway, and B.J. Cosby, 1984, Method of measuring alkalinity, in *The Integrated Lake-watershed Acidification Study*, Vol. 3, J.N. Galloway, E.R. Altwicker, M.R. Church, B.J. Cosby, A.O. Davis, G. Hendry, A.H. Johannes, K.D. Norstrom, N.E. Peters, C.L. Schofield, and J. Tokos, Eds., Electric Power Research Institute, Palo Alto, CA, 7-1.

M.R. Church, P.W. Shaffer, K.W. Thornton, D.L. Cassell, C.I. Liff, M.G. Johnson, D.A. Lammers, J.J. Lee, G.R. Holdren, J.S. Kern, L.H. Liegel, S.M. Pierson, D.L. Stevens, B.P. Rochelle, and R.S. Turner, 1992, Direct/delayed response project: future effects of long-term sulfur deposition on stream chemistry in the Mid-Appalachian region of the eastern United States, Environmental Protection Agency, EPA/600/R-92/186, Washington, DC.

M.R. Church, K.W. Thorton, P.W. Shaffer, D.L. Stevens, B.P. Rochelle, R.G. Holdren, M.G. Johnson, J.J. Lee, R.S. Turner, D.L. Cassell, D.A. Lammers, W.G. Campbell, C.I. Liff, C.C. Brandt, L.H. Liegel, G.D. Bishop, D.C. Mortenson, and S.M. Pierson. 1989, Future effects of long-term sulfur deposition on surface water chemistry in the northeast and southern Blue Ridge Province (results of the Direct/Delayed Response Project), Environmental Protection Agency Environmental Research Laboratory, Corvallis, OR.

W.E. Clark, R.H. Musgrove, C.G. Menke, and J.H. Cagle, Jr., 1964a, Water resources of Alachua, Bradford, Clay, and Union counties, Florida Bureau of Geology, Information Circular, Tallahassee, FL.

W.E. Clark, R.H. Musgrove, C.G. Menke, and J.H. Cagle, Jr., 1964b, Hydrology of Brooklyn Lake near Keystone Heights, Florida, Florida Geological Survey Report of Investigations, No. 33, Tallahassee, FL.

J.L. Clayton, 1998, Alkalinity generation in snowmelt and rain runoff during short distance flow over rock, Research Paper RMRS-RP-12, USDA Rocky Mountain Research Station, Ogden, UT.

A. Clemensson-Lindell and H. Persson, 1995, The effects of nitrogen addition and removal on Norway spruce fine-root vitality and distribution in three catchment areas at Gårdsjön, *For. Ecol. Manage.*, 71, 123.

D.C. Clow, N. Swoboda-Colberg, J.I. Drever, and F.S. Sanders, 1988, Chemistry of snowmelt, soil water and stream water at the West Glacier Lake Watershed, Wyoming, Abstract, *EOS*, 69, 1201.

D.W. Clow and M.A. Mast, 1995, Composition of precipitation, bulk deposition, and runoff at a granitic bedrock catchment in the Loch Vale watershed, Colorado, USA, in *Biogeochemistry of Seasonally Snow Covered Catchments*, K.A. Tonnessen, M.W. Williams, and M. Tranter, Eds., Proc. Boulder Symposium, July 1995, IAHS Pub. No. 228, 235.

D.W. Clow and M.A. Mast, 1999, Long-term trends in stream water and precipitation chemistry at five headwater basins in the northeastern United States, *Water Resour. Res.*, 35, 541.

D.W. Cole, H. Van Miegroet, and N.W. Foster, 1992, Retention or loss of N in IFS sites and evaluation of relative importance of processes, in *Atmospheric Deposition and Forest Nutrient Cycling*, D.W. Johnson and S.E. Lindberg, Eds., Ecological Studies 91, Springer-Verlag, New York, 196.

R.B. Cook and H.I. Jager, 1991, Upper Midwest: the effects of acidic deposition on lakes, in *Acidic Deposition and Aquatic Ecosystems: Regional Case Studies*, D.F. Charles, Ed., Springer-Verlag, New York.

R.B. Cook, J.W. Elwood, R.S. Turner, M.A. Bogle, P.J. Mulholland, and A.V. Palumbo, 1994, Acid-base chemistry of high-elevation streams in the Great Smoky Mountains, *Water Air Soil Pollut.*, 72, 331.

R.B. Cook, R.G. Kreis, Jr., J.C. Kingston, K.E. Camburn, S.A. Norton, M.J. Mitchell, B. Fry, and L.C.K. Shane, 1990, An acidic lake in northern Michigan, *J. Paleo.*, 3, 13.

R.B. Cook, K.A. Rose, A.L. Brenkert, and P.F. Ryan, 1992, Systematic comparison of ILWAS, MAGIC, and ETD watershed acidification models. 3. Mass balance budgets for acid neutralizing capacity, *Environ. Pollut.*, 77, 235.

P.S. Corn and F.A. Vertucci, 1992, Descriptive risk assessment of the effects of acidic deposition on Rocky Mountain amphibians, *J. Herpetol.*, 26, 361.

P.S. Corn, W. Stolzenburg, and R.B. Bury, 1989, Acid precipitation studies in Colorado and Wyoming: interim report of surveys of mountain amphibians and water chemistry, U.S. Fish and Wildlife Service Biological Report 80 (40.26).

B.J. Cosby, personal communication, Univeristy of Viginia, Charlottesville, VA.

B.J. Cosby and T.J. Sullivan, 1999, Application of the MAGIC Model to selected catchments: Phase I. Southern Appalachian Mountains Initiative (SAMI), final report submitted to Southern Appalachian Mountains Initiative, University of Virginia, Charlottesville, VA.

B.J. Cosby, R.C. Ferrier, A. Jenkins, B.A. Emmett, R.F. Wright, and A. Tietema, 1997, Modelling the ecosystem effects of nitrogen deposition: Model of Ecosystem Retention and Loss of Inorganic Nitrogen (MERLIN), *Hydrol. Earth Sys. Sci.*, 1, 137.

B.J. Cosby, G.M. Hornberger, J.N. Galloway, and R.F. Wright, 1985c, Time scales of catchment acidification: a quantitative model for estimating freshwater acidification, *Environ. Sci. Technol.*, 19, 1144.

B.J. Cosby, G.M. Hornberger, P.F. Ryan, and D.M. Wolock, 1989, MAGIC/DDRP Final Report, Vol. 1, Model, calibration, results, uncertainty analysis, QA/QC, Internal Report, EPA Environmental Research Laboratory-Corvallis, Corvallis, OR.

B.J. Cosby, A. Jenkins, R.C. Ferrier, J.D. Miller, and T.A.B. Walker, 1990, Modelling stream acidification in afforested catchments: long-term reconstruction at two sites in central Scotland, *J. Hydrol.*, 120, 143.

B.J. Cosby, S.A. Norton, and J.S. Kahl, 1996, Using a paired-catchment manipulation experiment to evaluate a catchment-scale biogeochemical model, *Sci. Tot. Environ.*, 183, 49.

B.J. Cosby, P.F. Ryan, J.R. Webb, G.M. Hornberger, and J.N. Galloway, 1991, Mountains of West Virginia, in *Acidic Deposition and Aquatic Ecosystems. Regional Case Studies*, D.F. Charles, Ed., Springer-Verlag, New York, 297.

B.J. Cosby, R.F. Wright, and E. Gjessing, 1995, An acidification model (MAGIC) with organic acids evaluated using whole-catchment manipulations in Norway, *J. Hydrol.*, 170, 101.

B.J. Cosby, R.F. Wright, G.M. Hornberger, and J.N. Galloway, 1985a, Modelling the effects of acid deposition: assessment of a lumped parameter model of soil water and streamwater chemistry, *Water Resour. Res.*, 21, 51.

B.J. Cosby, R.F. Wright, G.M. Hornberger, and J.N. Galloway, 1985b, Modelling the effects of acid deposition: estimation of long-term water quality responses in a small forested catchment, *Water Resour. Res.*, 21, 1591.

E.B. Cowling and L.S. Dochinger, 1980, Effects of acidic precipitation on health and productivity of forests, USDA Forest Service Technical Rep., PSW-43, 165.

R.G. Cress, M.W. Williams, and J. Sievering, 1995, Dry depositional loading of nitrogen to an alpine snowpack, Niwot Ridge, Colorado, in *Biogeochemistry of Seasonally Snow Covered Catchments*, K.A. Tonnessen, M.W. Williams, and M. Tranter, Eds., Proc. Boulder Symposium, July 1995, IAHS Pub. No. 228, 33.

T.L. Crisman, R.L. Schulze, P.L. Brezonik, and S.A. Bloom, 1980, Acid precipitation: the biotic response in Florida lakes, in *Ecological Impact of Acid Precipitation*, D. Drabløs and A. Tollan, Eds., Proc. Int. Conf., Sandefjord, Norway, 296.

C.S. Cronan and G.R. Aiken, 1985, Chemistry and transport of soluble humic substances in forested watersheds of the Adirondack Park, New York, *Geochem. Cosmochim. Acta*, 49, 1697.

C.S. Cronan and D.F. Grigal, 1995, Use of calcium/aluminum ratios as indicators of stress in forest ecosystems, *J. Environ. Qual.*, 24, 209.

C.S. Cronan and C.L. Schofield, 1979, Aluminum leaching response to acid precipitation: effects on high-elevation watersheds in the northeast, *Science*, 204, 304.

C.S. Cronan, W.J. Walker, and P.R. Bloom, 1986, Predicting aqueous aluminum concentrations in natural waters, Nature, 324, 140.

B.F. Cumming, K.A. Davey, J.P. Smol, and H.J.B. Birks, 1994, When did acid-sensitive Adirondack lakes (New York, USA) begin to acidify and are they still acidifying?, Can. J. Fish. Aquat. Sci., 51, 1550.

B.F. Cumming, J.P. Smol, J.C. Kingston, D.F. Charles, H.J.B. Birks, K.E. Camburn, S.S. Dixit, A.J. Uutala, and A.R. Selle, 1992, How much acidification has occurred in Adirondack region (New York, USA) lakes since pre-industrial times?, Can. J. Fish. Aquat. Sci., 49, 128.

R.A. Dahlgren, 1994, Soil acidification and nitrogen saturation from weathering of ammonium-bearing rock, Nature, 368, 838.

M. David, G. Vance, and J. Kahl, in press, Chemistry of dissolved organic carbon at Bear Brook Watershed, Maine: stream water response to $(NH_4)_2SO_4$ additions, Environ. Manage. Assess.

M.B. David and G.F. Vance, 1991, Chemical character and origin of organic acids in streams and seepage lakes of central Maine, Biogeochemistry, 12, 17.

M.B. David, G.F. Vance, and J.S. Kahl, 1992, Chemistry of dissolved organic carbon and organic acids in two streams draining forested watersheds, Water Resour. Res., 28, 389.

M.B. David, G.F. Vance and J.M. Rissing, 1989, Organic carbon fractions in extracts of O and B horizon solutions from a New England spodosol: effects of acid treatment, J. Environ. Qual., 18, 212.

R.B. Davis, D.S. Anderson, and F. Berge, 1985, Loss of organic matter, a fundamental process in lake acidification: paleolimnological evidence, Nature, 316, 436.

R.B. Davis, D.S. Anderson, D.F. Charles, and J.N. Galloway, 1988, Two-hundred-year pH history of Woods, Sagamore, and Panther Lakes in the Adirondack Mountains, New York state, in Aquatic Toxicology and Hazard Assessment, Vol. 10, W.J. Adams, G.A. Chapman, and W.G. Landis, Eds., ASTM STP 971, American Society for Testing and Materials, Philadelphia, 89.

R.B. Davis, D.S. Anderson, S.A. Norton, and M.C. Whiting, 1994, Acidity of twelve northern New England (U.S.A.) lakes in recent centuries, J. Paleolimnol., 12, 103.

E.S. Deevey, M.W. Binford, M. Brenner, and T.J. Whitmore, 1986, Sedimentary records of accelerated nutrient loading in Florida lakes, Hydrobiologia, 143, 49.

A.S. Denning, J. Baron, M.A. Mast, and M. Arthur, 1991, Hydrologic pathways and chemical composition of runoff during snowmelt in Loch Vale watershed, Rocky Mountain National Park, Colorado, USA, Water Air Soil Pollut., 59, 107.

T.E. Dennis and A.J. Bulger, 1995, Condition factor and whole-body sodium concentration in a freshwater fish: evidence of acidification stress and possible ionoregulatory over-compensation, Water Air Soil Pollut., 85, 377.

T.E. Dennis, S.E. MacAvoy, M.B. Steg, and A.J. Bulger, 1995, The association of water chemistry variables and fish condition in streams of Shenandoah National Park (USA), Water Air Soil Pollut., 85, 365.

W. de Vries, 1990, Philosophy, structure and application methodology of a soil acidification model for The Netherlands, in Impact Models to Assess Regional Acidification, J. Kämäri, Ed., Kluwer Academic Publishers, Dordrecht, 3.

W. de Vries, 1991, Methodologies for the assessment and mapping of critical loads and of the impact of abatement strategies on forest soils, Wagningen, the Winart Staring Center for Integrated Land, Soil and Water Research, Report 46.

W. de Vries and J. Kros, 1991, Assessment of critical loads and the impact of deposition scenarios by steady state and dynamic soil acidification models, in *Studies in Environmental Science*, Vol. 46, G.J. Heij, and T. Schneider, Eds., Acidification Research in the Netherlands, 569.

W.T. Dickson, 1978, Some effects of the acidification of Swedish lakes, *Verh. Int. Verein. Limnol.*, 20, 851.

W.T. Dickson, 1980, Properties of acidified water, in *Ecological Impact of Acid Precipitation*, D. Drabløs and A. Tollan, Eds., Proc. Int. Conf., Sandefjord, Norway, 75.

P.J. Dillon, R.A. Reid, and E. DeGrosbois, 1987, The rate of acidification of aquatic ecosystems in Ontario, Canada, *Nature*, 329, 45.

P.J. Dillon, R.A. Reid, and R. Girard, 1986, Changes in the chemistry of lakes near Sudbury, Ontario, following reduction of SO_2 emission, *Water Air Soil Pollut.*, 31, 59.

R.A. Dirks and B.E. Martner, 1982, The climate of Yellowstone and Grand Teton National Parks, National Park Service, Occasional Paper No. 6, Yellowstone National Park, WY.

N.B. Dise, 1984, A synoptic survey of headwater streams in Shenandoah National Park, Virginia, to evaluate sensitivity to acidification by acid deposition, M.S. thesis, Department of Environmental Sciences, University of Virginia, Charlottesville, VA.

N.B. Dise and R.F. Wright, 1995, Nitrogen leaching from European forests in relation to nitrogen deposition, *For. Ecol. Manage.*, 71, 153.

N.B. Dise and R.F. Wright, Eds., 1992, *The NITREX Project*, Commission of the European Communities; Ecosystems Research Rep. No. 2, Brussels.

N.B. Dise, E. Matzner, and P. Gundersen, 1998, Synthesis of nitrogen pools and fluxes from European forest ecosystems, *Water Air and Soil Pollut.*, 105, 143.

A.S. Dixit, S.S. Dixit, and J.P. Smol, 1992, Long-term trends in lake water pH and metal concentrations inferred from diatoms and chrysophytes in three lakes near Sudbury, Ontario, *Can. J. Fish. Aquat. Sci.*, 49, 1.

S.S. Dixit, B.F. Cumming, H.J.B. Birks, J.P. Smol, J.C. Kingston, A.J. Uutala, D.F. Charles, and K.E. Camburn, 1993, Diatom assemblages from Adirondack lakes (New York, USA) and the development of inference models for retrospective environmental assessment, *J. Paleolimnol.*, 8, 27.

S.S. Dixit, A.S. Dixit, and R.D. Evans, 1987, Paleolimnology evidence of recent acidification in two Sudbury (Canada) lakes, *Sci. Total Environ.*, 67, 53.

S.S. Dixit, A.S. Dixit, and J.P. Smol, 1989, Lake acidification recovery can be monitored using chrysophycean microfossils, *Can. J. Fish. Aquat. Sci.*, 46, 1309.

S.S. Dixit, A.S. Dixit, and J.P. Smol, 1991, Multivariable environmental inferences based on diatom assemblages from Sudbury (Canada) lakes, *Freshwater Biol.*, 26, 251.

J.E. Dobson, R.M. Rush, and R.W. Peplies, 1990, Forest blowdown and lake acidification, *Ann. Am. Assoc. Am. Geogr.*, 80, 343.

A.L. Donaldson, 1921, *A History of the Adirondacks*, Vol. II, The Century Co., NY.

J.I. Drever and D.R. Hurcomb, 1986, Neutralization of atmospheric acidity by chemical weathering in an alpine drainage basin in the North Cascade Mountains, *Geology*, 14, 221.

C.T. Driscoll, 1984, A procedure for the fractionation of aqueous aluminum in dilute acidic waters, *Int. J. Environ. Anal. Chem.*, 16, 267.

C.T. Driscoll and J.J. Bisogni, 1984, Weak acid/base systems in dilute acidified lakes and streams of the Adirondack region of New York State, in *Modeling of Total Acid Precipitation*, J.L. Schnoor, Ed., Butterworth Publishers, Boston, 53.

C.T. Driscoll and R.M. Newton, 1985, Chemical characteristics of Adirondack lakes, *Environ. Sci. Technol.*, 19, 1018.

C.T. Driscoll and G.D. Schafran, 1984, Short-term changes in the base neutralizing capacity of an acid Adirondack lake, New York, *Nature*, 310, 308.

C.T. Driscoll and R. van Dreason, 1993, Seasonal and long-term temporal patterns in the chemistry of Adirondack lakes, *Water Air Soil Pollut.*, 67, 319.

C.T. Driscoll, J.P. Baker, J.J. Bisogni, and C.L. Schofield, 1980, Effect of aluminum speciation on fish in dilute acidified waters, *Nature*, 284, 161.

C.T. Driscoll, R.D. Fuller, and W.D. Schecher, 1989c, The role of organic acids in the acidification of surface waters in the eastern U.S., *Water Air Soil Pollut.*, 43, 21.

C.T. Driscoll, M.D. Lehtinen, and T.J. Sullivan, 1994, Modeling the acid-base chemistry of organic solutes in Adirondack, New York, lakes, *Water Resour. Res.*, 30, 297.

C.T. Driscoll, G.E. Likens, L.O. Hedin, J.S. Eaton, and F.H. Bormann, 1989a, Changes in the chemistry of surface waters, *Environ. Sci. Technol.*, 23, 137.

C.T. Driscoll, R.M. Newton, C.P. Gubala, J.P. Baker, and S.W. Christensen, 1991, Adirondack Mountains, in *Acidic Deposition and Aquatic Ecosystems: Regional Case Studies*, D.F. Charles, Ed., Springer-Verlag, New York, 133.

C.T. Driscoll, K.M. Postek, W. Kretser, and D.J. Raynal, 1995, Long-term trends in the chemistry of precipitation and lake water in the Adirondack region of New York, USA, *Water Air Soil Pollut.*, 85, 583.

C.T. Driscoll, D.A. Schaefer, L.A. Molot, and P.J. Dillon, 1989b, Summary of North American Data, in *The Role of Nitrogen in the Acidification of Soils and Surface Waters*, J.L. Malanchuk and J. Nilsson, Eds., Nordic Council of Ministers, Copenhagen, 6.1.

C.T. Driscoll, B.J. Wykowski, C.C. Constentini, and M.E. Smith, 1987b, Processes regulating the temporal and longitudinal variations in chemistry of low-order woodland streams in the Adirondacks, *Biogeochemistry*, 3, 225.

C.T. Driscoll, C.P. Yatski, and F.J. Unangst, 1987a, Longitudinal and temporal trends in the water chemistry of the North Branch of the Moose River, *Biogeochemistry*, 3, 37.

L.E. Eary, E.A. Jenne, L.W. Vail, and D.C. Girvin, 1989, Numerical models for predicting watershed acidification, *Arch. Environ. Contam. Toxicol.*, 18, 29.

ECE, 1990, *Draft Manual for Mapping Critical Levels/Loads*, prepared by the Task Force on Mapping, Umweltbundesamt, Berlin.

J. Eilers, unpublished, E&S Environmental Chemistry Inc., Corvallis, OR.

J.M. Eilers and J.A. Bernert, 1989, An evaluation of National Lake Survey boundaries for the study of lake acidification, Report to the Environmental Protection Agency, E&S Environmental Chemistry, Inc., Corvallis, OR.

J.M. Eilers and J.A. Bernert, 1990, Umpqua Lakes base-line water quality inventory, E&S Report No. 90-12 to the Umpqua National Forest, E&S Environmental Chemistry, Inc., Corvallis, OR.

J.M. Eilers and S.S. Dixit, 1992, Diagnostic and feasibility study of Lake Notasha, Oregon, First Annual Report, E&S Environmental Chemistry, Inc., Corvallis, OR.

J.M. Eilers and A.R. Selle, 1991, Geographical overview, in *Acid Deposition and Aquatic Ecosystems: Regional Case Studies*, D.F. Charles, Ed., Springer-Verlag, New York, 107.

J.M. Eilers and K.B. Vaché, 1998, Lake response to atmospheric and watershed inputs in the Goat Rocks Wilderness, WA, final report to Weyerhaeuser Paper Co., Inc., Longview, WA, E&S Environmental Chemistry, Inc., Corvallis, OR.

J.M. Eilers, J.A. Bernert, and S.S. Dixit, 1994b, Lake Notasha, Oregon, A phase-I diagnostic and feasibility study, report submitted to Oregon Department of Environmental Quality, Portland, OR, E&S Environmental Chemistry, Inc., Corvallis, OR.

J.M. Eilers, D.F. Brakke, and D.H. Landers, 1988b, Chemical and physical characteristics of lakes in the Upper Midwest, United States, *Environ. Sci. Technol.*, 22, 164.

J.M. Eilers, D.F. Brakke, D.H. Landers, and P.E. Kellar, 1988a, Characteristics of lakes in mountainous areas of the western United States, *Vereh. Int. Verein. Limnol.*, 23, 144.

J.M. Eilers, D.F. Brakke, D.H. Landers, and W.S. Overton, 1989b, Chemistry of lakes in designated wilderness areas in the western United States, *Environ. Monitor. Assess.*, 12, 3.

J.M. Eilers, B.J. Cosby, and J.A. Bernert, 1991, Modeling lake response to acidic deposition in the northern Rocky Mountains, Report No. 91-02 to the USDA-Forest Service, Missoula, MT, E&S Environmental Chemistry, Inc., Corvallis, OR.

J.M. Eilers, G.E. Glass, A.K. Pollack, and J.A. Sorensen, 1989a, Changes in conductivity, alkalinity, calcium, and pH during a fifty year period in selected northern Wisconsin lakes, *Can. J. Fish. Aquat. Sci.*, 46, 1929.

J.M. Eilers, G.E. Glass, K.E. Webster, and J.A. Rogalla, 1983, Hydrologic control of lake susceptibility to acidification, *Can. J. Fish. Aquat. Sci.*, 40, 1896.

J.M. Eilers, P. Kanciruk, R.A. McCord, W.S. Overton, L. Hook, D.J. Blick, D.F. Brakke, P.E. Kellar, M.S. DeHaan, M.E. Silverstein, and D.H. Landers, 1987, *Characteristics of Lakes in the Western United States*, Vol. II, *Data Compendium for Selected Physical and Chemical Variables*, EPA-600/3-86/054b, Environmental Protection Agency, Washington, DC.

J.M. Eilers, D.H. Landers, and D.F. Brakke, 1988c, Chemical and physical characteristics of lakes in the southeastern United States, *Environ. Sci. Technol.*, 22, 172.

J.M. Eilers, G.J. Lien, and R.G. Berg, 1984, Aquatic organisms in acidic environments: a literature review, Technical Bulletin No. 150, Department of Natural Resources, Madison, WI.

J.M. Eilers, C.L. Rose, and T.J. Sullivan, 1994a, Status of air quality and effects of atmospheric pollutants on ecosystems in the Pacific Northwest Region of the National Park Service, Technical Report NPS/NRAQD/NRTR-94/160, National Park Service, Air Quality Division, Denver, CO.

J.M. Eilers, T.J. Sullivan, and K.C. Hurley, 1990, The most dilute lake in the world?, *Hydrobiologia*, 199, 1.

J.M. Eilers, P.R. Sweets, D.F. Charles, and K.B. Vaché, 1998, A diatom calibration set for the Cascade Mountain Ecoregion, submitted to PacifiCorp, Centralia, WA, E&S Environmental Chemistry, Inc., Corvallis, OR.

B.K. Ellis, J.A. Stanford, J.A. Craft, D.W. Chess, G.R. Gregory, and L.F. Marnell, 1992, Monitoring water quality of selected lakes in Glacier National Park, Montana: analysis of data collected 1984–1990, Open File Report 129-92 in conformance with Cooperative Agreement CA 1268-0-9001, Work Order 6, National Park Service, Glacier National Park, West Glacier, MO, Flathead Lake Biological Station, University of Montana, Polson, MT.

J.W. Elwood, M.J. Sale, P.R. Kaufmann, and G.F. Cada, 1991, The Southern Blue Ridge Province, in *Acidic Deposition and Aquatic Ecosystems, Regional Case Studies*, D.F. Charles, Ed., Springer-Verlag, New York, 319.

B.A. Emmett, D. Boxman, M. Bredemeier, P. Gundersen, O.J. Kjønaas, F. Moldan, P. Schleppi, A. Tietema, and R.F. Wright, 1998, Predicting the effects of atmospheric nitrogen deposition in conifer stands: evidence from the NITREX ecosystem-scale experiments, *Ecosystems*, 1, 352.

B.A. Emmett, S.A. Brittain, S. Hughes, J. Görres, V. Kennedy, D. Norris, R. Rafarel, B. Reynolds, and P.A. Stevens, 1995, Nitrogen additions ($NaNO_3$ and NH_4NO_3) at Aber forest, Wales. I. Response of throughfall and soil water chemistry, *For. Ecol. Manage.*, 71, 45.

B.A. Emmett, B.J. Cosby, R.C. Ferrier, A. Jenkins, A. Tietema, and R.F. Wright, 1997, Modelling the ecosystem effects of nitrogen deposition: simulation of nitrogen saturation in a Sitka spruce forest, Aber, Wales, UK, *Biogeochemistry*, 38, 129.

B.A. Emmett, B. Reynolds, P.A. Stevens, D.A. Norris, S. Hughes, J. Görres, and I. Lubrecht, 1993, Nitrate leaching from afforested Welsh catchments—Interactions between stand age and nitrogen deposition, *Ambio*, 22, 386.

D. Engle and J.M. Melack, 1995, Zooplankton of high elevation lakes of the Sierra Nevada, California: potential effects of chronic and episodic acidification, *Arch. Hydrobiol.*, 133, 1.

EPA, 1995a, Acid deposition standard feasibility study, a report to Congress, EPA 430-R-95-001A, Environmental Protection Agency, Washington, DC.

EPA, 1995b, National air pollution trends, 1990–1994, EPA/454/R-95/011, Environmental Protection Agency, Washington, DC.

E. Eriksson, 1981, Aluminum in groundwater, possible solution equilibria, *Nord. Hydrol.*, 12, 43.

J.W. Erisman, T. Brydges, K. Bull, E. Cowling, P. Grennfelt, L. Nordberg, K. Satake, T. Schneider, S. Smeulders, K. van der Hoek, J.R. Wisniewski, and J. Wisniewski, 1998, Nitrogen, the Confer-N-s, summary statement, First International Nitrogen Conference, March 23–27, 1998, Noordwijkerhout, The Netherlands.

K.N. Eshleman, 1988, Predicting regional episodic acidification of surface waters using empirical models, *Water Resour. Res.*, 34, 1118.

K.N. Eshleman and P.R. Kaufmann, 1987, Assessing the regional effects of sulfur deposition on surface water chemistry: the Southern Blue Ridge, *Environ. Sci. Technol.*, 22, 685.

K.N. Eshleman, T.D. Davies, M. Tranter, and P.J. Wigington, Jr., 1995, A two-component mixing model for predicting regional episodic acidification of surface waters during spring snowmelt periods, *Water Resour. Res.*, 31, 1011.

K.N. Eshleman, R.P. Morgan II, J.R. Webb, F.A. Deviney, and J.N. Galloway, 1998, Temporal patterns of nitrogen leakage from mid-Appalachian forested watersheds: role of insect defoliation, *Water Resour. Res.*, 34, 2005.

D. Fahey, G. Hubler, D. Parrish, E. Williams, R. Norton, B. Ridley, H. Singh, S. Liu, and F. Fehsenfeld, 1986, Reactive nitrogen species in the troposphere: measurements of NO, NO_2, HNO_3, particulate nitrate, peroxyacetyl nitrate (PAN), O_3, and total reactive nitrogen (NO_y) at Niwot Ridge, Colorado, *Geophys. Res.*, 91, 9781.

A.M. Farag, D.F. Woodward, E.E. Little, B. Steadman, and F.A. Vertucci, 1993, The effects of low pH and elevated aluminum on Yellowstone cutthroat trout (*Oncorhynchus clarki bouveri*), *Environ. Toxicol. Chem.*, 12, 719.

A.M. Farmer, 1990, The effects of lake acidification on aquatic macrophytes—a review, *Environ. Pollut.*, 65, 219.

FCG, 1986, Florida acid deposition study, Final report: a synthesis of the Florida Acid Deposition Study, Florida Electric Power Coordinating Group, Inc., Tampa, FL.

FDER, 1984, An analysis of acid deposition issues: the impacts of proposed national acid deposition control legislation on Florida, Florida Department of Environmental Regulation, Tallahassee, FL.

K.H. Feger, 1992, Nitrogen cycling in two Norway spruce (*Picea abies*) ecosystems and effects of a $(NH_4)_2SO_4$ addition, *Water Air Soil Pollut.*, 61, 295.

M.E. Fenn and A. Bytnerowicz, 1993, Dry deposition of nitrogen and sulfur to ponderosa and Jeffrey pine in the San Bernardino National Forest in the San Bernardino Mountains in southern California, *Environ. Pollut.*, 81, 277.

M.E. Fenn and M.A. Poth, 1999, Nitrogen deposition and cycling in Mediterranean forests—the new paradigm of nitrogen excess, in *Oxidant Air Pollution Impacts in the Montane Forests of Southern California: The San Bernardino Mountain Case Study*, P.R. Miller and J.R. McBride, Eds., Springer-Verlag, New York.

M.E. Fenn, M.A. Poth, J.D. Aber, J.S. Baron, B.T. Bormann, D.W. Johnson, A.D. Lemly, S.G. McNulty, D.F. Ryan, and R. Stottlemyer, 1998, Nitrogen excess in North American ecosystems: predisposing factors, ecosystem responses, and management strategies, *Ecol. Appl.*, 8, 706.

M.E. Fenn, M.A. Poth, and D.W. Johnson, 1996, Evidence for nitrogen saturation in the San Bernardino Mountains in southern California, *For. Ecol. Manage.*, 82, 211.

M. Ferm and H. Hultberg, 1998, Atmospheric deposition to the Gårdsjön research area, in *Experimental Reversal of Acid Rain Effects, The Gårdsjön Roof Project*, H. Hultberg and R. Skeffington, Eds., John Wiley & Sons, Chichester, UK, 71.

R.C. Ferrier, A. Jenkins, B.J. Cosby, R.C. Hall, R.F. Wright, and A.J. Bulger, 1995, Effects of future N deposition scenarios on the Galloway region of Scotland using a coupled sulphur and nitrogen model (MAGIC-WAND), *Water Air Soil Pollut.*, 85, 707.

R.J. Flower and R.W. Battarbee, 1983, Diatom evidence for recent acidification of two Scottish Lochs, *Nature*, 305, 130.

T. Flum and S.C. Nodvin, 1995, Factors affecting streamwater chemistry in the Great Smoky Mountains, USA, *Water Air Soil Pollut.*, 85, 1707.

N.W. Foster, J.D. Aber, J.M. Melillo, R.D. Bowden, and F.A. Bazazz, 1997, Forest response to disturbance and anthropogenic stress, *Bioscience*, 47, 437.

D.H. Freeman, 1987, *Applied Categorical Data Analysis*, Marcel Dekker, New York.

A.J. Friedland, E.K. Miller, J.J. Battles, and J.F. Thorne, 1991, Nitrogen deposition, distribution, and cycling in a subalpine spruce–fir forest in the Adirondacks, New York, USA, *Biogeochemistry*, 14, 31.

J.N. Galloway, 1984, Alkalinity as an indicator of sensitivity: relative importance of HNO_3 vs. H_2SO_4 and long-term acidification, in *The Acidic Deposition Phenomenon and Its Effects*. Vol. II. *Effects Sciences* 45, J.P. Altshuller and R.A. Linthurst, Eds., EPA-600/8-83-016 BF, Environmental Protection Agency, Washington, DC, 4-6.

J.N. Galloway, G.R. Hendrey, C.L. Schofield, N.E. Peters, and A.H. Johannes, 1987, Processes and causes of lake acidification during spring snowmelt in the west-central Adirondack Mountains, New York, *Can. J. Fish. Aquat. Sci.*, 44, 1595.

J.N. Galloway, G.E. Likens, and M.E. Hawley, 1984, Acid precipitation: natural versus anthropogenic components, *Science*, 226, 829.

J.N. Galloway, G.E. Likens, W.C. Keene, and J.M. Miller, 1982, The composition of precipitation in remote areas of the world, *J. Geophys. Res.*, 88, 10850.

S.A. Gherini, L. Mok, R.J. Hudson, G.F. Davis, C.W. Chen, and R.A. Goldstein, 1985, The ILWAS model: formulation and application, *Water Air Soil Pollut.*, 26, 425.

J.H. Gibson, J.N. Galloway, C. Schofield, W. McFee, R. Johnson, S. McCarley, N. Dise, and D. Herzog, 1983, Rocky Mountain acidification study, FWS/OBS-80/40.17, U.S. Fish and Wildlife Service, Division of Biological Services, Eastern Energy and Land Use Team.

F.S. Gilliam, M.B. Adams, and B.M. Yurish, 1996, Ecosystem nutrient responses to chronic nitrogen inputs at Fernow Experimental Forest, West Virginia, *Can. J. For. Res.*, 26, 196.

E. Gjessing, 1992, The HUMEX Project: experimental acidification of a catchment and its humic lake, *Environ. Int.*, 18, 535.

E.T. Gjessing, 1994, HUMEX (Humic Lake Acidification Experiment): chemistry, hydrology, and meteorology, *Environ. Internat.*, 20, 267.

W.H. Glaze, G. Abbt-Braun, A.M. Braun, R.F. Christman, H.G. Frimmel, W. Gasses, J.I. Giger, J.I. Hedges, M.J. Klug, A.H. Nehrkorn, E.M. Thurman, and D.C. White, 1990, What is the composition of organic acids in aquatic systems and how are they characterized?, in *Organic Acids in Aquatic Ecosystems*, E.M. Perdue and E.T. Gjessing, Eds., John Wiley & Sons, New York, 75.

F.R. Gobran, F. Courchesne, and A. Dufresne, 1998, Relationships between sulfate retention and release, solution pH and DOC in the Gårdsjön soils, in *Experimental Reversal of Acid Rain Effects. The Gårdsjön Roof Project*, H. Hultberg and R. Skeffington, Eds., John Wiley & Sons, Chichester, UK, 207.

R.A. Goldstein, S.A. Gherini, C.W. Chen, L. Mok, and R.J.M. Hudson, 1984, Integrated lake watershed acidification study (ILWAS): a mechanistic ecosystem analysis, *Phil. Trans. R. Soc. Lond. B*, 305, 409.

R.A. Goldstein, S.A. Gherini, C.T. Driscoll, R. April, C.L. Schofield, and C.W. Chen, 1987, Lake-watershed acidification in the North Branch of the Moose River, *Biogeochemistry*, 3, 5.

D.A. Graetz, C.D. Pollman, B. Roof, and E. Will, 1985, Effects of acidic treatments on soil chemistry and microbiology, Florida Acid Deposition Study, Phase IV Report, Vol. I, ESE No. 83-152-0106/0207/0307, Environmental Science and Engineering, Inc., Gainesville, FL, 4-5.

G. Gran, 1952, Determination of the equivalence point in potentiometric titrations, *Analyst*, 77, 661.

P. Grennfelt and H. Hultberg, 1986, Effects of nitrogen deposition on the acidification of terrestrial and aquatic ecosystems, *Water Air Soil Pollut.*, 30, 945.

R.W. Griffiths and W. Keller, 1992, Benthic macroinvertebrate changes in lakes near Sudbury, Ontario following a reduction in acid emissions, *Can. J. Fish. Aquat. Sci.*, 49 (Suppl. 1), 63.

G. Gschwandtner, K.C. Gschwandtner, and K. Elridge, 1985, Historic Emissions of Sulfur and Nitrogen Oxides in the United States from 1900–1980, Vol. I., EPA-600/7-5/009a, Environmental Protection Agency, Research Triangle Park, NC.

C.P. Gubala, C.T. Driscoll, R.M. Newton, and C.L. Schofield, 1991, The chemistry of a near-shore lake region during spring snowmelt, *Environ. Sci. Tech.*, 25, 2024.

D.D. Gulley and M. Parker, 1985, A limnological survey of 70 small lakes and ponds in Grand Teton National Park, Department of Zoology and Physiology, Univ. Wyoming, Laramie, WY.

P. Gundersen, 1992, Mass balance approaches for establishing critical loads for nitrogen in terrestrial ecosystems, background document for UN-ECE workshop, Critical Loads for Nitrogen, Lökeberg, Sweden, April 6–10, 1992.

P. Gundersen, 1998, Effects of enhanced nitrogen deposition in a spruce forest at Klosterhede, Denmark, examined by moderate NH_4NO_3 addition, *For. Ecol. Manage.*, 101, 251.

P. Gundersen and L. Rasmussen, 1990, Nitrification in forest soils: effects from nitrogen deposition on soil acidification and aluminum release, *Rev. Environ. Contam. Technol.*, 113, 1.

P. Gundersen and L. Rasmussen, 1995, Nitrogen mobility in a nitrogen limited forest at Klosterhede, Denmark, examined by NH_4NO_3 addition, *For. Ecol. Manage.*, 71, 75.

P. Gundersen, B.R. Andersen, C. Beier, and L. Rasmussen, 1995, Experimental manipulation of water and nutrient input to a Norway spruce plantation at Klosterhede, Denmark. 1. Unintended physical and chemical changes by roof experiments, *Plant Soil*, 168–169, 601.

P. Gundersen, B.A. Emmett, O.J. Kjønaas, C. Koopmans, and A. Tietema, 1998, Impact of nitrogen deposition on nitrogen cycling in forests: a synthesis of NITREX data, *For. Ecol. Manage.*, 101, 37.

J. Gunn, personal communication, Ontario Ministry of Natural Resources, Sudbury, Ontario, Canada.

J.M. Gunn, Ed., 1995, *Restoration and Recovery of an Industrial Region*, Springer-Verlag, New York.

J.M. Gunn and W. Keller, 1990, Biological recovery of an acid lake after reductions in industrial emissions of sulphur, *Nature*, 355, 431.

D. Haddow, R. Musselman, T. Blett, and R. Fisher, Technical Coordinators, 1998, Guidelines for evaluating air pollution impacts on wilderness within the Rocky Mountain Region: report of a workshop, 1990, USDA Forest Service, Rocky Mountain Research Station, Ft. Collins, CO.

T.A. Haines, J.J. Akielaszek, S.A. Norton, and R.B. Davis, 1983, Errors in pH measurement with colorimetric indicators in low alkalinity waters, *Hydrobiologia*, 107, 57.

K. Hansen, C. Beier, P. Gundersen, and L. Rasmussen, 1995, Experimental manipulation of water and nutrient input to a Norway spruce plantation at Klosterhede, Denmark. 3. Effects on throughfall, soil water chemistry, and decomposition, *Plant Soil*, 168–169, 623.

R. Harriman, T.E.H. Allott, R.W. Battarbee, C. Curtis, J. Hall, and K. Bull, 1995, Critical load maps for UK freshwaters, in *Critical Loads of Acid Deposition for UK Freshwaters*, DOE Report, 19.

D.V. Harris and E.P. Kiven, 1985, *The Geologic Story of the National Parks and Monuments*, John Wiley & Sons, New York.

J. Harte and E. Hoffman, 1989, Possible effects of acidic deposition on a Rocky Mountain population of the tiger salamander, *Ambystoma tigrinum*, *Conserv. Biol.*, 3, 149.

M. Hauhs, K. Rost-Siebert, G. Raben, T. Paces, and B. Vigerust, 1989, Summary of European data, in *The Role of Nitrogen in the Acidification of Soils and Surface Waters*, J.L. Malanchuk, and J. Nilsson, Eds., Miljørapport 1989, 10 (NORD 1989:92), Nordic Council of Ministers, Copenhagen, 5-1.

M. Havas, T.C. Hutchinson, and G.E. Likens, 1984, Red herrings in acid rain research, *Environ. Sci. Technol.*, 18, 176A.

M. Havas, D.G. Woodfine, P. Lutz, K. Yung, H.J. MacIsaac, and T.C. Hutchinson, 1995, Biological recovery of two previously acidified, metal-contaminated lakes near Sudbury, Ontario, Canada, *Water Air Soil Pollut.*, 85, 791.

H.G. Healy, 1975, Potentiometric surface and areas of artesian flow of the Florida aquifer in Florida, Map Series 73, Department of Natural Resources, Bureau of Geology, Tallahassee, FL.

L.O. Hedin, L. Granat, G.E. Likens, T.A. Bulshand, J.N. Galloway, T.J. Butler, and H. Rodhe, 1994, Steep declines in atmospheric base cations in regions of Europe and North America, *Nature*, 367, 351.

L.O. Hedin, G.E. Likens, and F.H. Bormann, 1987, Decrease in precipitation acidity resulting from decreased SO_4^{2-} concentration, *Nature*, 325, 244.

L.O. Hedin, G.E. Likens, K.M. Postek, and C.T. Driscoll, 1990, A field experiment to test whether organic acids buffer acid deposition, *Nature*, 345, 798.

D. Heinsdorf, 1993, The role of nitrogen in declining Scots pine forests (*Pinus sylvestris*) in the lowland of east Germany, *Water Air Soil Pollut.*, 69, 21.

M. Heit, Y.L. Tan, C. Klusek, and J.C. Burke, 1981, Anthropogenic trace elements and polycyclic aromatic hydrocarbon levels in sediment cores from two lakes in the Adirondack acid lake region, *Water Air Soil Pollut.*, 15, 441.

S. Heliwell, G.E. Batley, T.M. Florence, and B.G. Lumsden, 1983, Speciation and toxicity of aluminum in a model fresh water, *Environ. Technol. Lett.*, 4, 141.

H.F. Hemond, 1990, Wetlands as the source of dissolved organic carbon to surface waters, in *Organic Acids in Aquatic Ecosystems*, E.M. Perdue and E.T. Gjessing, Eds., John Wiley & Sons, New York, 301.

H.F. Hemond, 1994, Role of organic acids in acidification of fresh waters, in *Acidification of Freshwater Ecosystems. Implications for the Future*, C.E.W. Steinberg and R.F. Wright, Eds., John Wiley & Sons, Chichester, UK.

W.H. Hendershot, S. Savoie, and F. Courchesne, 1992, Simulation of stream-water chemistry with soil solution and groundwater flow contributions, *J. Hydrol.*, 136, 237.

C.D. Hendry, Jr. and P.L. Brezonik, 1984, Chemical composition of softwater Florida lakes and their sensitivity to acid precipitation, *Water Resour. Bull.*, 20, 75.

A. Henriksen, 1979, A simple approach for identifying and measuring acidification of freshwater, *Nature*, 278, 542.

A. Henriksen, 1980, Acidification of freshwaters—a large scale titration, in *Ecological Impact of Acid Precipitation*, D. Drabløs and A. Tollan, Eds., Proc. Int. Conf. Ecological Impact of Acid Precipitation, SNSF Project, Sandefjord, Norway, 68.

A. Henriksen, 1982, Changes in base cation concentrations due to freshwater acidification, Acid Rain Research Report, Norwegian Institute for Water Research, Oslo, Norway.

A. Henriksen, 1984, Changes in base cation concentrations due to freshwater acidification, *Verh. Int. Verein. Limnol.*, 22, 692.

A. Henriksen and D.F. Brakke, 1988, Increasing contributions of nitrogen to the acidity of surface waters in Norway, *Water Air Soil Pollut.*, 42, 183.

A. Henriksen, J. Kämäri, M. Posch, G. Lövblad, M. Forsius, and A. Wilander, 1990b, Critical loads to surface waters in Fennoscandia: intra- and inter-grid variability of critical loads and their exceedance, Miljørapport (Environmental Report) 17, Nordic Council of Ministers, Copenhagen.

A. Henriksen, J. Kämäri, M. Posch, and A. Wilander, 1992, Critical loads of acidity: Nordic surface waters, *Ambio*, 21, 356.

A. Henriksen, L. Lien, and T.S. Traaen, 1990a, Critical loads for surface waters—chemical criteria for inputs of strong acids, Acid Rain Research Report 22, Norwegian Institute for Water Research, Oslo.

A.L. Herczog, W.S. Broeker, R.F. Anderson, S.L. Schiff, and D.W. Schindler, 1985, A new method for monitoring temporal trends in the acidity of fresh waters, *Nature*, 315, 133.

A.T. Herlihy, P.R. Kaufmann, M.R. Church, P.J. Wigington, Jr., J.R. Webb, and M.J. Sale, 1993, The effects of acid deposition on streams in the Appalachian Mountain and Piedmont region of the mid-Atlantic United States, *Water Resour. Res.*, 29, 2687.

A.T. Herlihy, P.R. Kaufmann, M.E. Mitch, and D.D. Brown, 1990, Regional estimates of acid mine drainage impact on streams in the mid-Atlantic and Southeastern United States, *Water Air Soil Pollut.*, 50, 91.

A.T. Herlihy, P.R. Kaufmann, J.L. Stoddard, K.N. Eshleman, and A.J. Bulger, 1996, Effects of acidic deposition on aquatic resources in the Southern Appalachians with a special focus on Class I wilderness areas, report prepared for the Southern Appalachian Mountains Initiative (SAMI), Asheville, NC.

A.T. Herlihy, D.H. Landers, R.F. Cusimano, W.S. Overton, P.J. Wigington, Jr., A.K. Pollack, and T.E. Mitchell-Hall, 1991, Temporal variability in lakewater chemistry in the northeastern United States: results of phase II of the Eastern Lake Survey, EPA/600/3-91/012, Environmental Protection Agency, Corvallis, OR.

D. Hessen, O. Vadstein, and J. Magnusson, 1992, Nitrogen to marine areas, on the application of a critical load concept, background document for workshop on critical loads of nitrogen, April 6–10, 1992, in Lökeberg, Sweden.

M.J. Hinton, S.L. Schiff, and M.C. English, 1994, Examining the contributions of glacial till water to storm runoff using two- and three-compartment hydrograph separations, *Water Resour. Res.*, 30, 983.

R.S. Holmes, M.L. Whiting, and J.L. Stoddard, 1989, Changes in diatom-inferred pH and acid neutralizing capacity in a dilute, high elevation, Sierra Nevada lake since A.D. 1825, *Freshwater Biol.*, 21, 295.

R.P. Hooper and N. Christophersen, 1992, Predicting episodic stream acidification in the southeastern United States: combining a long-term acidification model and the end-member mixing concept, *Water Resour. Res.*, 28, 1983.

R.P. Hooper and C.A. Shoemaker, 1985, Aluminum mobilization in an acidic headwater stream: temporal variation and mineral dissolution disequilibria, *Science*, 229, 463.

R.P. Hooper and C.A. Shoemaker, 1986, A comparison of chemical and isotopic hydrograph separation, *Water Resour. Res.*, 22, 1444.

M. Hornung, K.R. Bull, M. Cresser, J. Hall, P.J. Loveland, S.J. Langan, B. Reynolds, and W.H. Robertson, 1995, Mapping critical loads for soils in Great Britain, in *Acid Rain and Its Impacts: The Critical Loads Debate*, Battarbee, R.W., Ed., ENSIS Publishing, London, 43.

G. Howells, T.R.K. Falziel, J.P. Reader, and J.F. Solbe, 1990, EIFAC water quality criteria for European freshwater fish: report on aluminum, *Chem. Ecol.*, 4, 117.

G.H. Hughes, 1967, Analysis of the water-level fluctuations of Lake Jackson near Tallahassee, Florida, Florida Bureau of Geology Report of Investigations No. 48, Tallahassee, FL.

H. Hultberg, 1992. Biochemical cycling of sulfur and its effects on surface chemistry in an acid catchment in SW Sweden, in *Responses of Forest Ecosystems to Environmental Changes*, A. Teller, P. Mathy, and J.N.R. Jeffers, Eds., Elsevier Applied Science, Amsterdam, 677.

H. Hultberg and R. Skeffington, Eds., 1998, *Experimental Reversal of Acid Rain Effects. The Gårdsjön Roof Project*, John Wiley & Sons, Chichester, UK.

H. Hultberg, F. Moldan, B.I. Anderson, and R.A. Skeffington, 1998, Recovery from acidification in the forested covered catchment experiment at Gårdsjön: effects on biogeochemical output fluxes and concentrations, in *Experimental Reversal of Acid Rain Effects. The Gårdsjön Roof Project*, H. Hultberg and R. Skeffington, Eds., John Wiley & Sons, Chichester, UK, 157.

C.T. Hunsaker, J.L. Malanchuk, R.J. Olson, S.W. Christensen, and R.S. Turner, 1986a, Adirondack headwater lake chemistry relationships with watershed characteristics, *Water Air Soil Pollut.*, 31, 79.

C.T. Hunsaker, R.J. Olson, S.W. Christensen, R.S. Turner, and J.J. Beauchamp, 1986b, Empirical relationships between watershed attributes and headwater lake chemistry in the Adirondack region, Report ORNL/TM-9838, Oak Ridge National Laboratory, Oak Ridge, TN.

R.B. Husar, T.J. Sullivan, and D.F. Charles, 1991, Historical trends in atmospheric sulfur deposition and methods for assessing long-term trends in surface water chemistry, in *Acidic Deposition and Aquatic Ecosystems. Regional Case Studies*, D.F. Charles, Ed., Springer-Verlag, New York, 65.

G.E. Hutchinson, 1957, *A Treatise on Limnology*, Vol. I, *Geography, Physics, and Chemistry*, John Wiley & Sons, New York.

N.J. Hutchinson, K.E. Holtze, J.R. Munro, and T.W. Pawson, 1989, Modifying effect of life stage, ionic strength and post-exposure mortality on lethality of H^+ and Al to lake trout and brook trout, *Aquat. Toxicol.*, 15, 1.

T.C. Hutchinson and M. Havas, 1986, Recovery of previously acidified lakes near Coniston, Canada following reductions in atmospheric sulphur and metal emissions, *Water Air Soil Pollut.*, 28, 319.

K.E. Hyer, J.R. Webb, and K.N. Eshleman, 1995, Episodic acidification of three streams in Shenandoah National Park, Virginia (U.S.A.), *Water Air Soil Pollut.*, 85, 523.

G. Ingersoll, personal communication, U.S. Geological Survey, Denver, CO.

G. Ingersoll, unpublished, U.S. Geological Survey, Denver, CO.

D.A. Jackson and H.H. Harvey, 1995, Gradual reduction and extinction of fish populations in acid lakes, *Water Air Soil Pollut.*, 85, 389.

D.S. Jeffries, Ed., 1997, *1997 Canadian Acid Rain Assessment*, Vol. 3, *The Effects on Canada's Lakes, Rivers, and Wetlands*, Environment Canada, Ottawa.

A. Jenkins, R.C. Ferrier, and B.J. Cosby, 1997b, A dynamic model for assessing the impact of coupled sulphur and nitrogen deposition scenarios on surface water acidification, *J. Hydrol.*, 197, 111.

A. Jenkins, R.C. Helliwell, P.J. Swingewood, C. Seftron, M. Renshaw, and R.C. Ferrier, 1998, Will reduced sulphur emissions under the Second Sulphur Protocol lead to recovery of acid sensitive sites in UK?, *Environ. Pollut.*, 99, 309.

A. Jenkins, M. Renshaw, R. Helliwell, C. Sefton, R. Ferrier, and P. Swingewood, 1997a, Modelling surface water acidification in the UK, Report No. 131, Institute of Hydrology, Oxfordshire.

A. Jenkins, P.G. Whitehead, B.J. Cosby, and H.J.B. Birks, 1990, Modelling long term acidification: a comparison with diatom reconstructions and the implications for reversibility, *Phil. Trans. R. Soc. Lond. B.*, 327, 435.

A. Jenkins, R.F. Wright, N. van Breemen, E.D. Schulze, F. Berendse, F.I. Woodward, and L. Brussard, 1992, The CLIMEX Project—Climate change experiment, International Symposium on Experimental Manipulations of Biota and Biogeochemical Cycling in Ecosystems—Approach, Methodologies, Findings, May 18–20, 1992, Copenhagen.

M. Johannessen and A. Henriksen, 1978, Chemistry of snow meltwater. Changes in concentrations during melting, *Water Resour. Res.*, 14, 615.

A.H. Johnson, S.B. Anderson, and T.G. Siccama, 1994, Acid rain and soils of the Adirondacks. I. Changes in pH and available calcium, 1930–1984, *Can. J. For. Res.*, 24, 39.

D.W. Johnson and D.W. Cole, 1980, Mobility in soils: relevance to nutrient transport from forest ecosystems, *Environ. Int.*, 3, 79.

D.W. Johnson and S.W. Lindberg, Eds., 1991, *Atmospheric Deposition and Forest Nutrient Cycling*, Ecological Series 91, Springer-Verlag, New York.

D.W. Johnson and M.J. Mitchell, 1998, Responses of forest ecosystems to changing sulfur inputs, in *Sulfur in the Environment*, D. Maynard, Ed., Marcel Dekker, New York, 219.

D.W. Johnson, D. Binkley, and P. Conklin, 1995, Simulated effects of atmospheric deposition, harvesting, and species change on nutrient cycling in a loblolly pine forest, *For. Ecol. Manage.*, 76, 29.

D.W. Johnson, W.T. Swank, and J.M. Vose, 1993, Simulated effects of atmospheric sulfur deposition on nutrient cycling in a mixed deciduous forest, *Biogeochemistry*, 23, 169.

D.W. Johnson, H. Van Miegroet, S.E. Lindberg, R.B. Harrison, and D.E. Todd, 1991, Nutrient cycling in red spruce forests of the Great Smoky Mountains, *Can. J. For. Res.*, 21, 769.

N.M. Johnson, C.T. Driscoll, J.S. Eaton, G.E. Likens, and W.H. McDowell, 1981, "Acid rain," dissolved aluminum and chemical weathering at the Hubbard Brook Experimental Forest, New Hampshire, *Geochim. Cosmochim. Acta*, 45, 1421.

J. Kahl, S. Norton, I. Fernandez, L. Rustad, and M. Handley, in press, Nitrogen and sulfur input–output budgets in the experimental and reference watersheds, Bear Brook watershed in Maine (BBWM), *Environ. Manage. Assess.*

J.S. Kahl, personal communication, University of Maine, Orono, ME.

J.S. Kahl, T.A. Haines, S.A. Norton, and R.B. Davis, 1993b, Recent temporal trends in the acid–base chemistry of surface waters in Maine, USA, *Water Air Soil Pollut.*, 67, 281.

J.S. Kahl, S.A. Norton, I.J. Fernandez, K.J. Nadelhoffer, C.T. Driscoll, and J.D. Aber, 1993a, Experimental inducement of nitrogen saturation at the watershed scale, *Environ. Sci. Technol.*, 27, 565.

J.S. Kahl, S.A. Norton, T.A. Haines, E.A. Rochette, R.C. Heath, and S.C. Nodvin, 1992, Mechanisms of episodic acidification in low-order streams in Maine, USA, *Environ. Pollut.*, 78, 37.

J. Kämäri, M. Amann, Y.-W Brodin, M.J. Chadwick, A. Henriksen, J.P. Hettelingh, J.C.I. Kuylenstierna, M. Posch, and H. Sverdrup, 1992, The use of critical loads for the assessment of future alternatives to acidification, *Ambio*, 21, 377.

J. Kämäri, M. Forsius, and M. Posch, 1993, Critical loads of sulfur and nitrogen for lakes II: Regional extent and variability in Finland, *Water Air Soil Pollut.*, 66, 77.

P. Kanciruk, J.M. Eilers, R.A. McCord, D.H. Landers, D.F. Brakke, and R.A. Linthurst, 1986, *Characteristics of Lakes in the Eastern United States*, Vol. III. *Data Compendium of Site Characteristics and Chemical Variables*, EPA-600/4-86/007c, Environmental Protection Agency, Washington, DC.

T. Katoh, T. Konno, H. Koyama, H. Tsuruta, and H. Makino, 1990, Acidic precipitation in Japan, in *Acidic Precipitation*, Vol. 5, *International Overview and Assessment*, A.H.M. Bresser and W. Salomons, Eds., Springer-Verlag, New York, 41.

R. Kattelmann and K. Elder, 1991, Hydrologic characteristics and water balance of an alpine basin in the Sierra Nevada, *Water Resour. Res.*, 27, 1553.

P.R. Kaufmann, A.T. Herlihy, J.W. Elwood, M.E. Mitch, W.S. Overton, M.J. Sale, J.J. Messer, K.A. Cougar, D.V. Peck, K.H. Reckhow, A.J. Kinney, S.J. Christie, D.D. Brown, C.A. Hagley, and H.I. Jager, 1988, *Chemical Characteristics of Streams in the Mid-Atlantic and Southeastern United States*, Vol. I, *Population Descriptions and Physico–Chemical Relationships*, EPA/600/3-88/021a, Environmental Protection Agency, Washington, DC.

P.E. Kauppi, K. Mielikäinen, and K. Kuusela, 1992, Biomass and carbon budget of European forests, 1971 to 1990, *Science*, 256, 70.

W. Keller, 1992, Introduction and overview to aquatic acidification studies in the Sudbury, Ontario, Canada area, *Can. J. Fish. Aquat. Sci.*, 49 (Suppl. 1), 3.

W. Keller and J.R. Pitblado, 1986, Water quality changes in Sudbury area lakes: a comparison of synoptic surveys in 1974–1976 and 1981–1983, *Water Air Soil Pollut.*, 29, 285.

W. Keller and N. Yan, 1991, Recovery of crustacean zooplankton species richness in Sudbury area lakes following water quality improvements, *Can. J. Fish. Aquat. Sci.*, 48, 1635.

W. Keller, J.R. Pitblado, and N.I. Conroy, 1986, Water quality improvements in the Sudbury, Ontario, Canada area related to reduced smelter emissions, *Water Air Soil Pollut.*, 31, 765.

T.J. Kelley and D.H. Stedman, 1980, Effects of urban sources on acid precipitation in the western United States, *Science*, 210, 1043.

C.A. Kelly, J.W.M. Rudd, R.H. Hesslein, C.T. Driscoll, S.A. Gherini, and R.E. Hecky, 1987, Prediction of biological acid neutralization in acid–sensitive lakes, *Biogeochemistry*, 3, 129.

C. Kendall, D.H. Campbell, D.A. Burns, J.B. Shanley, S.R. Silva, and C.C.Y. Chang, 1995, Tracing sources of nitrate in snowmelt runoff using the oxygen and nitrogen isotopic compositions of nitrate, in *Biogeochemistry of Seasonally Snow-Covered Catchments*, K. Tonnessen, M. Williams, and M. Trantner, Eds., International Association of Hydrological Sciences Proceedings, July 13–14, 1995, Boulder, CO, 339.

G.J. Kenoyer, 1986, Groundwater/lake dynamics and chemical evolution in a sandy silicate aquifer in northern Wisconsin, Ph.D. thesis, University of Wisconsin–Madison, WI.

G.J. Kenoyer and M.P. Anderson, 1989, Groundwater's dynamic role in regulating acidity and chemistry in a precipitation-dominated lake, *J. Hydrol.*, 109, 287.

E.H. Ketchledge, 1965, Changes in the forests of New York, *The Conservationist*, 19, 29.

J.M. Kiesecker, 1991, Acidification and its effects on amphibians breeding in temporary ponds in montane Colorado, unpublished M.A. thesis, Univ. Northern Colorado, Greeley, CO.

J.C. Kingston and H.J.B. Birks, 1990, Dissolved organic carbon reconstruction from diatom assemblages in PIRLA project lakes, North America, *Phil. Trans. R. Soc. Lond. B.*, 327, 279.

J.C. Kingston, H.J.B. Birks, A.J. Uutala, B.F. Cumming, and J.P. Smol, 1992, Assessing trends in fishery resources and lake water aluminum from paleolimnological analyses of siliceous algae, *Can. J. Fish. Aquat. Sci.*, 49, 116.

J.C. Kingston, R.B. Cook, R.G. Kreis, K.E. Camburn, S.A. Norton, P.R. Sweets, M.W. Binford, M.J. Mitchell, S.C. Schindler, L. Shane, and G. King, 1990, Paleoecological investigation of recent lake acidification in the Northern Great Lakes States, in *Paleoecological Investigation of Recent Lake Acidification*, D.F. Charles and D.W. Whitehead, Eds., Kluwer Academic Press, Dordrecht, The Netherlands.

J.W. Kirchner and E. Lydersen, 1995, Base cation depletion and potential long-term acidification of Norwegian catchments, *Environ. Sci. Technol.*, 29, 1953.

J.V. Klump, C.C. Remsen, and J.L. Kaster, 1988, The presence and potential impact of geothermal activity on the chemistry and biology of Yellowstone Lake, Wyoming, *NOAA Symp. Series for Undersea Research*, Vol. 6(2).

P. Kortelainen, M.B. David, T. Roila, and I. Mäkinen, 1992, Acid–base characteristics of organic carbon in the Humex Lake Skjervatjern, *Environ. Int.*, 18, 621.

J.R. Kramer and S.S. Davies, 1988, Estimation of non-carbonate protolytes for selected lakes of the Eastern Lakes Survey, *Environ. Sci. Technol.*, 22, 182.

J.R. Kramer and A. Tessier, 1982, Acidification of aquatic systems: a critique of chemical approaches, *Environ. Sci. Technol.*, 16, 606.

J.R. Kramer, A.W. Andren, R.A. Smith, A.H. Johnson, R.B. Alexander, and G. Oehlert, 1986, Streams and lakes, in *Acid Deposition: Long-Term Trends*, Committee on Monitoring and Assessment of Trends in Acid Deposition, National Academy Press, Washington, DC, 231.

K.W. Kratz, S.D. Cooper, and J.M. Melack, 1994, Effects of single and repeated experimental acid pulses on invertebrates in a high altitude Sierra Nevada stream, *Freshwater Biol.*, 32, 161.

W. Kretser, J. Gallagher, and J. Nicolette, 1989, *An Evaluation of Fish Communities and Water Chemistry*, Adirondack Lakes Survey Corporation, Ray Brook, New York.

E.C. Krug, 1989, Assessment of the theory and hypotheses of the acidification of watersheds, Illinois State Water Survey Contract Report 457, Champaign, IL.

E.C. Krug, 1991a, Geographic relationships between soil and water acidity, soil-forming factors and acid rain, in *Plant-Soil Interactions at Low pH*, R.J. Wright et al., Eds., Kluwer Academic Publishers, The Netherlands, 123.

E.C. Krug, 1991b, Review of acid-deposition-catchment interactions and comments on future research needs, *J. Hydrol.*, 128, 1.

E.C. Krug and C.R. Frink, 1983, Acid rain on acid soil: a new perspective, *Science*, 221, 520.

E.C. Krug, P.J. Isaacson, and C.R. Frink, 1985, Appraisal of some current hypotheses describing acidification of watersheds, *J. Air Pollut. Contr. Assoc.*, 35, 109.

D.C.L. Lam, A.G. Bobba, R.A. Bourbonniere, G.D. Howell, and M.E. Thompson, 1989, Modeling organic and inorganic acidity in two Nova Scotia rivers, *Water Air Soil Pollut.*, 46, 277.

D.C.L. Lam, C.I. Mayfield, D.A. Swayne, and K. Hopkins, 1994, A prototype information system for watershed management and planning, *J. Biol. Sys.*, 2, 499.

J. Lancaster, M. Real, S. Juggins, D.T. Moneeith, R.J. Flower, and W.R.C. Beaumont, 1996, Monitoring temporal changes in the biology of acid waters, *Freshwater Biol.*, 36, 179.

D.H. Landers, J.M. Eilers, D.F. Brakke, W.S. Overton, P.E. Kellar, M.E. Silverstein, R.D. Schonbrod, R.E. Crowe, R.A. Linthurst, J.M. Omernik, S.A. Teague, and E.P. Meier, 1987, *Characteristics of Lakes in the Western United States*, Vol. I, *Population Descriptions and Physico–Chemical Relationships*, EPA/600/3-86/054a, Environmental Protection Agency, Washington, DC.

G. Landmann, Ed., 1991, *French Research into Forest Decline*, Ecole Nationale du Génie Rural, des eaux et des Foréts, Nancy, France.

A.O. Langford and F.C. Fehsenfeld, 1992, Natural vegetation as a source or sink for atmospheric ammonia: a case study, *Science*, 255, 581.

G.B. Lawrence and T.G. Huntington, 1999, Soil-calcium depletion linked to acid rain and forest growth in the eastern United States, U.S. Geological Survey WRIR 98-4267.

G.B. Lawrence, M.B. David, G.M. Lovett, P.S. Murdoch, D.A. Burns, J.L. Stoddard, B.P. Baldigo, J.H. Porter, and A.W. Thompson, in press, Soil-calcium status and the response of stream chemistry to changing acidic deposition rates, *Ecol. Appl.*

G.B. Lawrence, M.B. David, and W.C. Shortle, 1995, A new mechanism for calcium loss in forest-floor soils, *Nature*, 378, 162.

B.D. LaZerte, 1984, Forms of aqueous aluminum in acidified catchments of central Ontario: a methodological analysis, *Can. J. Fish. Aquat. Sci.*, 41, 766.

B.D. LaZerte, 1986, Metals and acidification: an overview, *Water Air Soil Pollut.*, 31, 569.

B.D. LaZerte, 1993, The impact of drought and acidification on the chemical exports from a minerotrophic conifer swamp, *Biogeochemistry*, 18, 153.

J.A. Leenheer, 1981, Comprehensive approach to preparative isolation and fractionation of dissolved organic carbon from natural waters and wastewaters, *Environ. Sci. Technol.*, 15, 578.

H. Leivestad, J.E. Jensen, J. Kjartansson, and L. Xingfu, 1987, Aqueous speciation of aluminum and toxic effects on Atlantic salmon, *Annis. Soc. R. Zool.*, 117 (supple 1), 387.

G.E. Likens and F.H. Bormann, 1995, *Biogeochemistry of a Forested Ecosystem*, 2nd ed., Springer-Verlag, New York.

G.E. Likens, F.H. Bormann, L.O. Hedin, and C.T. Driscoll, 1990, Dry deposition of sulfur: a 23-year record for the Hubbard Brook Forest ecosystem, *Tellus*, 42B, 319.

G.E. Likens, F.H. Bormann, R.J. Pierce, and R.E. Munn, 1984, Long-term trends in precipitation chemistry at Hubbard Brook, New Hampshire, *Atmos. Environ.*, 18, 2641.

G.E. Likens, C.T. Driscoll, and D.C. Buso, 1996, Long-term effects of acid rain: response and recovery of a forest ecosystem, *Science*, 272, 244.

J.C. Lin and J.L. Schnoor, 1986, Acid precipitation model for seepage lakes, *J. Environ. Eng.*, 112, 677.

C.J. Lind and J.D. Hem, 1975, Effects of organic solutes on chemical reactions of aluminum, U.S. Geological Survey Water Supply Paper 1827-G.

R.A. Linthurst, D.H. Landers, J.M. Eilers, D.F. Brakke, W.S. Overton, E.P. Meier, and R.E. Crowe, 1986, *Characteristics of Lakes in the Eastern United States*, Vol. I, *Population Descriptions and Physico–Chemical Relationships*, EPA/600/4-86/007a, Environmental Protection Agency, Washington, DC.

S. Liu, R. Munson, D. Johnson, S. Gherini, K. Summers, R. Hudson, K. Wilkinson, and L. Pitelka, 1991, Application of a nutrient cycling model (NuCM) to northern mixed hardwood and southern coniferous forest, *Tree Physiol.*, 9, 173.

A. Locke, 1992, Factors influencing community structure along stress gradients: zooplankton responses to acidification, *Ecology*, 73, 903.

T.J. Loranger and D.F. Brakke, 1988, The extent of snowpack influence on water chemistry in a North Cascade lake, *Water Resour. Res.*, 24, 723.

T.J. Loranger, D.F. Brakke, M.B. Bonoff, and B.F. Gall, 1986, Temporal variability of lake waters in the North Cascade Mountains (Washington, USA), *Water Air Soil Pollut.*, 31, 123.

G. Lövblad and J.W. Erisman, 1992, Deposition of nitrogen species on a small scale in Europe, Report of Working Group 5, Workshop on Critical Loads for Nitrogen.

G.M. Lovett, 1994, Atmospheric deposition of nutrients and pollutants in North America: an ecological perspective, *Ecol. Appl.*, 4, 629.

E. Lyderson, E. Fjeld, and E.T. Gjessing, 1996, The Humic Lake Acidification Experiment (HUMEX): main physico–chemical results after five years of artificial acidification, *Environ. Int.*, 22, 591.

E. Lydersen, A.B.S. Polèo, M. Nandrup Pettersen, G. Riise, B. Salbu, F. Kroglund, and
    B.O. Rosseland, 1994, The importance of *"in situ"* measurements to relate toxicity
    and chemistry in dynamic aluminum freshwater systems, *J. Ecol. Chem.*, 3(3), 357.
E. Lydersen, A.B.S. Polèo, I.P. Muniz, B. Salbu, and H.J. Bjørnstad, 1990, The effects
    of naturally occurring high and low molecular weight inorganic and organic
    species on the yolk–sack larvae of Atlantic salmon (*Salmo salar* L.) exposed to
    acidic aluminum-rich lake water, *Aquat. Toxicol.*, 18, 219.
J.A. Lynch, V.C. Bowersox, and J.W. Grimm, 1996, Trends in precipitation chemistry
    in the United States, 1983–94—an analysis of the effects in 1995 of phase I of the
    Clean Air Act Amendments of 1990, Title IV, U.S. Geological Survey, Open File
    Report 96-0346, Washington, DC.
S.E. MacAvoy and A.J. Bulger, 1995, Survival of brook trout (*Salvelinus fontinalis*)
    embryos and fry in streams of different acid sensitivity in Shenandoah National
    Park, USA, *Water Air Soil Pollut.*, 85, 439.
A.H. Magill, M.R. Downs, K.J. Nadelhoffer, R.A. Hallett, and J.D. Aber, 1996, Forest
    ecosystem response to four years of chronic nitrate and sulfate additions at Bear
    Brook Watershed, Maine, USA, *For. Ecol. Manage.*, 84, 29.
M.D. Marcus, B.R. Parkhurst, R.W. Brocksen, and F.E. Payne, 1983, An assessment of
    the relationship among acidifying deposition, surface water acidification, and
    fish populations in North America, Vol. 1, Electric Power Research Institute, Final
    Report No. EA-3127.
D.R. Marmorek, D.P. Bernard, M.L. Jones, L.P. Rattie, and T.J. Sullivan, 1988, The
    effects of mineral acid deposition on concentrations of dissolved organic acids
    in surface waters, ERL-COR-500 AP, Environmental Protection Agency, Environ-
    mental Research Laboratory, Corvallis, OR.
L. Marquez, 1994, Atmospheric loading of nitrogen to alpine tundra at Niwot Ridge,
    Colorado, Masters Project, Center for Environmental Sciences, University of
    Colorado at Denver.
C.J. Martinka, 1992, Conserving the natural integrity of mountain parks: lessons from
    Glacier National Park, Montana, *Oecologia*, 1, 41.
J. Mason and H.M. Seip, 1985, The current status of knowledge on acidification of
    surface waters and guidelines for further research, *Ambio*, 14, 45.
M.A. Mast, J.I. Drever, and J. Baron, 1990, Chemical weathering in the Loch Vale
    watershed, Rocky Mountain National Park, Colorado, *Water Resour. Res.*, 26, 2971.
M.A. Mast, C. Kendall, D.H. Campbell, D.W. Clow, and J. Back, 1995, Determination
    of hydrologic pathways in alpine subalpine basin using isotopic and chemical
    tracers, in *Biogeochemistry of Seasonally Snow Covered Catchments*, K. Tonnessen,
    M.W. Williams, and M. Tranter, Eds., Proc. Boulder Symp., July 1995, IAHS Publ.
    No. 228.
E. Matzner and D. Murach, 1995, Soil changes induced by air pollutant deposition
    and their implication for forests in central Europe, *Water Air Soil Pollut.*, 85, 63.
C.P. McCahon, A.F. Brown, M.J. Poulton, and D. Pascoe, 1989, Effects of acid, alumi-
    num and lime additions on fish and invertebrates in a chronically acidic Welch
    stream, *Water Air Soil Pollut.*, 45, 345.
J.G. McColl, 1981, Effects of acid rain on plants and soils in California, Final Report,
    Contract A7-169-30, California Air Resources Board, Sacramento, CA.
D.M. McKnight, E.M. Thurman, R.L. Wershaw, and H.F. Hemond, 1985, Biogeochem-
    istry of aquatic humic substances in Thoreau's Bog, Concord, Massachusetts,
    Ecology, 66, 1339.
B. McMartin, 1994, *The Great Forest of the Adirondacks*, North Country Books, Utica, NY.

D.K. McNichol, B.E. Bendell, and M.L. Mallory, 1995, Evaluating macroinvertebrate responses to recovery from acidification in small lakes in Ontario, Canada, *Water Air Soil Pollut.*, 85, 451.

S.G. McNulty, J.D. Aber, and R.D. Boone, 1991, Spatial changes in forest floor and foliar chemistry of spruce-fir forests across New England, *Biogeochemistry*, 14, 13.

S.G. McNulty, J.D. Aber, and S.D. Newman, 1996, Nitrogen saturation in a high elevation spruce-fir stand, *For. Ecol. Manage.*, 84, 109.

J.M. Melack and J.L. Stoddard, 1991, Sierra Nevada: unacidified, very dilute waters and mildly acidic atmospheric deposition, in *Acidic Deposition and Aquatic Ecosystems: Regional Case Studies*, D.F. Charles, Ed., Springer-Verlag, New York, 503.

J.M. Melack, S.C. Cooper, T.M. Jenkins, L. Barmuta, Jr., S. Hamilton, K. Kratz, J. Sickman, and C. Soiseth, 1989, Chemical and biological characteristics of Emerald Lake and the streams in its watershed, and the response of the lake and streams to acidic deposition, final report, Contract A6-184-32, California Air Resources Board, Sacramento, CA.

J.M. Melack, S.D. Cooper, R.W. Holmes, J.O. Sickman, K. Kratz, P. Hopkins, H. Hardenbergh, M. Thieme, and L. Meeker, 1987, Chemical and biological survey of lakes and streams located in the Emerald Lake watershed, Sequoia National Park, final report, Contract A3-096-32, California Air Resources Board.

J.M. Melack, J.O. Sickman, and A. Leydecker, 1998, Comparative analyses of high-altitude lakes and catchments in the Sierra Nevada: susceptibility to acidification, final report prepared for the California Air Resources Board, Contract A032-188, Marine Science Institute and Institute for Computational Earth System Science, Univ. California, Santa Barbara.

J.M. Melack, J.O. Sickman, F. Setaro, and D. Dawson, 1997, Monitoring of wet deposition in alpine areas in the Sierra Nevada, final report, Contract A932-081, California Air Resources Board.

J.M. Melack, J.O. Sickman, F.V. Setaro, and D. Engle, 1993, Long-term studies of lakes and watersheds in the Sierra Nevada, patterns and processes of surface-water acidification, final report, Contract No. A932-060, California Air Resources Board, Marine Science Institute, Univ. of California, Santa Barbara.

J.M. Melack, J.L. Stoddard, and C.A. Ochs, 1985, Major ion chemistry and sensitivity to acid precipitation of Sierra Nevada lakes, *Water Resour. Res.*, 21, 27.

S.E. Metcalf and J.D. Whyatt, 1995, Modeling future acid deposition with HARM, in *Acid Rain and Its Impact: The Critical Loads Debate*, R. Battarbee, Ed., ENSIS Publishing, London.

D.C. Middleton, 1904, Ninth annual report of the forest, fish, and game commission, J.B. Lyon Co., Albany, NY.

W. Miller and M. Bellini, 1996, Trophic state evaluation of selected lakes in Grand Teton National Park, Brigham Young University, Provo, UT.

M.J. Mitchell, M.B. David, and R.B. Harrison, 1992, Sulfur dynamics of forest ecosystems, in *Sulfur Cycling on the Continents*, R.W. Howarth, J.W.B. Stewart, and M.V. Ivanov, Eds., SCOPE Vol. 48, John Wiley & Sons, New York, chap. 9, 215.

M.J. Mitchell, G. Iwatsubo, K. Ohrui, and Y. Nakagawa, 1997, Nitrogen saturation in Japanese forests: an evaluation, *For. Ecol. Manage.*, 97, 39.

M.J. Mitchell, C.R. Krouse, B. Mayer, A.C. Stam, and Y. Zhang, 1998, Use of stable isotopes in evaluating sulfur biogeochemistry of forest ecosystems, in *Isotope Tracers in Catchment Hydrology*, C. Kendall and J. McDonnell, Eds., Elsevier Science, The Netherlands.

M.J. Mitchell, D.J. Raynal, and C.T. Driscoll, 1996, Biogeochemistry of a forested watershed in the central Adirondack Mountains, Temporal changes and mass balances, *Water Air Soil Pollut.*, 88, 355.

F. Moldan and R.F. Wright, 1998a, Episodic behavior of nitrate in runoff during six years of nitrogen addition to the NITREX catchment at Gårdsjön, Sweden, *Environ. Pollut.*, 102, 439.

F. Moldan and R.F. Wright, 1998b, Changes in runoff chemistry after five years of N addition to a forested catchment at Gårdsjön, Sweden, *For. Ecol. Manage.*, 101, 187.

F. Moldan, H. Hultberg, U. Nyström, and R.F. Wright, 1995, Nitrogen saturation at Gårdsjön, southwest Sweden, induced by experimental addition of ammonium nitrate, *For. Ecol. Manage.*, 71, 89.

B. Momen and J.P. Zehr, 1998, Watershed classification by discriminate analyses of lakewater-chemistry and terrestrial characteristics, *Ecol. Appl.*, 8, 497.

MPCA, 1985, Statement of need and reasonableness: proposed acid deposition standard and control plan, Minnesota Pollution Control Agency, St. Paul, MN.

I.P. Muniz and H. Leivestad, 1980, Acidification effects on freshwater fish, in *Ecological Impact of Acid Precipitation*, D. Drabløs and A. Tollan, Eds., Proc. Int. Conf. Sandefjord, Norway, SNSF-project, Oslo, Norway, 84.

R.K. Munson and S.A. Gherini, 1991, Processes influencing the acid–base chemistry of surface waters, in *Acidic Deposition and Aquatic Ecosystems. Regional Case Studies*, D.F. Charles, Ed., Springer-Verlag, New York, 9.

R.K. Munson and S.A. Gherini, 1993, Influence of organic acids on the pH and ANC of Adirondack Lakes, *Water Resour. Res.*, 29, 891.

P.S. Murdoch and J.L. Stoddard, 1992, The role of nitrate in the acidification of streams in the Catskill Mountains of New York, *Water Resour. Res.*, 28, 2707.

P.S. Murdoch and J.L. Stoddard, 1993, Chemical characteristics and temporal trends in eight streams of the Catskill Mountains, New York, *Water Air Soil Pollut.*, 67, 367.

P.S. Murdoch, D.A. Burns, and G.B. Lawrence, 1998, Relation of climate change to the acidification of surface waters by nitrogen deposition, *Environ. Sci. Technol.*, 32, 1642.

R.C. Musselman, L. Hudnell, M.W. Williams, and R.A. Sommerfeld, 1996, Water chemistry of Rocky Mountain Front Range aquatic ecosystems, Research Paper RM-RP-325, USDA Rocky Mountain Forest and Range Experiment Station, Ft. Collins, CO.

K. Nadelhoffer, M. Downs, B. Fry, A. Magill, and J. Aber, in press, Controls on N retention and exports in a forested watershed, *Environ. Manage. Assess.*

NAPAP, 1991, Integrated assessment report, National Acid Precipitation Assessment Program, Washington, DC.

NAPAP, 1998, Biennial report to Congress: an integrated assessment, National Acid Precipitation Assessment Program, Silver Spring, MD.

National Research Council, 1986, *Acid Deposition: Long-Term Trends*, National Academy Press, Washington, DC.

C. Neal, P. Whitehead, R. Neale, and B.J. Cosby, 1986, Modelling the effects of acidic deposition and conifer afforestation on stream acidity in the British uplands, *J. Hydrol.*, 86, 15.

B.P. Neary and P.J. Dillon, 1988, Effects of sulfur deposition on lake-water chemistry in Ontario, Canada, *Nature*, 333, 340.

P.O. Nelson, 1991, Cascade Mountains, in *Acid Deposition and Aquatic Ecosystems, Regional Case Studies*, D.F. Charles, Ed., Springer-Verlag, New York, 531.

A. Newell, personal communication, Oregon Department of Environmental Quality, Portland, OR.

A.D. Newell, 1993, Inter-regional comparison of patterns and trends in surface water acidification across the United States, *Water Air Soil Pollut.*, 67, 257.

R.M. Newton and C.T. Driscoll, 1990, Classifications of ALSC lakes, in *Adirondack Lakes Survey: An Interpretive Analysis of Fish Communities and Water Chemistry, 1984–1987*, J.P. Baker, S.A. Gherini, S.W. Christensen, C.T. Driscoll, J. Gallagher, R.K. Munson, R.M. Newton, K.H. Reckhow, and C.L. Schofield, Eds., Adirondack Lakes Survey Corporation, Ray Brook, NY, 2-70.

R.M. Newton, J. Weintraub, and R. April, 1987, The relationship between surface water chemistry and geology in the North Branch of the Moose River, *Biogeochemistry*, 3, 21.

D.S. Nichols and R.E. McRoberts, 1986, Relations between lake acidification and sulfate deposition in northern Minnesota, Wisconsin, and Michigan, *Water Air Soil Pollut.*, 31, 197.

J. Nilsson, Ed., 1986, *Critical Loads for Sulphur and Nitrogen*, Miljørapport 1986:11, Nordic Council of Ministers, Copenhagen.

J. Nilsson and P. Grennfelt, Eds., 1988, Critical loads for sulphur and N, report from a workshop held at Skokloster, Sweden, March 19–24, 1988, NORD Miljørapport 1988:15, Nordic Council of Ministers, Copenhagen, 225.

S.I. Nilsson, 1993, Acidification of Swedish oligotrophic lakes—interactions between deposition, forest growth, and effects on lake–water quality, *Ambio*, 22, 272.

S.I. Nilsson, H.G. Miller, and J.D. Miller, 1982, Forest growth as a possible cause of soil and water acidification: an examination of the concepts, *Oikos*, 39, 40.

S.C. Nodvin, H. Van Miegroet, S.E. Lindberg, N.S. Nicholas, and D.W. Johnson, 1995, Acidic deposition, ecosystem processes, and nitrogen saturation in a high elevation Southern Appalachian watershed, *Water Air Soil Pollut.*, 85, 1647.

S.A. Norton and A. Henriksen, 1983, The importance of CO_2 in evaluation of effects of acidic depositions, *Vatten*, 39, 346.

S.A. Norton and J.S. Kahl, in press, Impacts of marine aerosols on surface water chemistry at Bear Brook Watershed, Maine, *Verh. Int. Ver. Limnol.*

S.A. Norton, J.J. Akielaszek, T.A. Haines, K.J. Stromborg, and J.R. Longcore, 1982, in *Bedrock Geologic Control of Sensitivity of Aquatic Ecosystems in the United States to Acidic Deposition*, National Atmospheric Deposition Program, Fort Collins, CO, 1.

S.A. Norton, J.S. Kahl, D.F. Brakke, G.F. Brewer, T.A. Haines, and S.C. Nodvin, 1988, Regional patterns and local variability of dry and occult deposition strongly influence sulfate concentrations in Maine lakes, *Sci. Total Environ.*, 72, 183.

S.A. Norton, J.S. Kahl, I. Fernandez, T. Haines, L. Rustad, S. Nodvin, J. Schofield, T. Strickland, H. Erickson, P. Wigington, Jr., and J. Lee, 1999, The Bear Brook Watershed, Maine (BBWM), USA, *Environ. Manage. Assess.*, 55, 7.

S.A. Norton, J.S. Kahl, I.J. Fernandez, L.E. Rustad, J.P. Schofield, and T.A. Haines, 1994, Response of the West Bear Brook Watershed, Maine, USA, to the addition of $(NH_4)_2SO_4$: 3-year results, *For. Ecol. Manage.*, 68, 61.

S.A. Norton, J.S. Kahl, I.J. Fernandez, J.P. Schofield, L.E. Rustad, T.A. Haines, and J. Lee, 1993, The watershed manipulation project: two-year results at the Bear Brook Watershed in Maine (BBWM), in *Experimental Manipulations of Biota and Biogeochemical Cycling in Ecosystems*, L. Rasmussen, T. Brydges, and P. Mathy, Eds., ECSC-EEC-EAEC, Brussels, 55.

S.A. Norton, R.F. Wright, J.S. Kahl, and J.P. Schofield, 1992, The MAGIC simulation of surface water at, and first year results from, the Bear Brook Watershed Manipulation, Maine, USA, *Environ. Pollut.*, 77, 279.

K. Ohrui and M.J. Mitchell, 1997, Nitrogen saturation in Japanese forested watersheds, *Ecol. Appl.*, 7, 391.

K. Ohrui and M.J. Mitchell, 1998a, Spatial patterns of soil nitrate in Japanese forested watersheds: importance of the near-stream zone as a source of nitrate in stream water, *Hydrol. Proc.*, 12, 1433.

K. Ohrui and M.J. Mitchell, 1998b, Effects of nitrogen fertilization on stream chemistry of Japanese forested watersheds, *Water Air Soil Pollut.*, 107, 219.

J. Økland and K.A. Økland, 1986, The effects of acid deposition on benthic animals in lakes and streams, *Experientia*, 42, 471.

H. Olem, 1990, Liming acidic surface waters, State of the Science, SOS/T 15, National Acid Precipitation Assessment Program, Washington, DC.

B.G. Oliver, E.M. Thurman, and R.L. Malcolm, 1983, The contribution of humic substances to the acidity of colored natural waters, *Geochim. Cosmochim. Acta*, 47, 2031.

S.V. Ollinger, J.D. Aber, G.M. Lovett, S.E. Millham, R.G. Lathrop, and J.M. Ellis, 1993, A spatial model of atmospheric deposition for the northeastern U.S., *Ecol. Appl.*, 3, 459.

J.M. Omernik and C.S. Powers, 1982, *Total Alkalinity of Surface Waters—National Map*, EPA Environmental Research Laboratory, Corvallis, OR.

N. Oreskes, K. Shrader-Frechette, and K. Belitz, 1994, Verification, validation, and confirmation of numerical models in the earth sciences, *Science*, 263, 641.

S.J. Ormerod and S.J. Tyler, 1991, Exploitation of prey by a river bird, the dipper *Cinclus cinclus* (L.), along acidic and circumneutral streams in upland Wales, *Freshwater Biol.*, 25, 105.

S.J. Ormerod, G.W. Mawle, and R.W. Edwards, 1989, The influence of plantation forestry on the pH and aluminum concentrations of upland Welsh streams, A re-examination, *Environ. Pollut.*, 62, 47.

E.J. Orr, 1993, 1992 Compliance Report, Minnesota Wet Sulfate Deposition Standard, Minnesota Pollution Control Agency, St. Paul, MN.

OTA, 1984, Acid rain and transported air pollutants: implications for public policy, U.S. Congress Office of Technology Assessment, Washington, DC.

D.R. Parker, L.W. Zelazny, and T.B. Kinraide, 1989, Chemical speciation and plant toxicity of aqueous aluminum, in *Environmental Chemistry and Toxicology of Aluminum*, T.E. Lewis, Ed., American Chemical Society, 117.

D. Parrish, R. Norton, M. Bollinger, S. Liu, P. Murphy, D. Albritton, and F. Fehsenfeld, 1986, Measurements of HNO_3 and NO_3^- particulates at a rural site in the Colorado Mountains, *J. Geophys. Res.*, 91, 5379.

W.J. Parton, J.M.O. Scurlock, D.S. Ojima, T.G. Gilmanov, R.J. Scholes, D.S. Schimel, T. Kirchner, J-C. Menaut, T. Seastedt, M.E. Garcia, A. Kamnalrut, and J.I. Kinyamario, 1993, Observations and modeling of biomass on soil organic matter dynamics for the grassland biome worldwide, *Global Biogeochemical Cycles*, 7, 785.

R.K. Peet, 1992, Community structure and ecosystem function, in *Plant Succession: Theory and Prediction*, D.L. Glenn-Lewis, R.K. Peet, and T.T. Veblen, Eds., Chapman and Hall, London, 103.

E.M. Perdue, J.H. Reuter, and R.S. Parrish, 1984, A statistical model of proton binding by humus, *Geochim. Cosmochim. Acta*, 48, 1257.

W.T. Peterjohn, M.B. Adams, and F.S. Gilliam, 1996, Symptoms of nitrogen saturation in two central Appalachian hardwood forest ecosystems, *Biogeochemistry,* 35, 507.

D.L. Peterson and T.J. Sullivan, 1998, Assessment of air quality and air pollutant impacts in National Parks of the Rocky Mountains and Northern Great Plains, NPS D-657, U.S. Department of the Interior, National Park Service, Air Resources Division.

K. Phillips, 1990, Where have all the frogs and toads gone?, *Bioscience,* 40, 422.

M. Placet, R.E. Battye, F.C. Fehsenfeld, and G.W. Bassett, 1990, Emissions involved in acidic deposition processes, State of Science and Technology Report 1, National Acid Precipitation Assessment Program, Washington, DC.

A.B.S. Poléo, 1995, Aluminum polymerization—a mechanism of acute toxicity of aqueous aluminum to fish, *Aquat. Toxicol.,* 31, 347.

A.B.S. Poléo and I.P. Muniz, 1993, The effect of aluminum in soft water at low pH and different temperatures on mortality, ventilation frequency and water balance in smoltifying Atlantic salmon, *Salmo salar., Environ. Biol. Fish.,* 36, 193.

A.B.S. Poléo, E. Lydersen, B.O. Rosseland, F. Kroglund, B. Salbu, R. Vogt, and A. Kvellestad, 1994, Increased mortality of fish due to changing Al-chemistry of mixing zones between limed streams and acidic tributaries, *Water Air Soil Pollut.,* 75, 339.

A.K. Pollack, A.B. Hudischewskyj, T.S. Stoeckenius, and P. Guttorp, 1989, Analysis of variability of UAPSP precipitation chemistry measurements, draft final report, Utility Acid Precipitation Study Program (UAPSP 118, Contract U101-06), Washington, DC.

C.D. Pollman and D.E. Canfield, 1991, Florida: a case study in hydrologic and biogeochemical controls on seepage lake chemistry, in *Acidic Deposition and Aquatic Ecosystems: Regional Case Studies,* D.F. Charles, Ed., Springer-Verlag, New York.

C.D. Pollman and P.R. Sweets, 1990, Hindcasting of pre-cultural ANC and pH of seepage lakes in the Adirondacks, Upper Midwest, and Florida, EPA Report 89007B1/RES-1, Corvallis, OR.

C.D. Pollman, T.M. Lee, W.J. Andrews, L.A. Sacks, S.A. Gherini, and R.K. Munson, 1991, Preliminary analysis of the hydrologic and geochemical controls on acid-neutralizing capacity in two acidic seepage lakes in Florida, *Water Resour. Res.,* 27, 2321.

C.D. Pollman, H.S. Prentice, and M.H. Robbins, 1990, Florida softwater lake survey, final report to Florida Department of Environmental Regulation, Tallahassee, FL, KBN Engineering and Applied Sciences, Inc., Gainesville, FL.

M. Posch, J.-P. Hettelingh, P.A.M. de Smet, and R.J. Downing, 1997, Calculation and mapping of critical thresholds in Europe: CCE status report 1997, RIVM Report 259101007, National Institute for Public Health and the Environment, Bilthoven, The Netherlands.

K.M. Postek, C.T. Driscoll, J.D. Aber, and R.C. Santore, 1995, Application of PnET-CN/CHESS to a spruce stand in Solling, Germany, *Ecol. Model.,* 83, 163.

M.L. Ralston and R.I. Jennrich, 1978, Dud, a derivative-free algorithm for nonlinear least squares, *Technometrics,* 20, 7.

C.M. Rascher, C.T. Driscoll, and N.E. Peters, 1987, Concentration and flux of solutes from snow and forest floor during snowmelt in the west central Adirondack region of New York, *Biogeochemistry,* 3, 209.

L. Rasmussen, 1990, Study on acid deposition effects by manipulating forest ecosystems (EXMAN), Air Pollution Research Rep. 24, Commission of the European Communities, Brussels.

L. Rasmussen, C. Beier, N. van Breemen, P. de Visser, K. Kreutzer, R. Schierl, E. Matzner, and E.P. Farrell, 1990, EXMAN—EXperimental MANipulation of forest ecosystems in Europe, Air Pollution Research Rep. 24, Commission of the European Communities, Brussels.

M.M. Reddy and N. Caine, 1988, Chemical budget for a small alpine basin in Colorado, Abstract, *EOS*, 69, 1214.

I. Renberg and H. Hultberg, 1992, A paleolimnological assessment of acidification and liming effects on diatom assemblages in a Swedish lake, *Can. J. Fish. Aquat. Sci.*, 49, 65.

I. Renberg, T. Korsman, and H.J.B. Birks, 1993, Prehistoric increases in the pH of acid-sensitive Swedish lakes caused by land-use changes, *Nature*, 362, 824.

J.O. Reuss and D.W. Johnson, 1985, Effect of soil processes on the acidification of water by acid deposition, *J. Environ. Qual.*, 14, 26.

J.O. Reuss, N. Christopherson, and H.M. Seip, 1986, A critique of models for freshwater and soil acidification, *Water Air Soil Pollut.*, 30, 909.

J.O. Reuss, F.A. Vertucci, R.C. Musselman, and R.A. Sommerfeld, 1995, Chemical fluxes and sensitivity to acidification of two high-elevation catchments in southern Wyoming, *J. Hydrol.*, 173, 165.

B. Reynolds, S.J. Ormerod, and A.S. Gee, 1994, Spatial patterns in stream nitrate concentrations in upland Wales in relation to catchment forest cover and forest age, *Environ. Pollut.*, 84, 27.

B. Reynolds, P.A. Stevens, J.K. Adamson, S. Hughes, and J.D. Roberts, 1992, Effects of clearfelling on stream and soil water aluminum chemistry in three UK forests, *Environ. Pollut.*, 77, 157.

P.J. Riggan, R.N. Lockwood, P.M. Jacks, C.G. Colver, F. Weirich, L.F. Debano, and J.A. Brass, 1994, Effects of fire severity on nitrate mobilization in watersheds subject to chronic atmospheric deposition, *Environ. Sci. Technol.*, 28, 369.

P.J. Riggan, R.N. Lockwood, and E.N. Lopez, 1985, Deposition and processing of airborne nitrogen pollutants in Mediterranean-type ecosystems of southern California, *Environ. Sci. Technol.*, 19, 781.

L. Rochefort, D.H. Vitt, and S.E. Bayley, 1990, Growth, production, and decomposition dynamics of Sphagnum under natural and experimentally acidified conditions, *Ecology*, 71, 1986.

J.A. Rogalla, P.L. Brezonik, and G.E. Glass, 1986, Empirical models for lake acidification in the Upper Great Lakes region, *Water Air Soil Pollut.*, 31, 95.

E.J.S. Røgeberg and A. Henriksen, 1985, An automated method for fractionation and determination of aluminum species in freshwaters, *Vatten.*, 41, 48.

A.D. Rosemond, S.R. Reice, J.W. Elwood, and P.J. Mulholland, 1992, The effects of stream acidity on benthic invertebrate communities in the southeastern United States, *Freshwater Biol.*, 27, 193.

K. Rosén, P. Gundersen, L. Tagnhammar, M. Johansson, and T. Frogner, 1992, Nitrogen enrichment of Nordic forest ecosystems, The concept of critical loads, *Ambio*, 21, 364.

H.H. Rosenbrock, 1960, An automatic method for finding the greatest or least value of a function, *Computer J.*, 3, 175.

B.D. Rosenlund and D.R. Stevens, 1988, Fisheries and aquatic management, Rocky Mountain National Park, U.S. Fish and Wildlife Service, Colorado Fish and Wildlife Assistance Office, Golden, CO.

I. Rosenqvist, 1978, Acid precipitation and other possible sources for acidification of rivers and lakes, *Sci. Total Environ.*, 10, 271.

B.O. Rosseland, I.A. Blakar, A. Bulger, F. Kroglund, A. Kvellestad, E. Lydersen, D.H. Oughton, B. Salbu, M. Stuarnes, and R. Vogt, 1992, The mixing zone between limed and acidic river waters: complex aluminum chemistry and extreme toxicity for salmonids, *Environ. Pollut.*, 78, 3.

J.W.M. Rudd, C.A. Kelly, and A. Furutani, 1986, The role of sulfate reduction in long term accumulation of organic and inorganic sulfur in lake sediments, *Limnol. Oceanogr.*, 31, 1281.

S.D. Rundle and A.G. Hildrew, 1990, The distribution of micro-anthropods in some southern English streams: the influence of physicochemistry, *Freshwater Biol.*, 23, 411.

D. Rusch and H. Sievering, 1995, Variation in ambient air nitrogen concentration and total annual atmospheric deposition at Niwot Ridge, Colorado, in *Biogeochemistry of Seasonally Snow Covered Catchments*, K.A. Tonnessen, M.W. Williams, and M. Tranter, Eds., Proc. Boulder Symposium July 1995, IAHS Publ. No. 228, 23.

L.E. Rustad, I.J. Fernandez, M.J. Mitchell, K.J. Nadelhoffer, M.B. David, and R.B. Fuller, 1996, Experimental soil acidification and recovery at the Bear Brook Watershed in Maine, *Soil Sci. Soc. Amer. J.*, 60, 1933.

L.E. Rustad, J.S. Kahl, S.A. Norton, and I.J. Fernandez, 1995, Underestimation of dry deposition by throughfall in mixed northern hardwood forest, *J. Hydrol.*, 162, 319.

S. Rustad, N. Christophersen, H.M. Seip, and D. Dillon, 1986, Model for streamwater chemistry of a tributary to Harp Lake, Ontario, *Can. J. Fish. Aquat. Sci.*, 43, 625.

P.F. Ryan, J.N. Galloway, B.J. Cosby, G.M. Hornberger, and J.R. Webb, 1989, Changes in the chemical composition of streamwater in two catchments in the Shenandoah National Park, Virginia, in response to atmospheric deposition of sulfur, *Water Resour. Res.*, 25, 2091.

C.J. Sampson, P.L. Brezonik, and E.P. Weir, 1994, Effects of acidification on chemical composition and chemical cycles in a seepage lake: inferences from a whole-lake experiment, in *Environmental Chemistry of Lakes and Reservoirs*, L.A. Baker, Ed., Adv. Chem. Ser., American Chemical Society, Washington, DC.

D.A. Schaefer and C.T. Driscoll, 1993, Identifying sources of snowmelt acidification with a watershed mixing model, *Water Air Soil Pollut.*, 67, 345.

D.A. Schaefer, C.T. Driscoll, R.S. Van Dreason, and C.P. Yatsko, 1990, The episodic acidification of Adirondack Lakes during snowmelt, *Water Resour. Res.*, 26, 1639.

W.D. Schecher and C.T. Driscoll, 1987, An evaluation of uncertainty associated with aluminum equilibrium calculations, *Water Resour. Res.*, 23, 525.

W.D. Schecher and C.T. Driscoll, 1994, ALCHEMI: a chemical equilibrium model to assess the acid–base chemistry and speciation of aluminum in dilute solutions, in *Chemical Equilibrium and Reaction Models*, R. Loeppert, A.P. Schwab, and S. Goldberg, Eds., Soil Sci. Soc. Am., Madison, WI.

J.P. Schimel, K. Kielland, and F.S. Chapin, III, 1995, Nutrient availability and uptake by tundra plants, in *Landscape Function: Implications for Ecosystem Response to Disturbance: A Case Study in Arctic Tundra*, J.F. Reynold and J.D. Tenhunen, Eds., Springer-Verlag, New York.

D.W. Schindler, 1988, The effects of acid rain on freshwater ecosystems, *Science*, 239, 149.

D.W. Schindler, 1990, Experimental perturbations of whole lakes as tests of hypotheses concerning ecosystem structure and function, *Oikos*, 57, 25.

D.W. Schindler, 1998, A dim future for boreal waters and landscapes, *Bioscience*, 48, 157.

D.W. Schindler and M.A. Turner, 1982, Biological, chemical and physical responses of lakes to experimental acidification, *Water Air Soil Pollut.*, 18, 259.

D.W. Schindler, M.A. Turner, M.P. Stainton, and G.A. Linsey, 1986, Natural sources of acid neutralizing capacity in low-alkalinity lakes of the Precambrian Shield, *Science*, 232, 844.

W. Schmidt and M.W. Clark, 1980, Geology of Bay County, Florida, Florida Geological Survey Bulletin No. 57, Tallahassee, FL.

M. Schnitzer and S.I.M. Skinner, 1963, Organo-metallic interactions in soils. 2. Reactions between different forms of iron and aluminum and the organic matter of a podzol Bh horizon, *Soil Sci.*, 96, 181.

J.L. Schnoor, N.P. Nikolaidis, and G.E. Glass, 1986, Lake resources at risk to acidic deposition in the Upper Midwest, *J. Water Pollut. Control Fed.*, 58, 139.

C.L. Schofield, 1976, Acidification of Adirondack lakes by atmospheric precipitation. Extent and magnitude of the problem, final report, Project F-28-R, New York Department of Environmental Conservation, Albany, NY.

C.L. Schofield, 1982, Historical fisheries changes in the United States related to decreases in surface water pH, in *Acid Rain/Fisheries*, R.E. Johnson, Ed., American Fisheries Society, Bethesda, MD.

C.L. Schofield and C. Keleher, 1996, Comparison of brook trout reproductive success and recruitment in an acidic Adirondack lake following whole lake liming and watershed liming, *Biogeochemistry*, 32, 323.

C.L. Schofield and J.R. Trojnar, 1980, Aluminum toxicity to brook trout (*Salvelinus fontinalis*) in acidified waters, in *Polluted Rain*, T.Y. Toribara, M.W. Miller, and P.E. Morrows, Eds., Plenum Press, 341.

C.L. Schofield, J.N. Galloway, and G.R. Hendry, 1985, Surface water chemistry in the ILWAS basin, *Water Air Soil Pollut.*, 26, 403.

E.-D. Schulze, 1989, Air pollution and forest decline in a spruce (*Picea abies*) forest, *Science*, 244, 776.

H.M. Seip, 1980, Acidification of freshwater—sources and mechanisms, in *Ecological Impacts of Acid Precipitation*, D. Drabløs and A. Tollan, Eds., Proc. Int. Conf., SNSF Project, Sandefjord, Norway, Oslo, Norway, 358.

H.M. Seip, N. Christophersen, and T.J. Sullivan, 1989, Episodic variations in streamwater aluminum chemistry at Birkenes, southernmost Norway, in *Environmental Chemistry and Toxicology of Aluminum*, T.E. Lewis, Ed., Lewis Publishers, Chelsea, MI, 159.

H.M. Seip, L. Müller, and A. Naas, 1984, Aluminum speciation: comparison of two spectrophotometric analytical methods and observed concentrations in some acidic aquatic systems in southern Norway, *Water Air Soil Pollut.*, 23, 81.

SFT, 1987, Thousand Lake survey, Norwegian State Pollution Control Authority, Norwegian Institute for Water Research (NIVA), Oslo, Norway.

SFT, 1988, Thousand Lake survey 1986—Fish Status, Norwegian State Pollution Control Authority, Norwegian Institute for Water Research (NIVA), Oslo, Norway.

E.E. Shannon, 1970, Eutrophication factors in north-central Florida lakes, Ph.D. dissertation, University of Florida, Gainesville, FL.

W.C. Shortle and E.A. Bondietti, 1992, Timing, magnitude, and impact of acidic deposition on sensitive forest sites, *Water Air Soil Pollut.*, 61, 253.

W.C. Shortle, K.T. Smith, R. Minocha, G.B. Lawrence, and M.B. David, 1997, Acidic deposition, cation mobilization, and biochemical indicators of stress in healthy red spruce, *J. Environ. Qual.*, 26, 871.

J.O. Sickman and J.M. Melack, 1989, Characterization of year-round sensitivity of California's montane lakes to acidic deposition, final report, Contrib. A5-203-32, California Air Resources Board, Sacramento, CA.

H. Sievering, J. Braus, and J. Caine, 1989, Dry deposition of nitrate and sulfate to coniferous canopies in the Rocky Mountains, in *Transactions Effects of Air Pollution on Western Forests*, R.K. Olson and A.S. Lefohn, Eds., Air and Waste Management Association, Pittsburgh, PA, 171.

H.A. Simonin, W.A. Kretser, D.W. Bath, M. Olson, and J. Gallagher, 1993, *In situ* bioassays of brook trout (*Salvelinus fontinalis*) and blacknose dace (*Rhinichthys atratulus*) in Adirondack streams affected by episodic acidification, *Can. J. Fish. Aquat. Sci.*, 50, 902.

R. Sinha, M.J. Small, P.F. Ryan, T.J. Sullivan, and B.J. Cosby, 1998, Reduced-form modeling of surface water and soil chemistry for the tracking and analysis framework, *Water Air Soil Pollut.*, 105, 617.

D.L. Sisterson, V.C. Bowersox, T.P. Meyers, A.R. Olsen, and R.J. Vong, 1990, Deposition monitoring: methods and results, State of Science and Technology Report 6, National Acid Precipitation Assessment Program, Washington, DC.

R.A. Skeffington and H. Hultberg, 1998, Summary and conclusions from the Gårdsjön covered catchment experiment—and what remains to be discovered, in *Experimental Reversal of Acid Rain Effects, The Gårdsjön Roof Project*, H. Hultberg and R. Skeffington, Eds., John Wiley & Sons, Chichester, UK, 433.

R.A. Skeffington and E.J. Wilson, 1988, Excess nitrogen deposition: issues for consideration, *Environ. Pollut.*, 54, 159.

B.L. Skjelvåle, R.F. Wright, and A. Henriksen, 1998, Norwegian lakes show widespread recovery from acidification; results from national surveys of lakewater chemistry 1986–1997, *Hydrol. Earth Syst. Sci.*, 2, 555.

M.J. Small and M.C. Sutton, 1986a, A regional pH-alkalinity relationship, *Water Res.*, 20, 335.

M.J. Small and M.C. Sutton, 1986b, A direct distribution model for regional aquatic acidification, *Water Resour. Res.*, 22, 1749.

R.A. Smith, E.B. Alexander, and M.G. Wolman, 1987, Water-quality trends in the nation's rivers, *Science*, 235, 1607.

J.P. Smol, R.W. Battarbee, R.B. Davis, and J. Meriläinen, Eds., 1986, *Diatoms and Lake Acidity*, Dr. W. Junk. Dordrecht, The Netherlands.

J.P. Smol, D.F. Charles, and D.R. Whitehead, 1984, Mallomonadacean (Chrysophyceae) assemblages and their relationships with limnological characteristics in 38 Adirondack (New York) lakes, *Can. J. Bot.*, 62, 911.

J.P. Smol, B.F. Cumming, A.S. Dixit, and S.S. Dixit, 1998, Tracking recovery patterns in acidified lakes: a paleolimnological perspective, *Restor. Ecol.*, 6, 318.

C.R. Soiseth, 1992, The pH and acid neutralizing capacity of ponds containing *Pseudacris regilla* larvae in an alpine basin of the Sierra Nevada, *Calif. Fish Game*, 78, 11.

P. Sollins, C.C. Grier, F.M. McCorison, K. Cromack, Jr., R. Fogel, and R.L. Fredriksen, 1980, The internal element cycles of an old-growth Douglas-fir ecosystem in western Oregon, *Ecol. Monogr.*, 50, 261.

R.A. Sommerfeld, A.R. Mosier, and R.C. Musselman, 1993, CO_2, CH_4, and N_2O flux through a Wyoming snowpack, *Nature*, 361, 140.

V.L. St. Louis, L. Breebaart and J.C. Barlow, 1990, Foraging behaviour of tree swallows over acidified and nonacidic lakes, *Can. J. Zool.*, 68, 2385.

A.C. Stam, M.J. Mitchell, H.R. Krouse and J.S. Kahl, 1992, Stable sulfur isotopes of sulfate in precipitation and stream solutions in a northern hardwood watershed, *Water Resour. Res.*, 28, 231.

R.E. Stauffer, 1990, Granite weathering and the sensitivity of alpine lakes to acid deposition, *Limnol. Oceanogr.*, 35, 1112.

R.E. Stauffer and B.D. Wittchen, 1991, Effects of silicate weathering on water chemistry in forested, upland, felsic terrain of the USA, *Geochim. Cosmochim. Acta*, 55, 3253.

A.C. Stevenson, H.J.B. Birks, R.J. Flower, R.W. Battarbee, 1989, Diatom-based pH reconstruction of lake acidification using canonical correspondence analysis, *Ambio*, 18, 228.

J.L. Stoddard, 1994, Long-term changes in watershed retention of nitrogen: its causes and aquatic consequences, in *Environmental Chemistry of Lakes and Reservoirs*, L.A. Baker, Ed., Advances in Chemistry Series, No. 237, American Chemical Society, Washington, DC, 223.

J.L. Stoddard, 1995, Episodic acidification during snowmelt of high elevation lakes in the Sierra Nevada Mountains of California, *Water Air Soil Pollut.*, 85, 353.

J.L. Stoddard and J.H. Kellogg, 1993, Trends and patterns in lake acidification in the state of Vermont: evidence from the long-term monitoring project, *Water Air Soil Pollut.*, 67, 301.

J.L. Stoddard and P.S. Murdoch, 1991, Catskill Mountains: an overview of the impact of acidifying pollutants on aquatic resources, in *Acidic Deposition and Aquatic Ecosystems: Regional Case Studies*, D.F. Charles, Ed., Springer-Verlag, New York.

J.L. Stoddard, C.T. Driscoll, J.S. Kahl, and J.H. Kellogg, 1998, Can site-specific trends be extrapolated to a region? An acidification example for the Northeast, *Ecol. Appl.*, 8, 288.

A. Stone and H.M. Seip, 1989, Mathematical models and their role in understanding water acidification: an evaluation using the Birkenes model as an example, *Ambio*, 18, 192.

A. Stone and H.M. Seip, 1990, Are mathematical models useful for understanding water acidification?, *Sci. Total Environ.*, 96, 159.

W. Stumm and J.J. Morgan, 1981, *Aquatic Chemistry*, Wiley-Interscience, New York.

J.K. Sueker, 1995, Chemical hydrograph separation during snowmelt for three headwater basins in Rocky Mountain National Park, Colorado, in *Biogeochemistry of Seasonally Snow Covered Catchments*, Proc. Boulder Symposium, July 1995, IAHS Publ. No. 228.

T. Sullivan, unpublished, E&S Enviromental Chemistry, Inc., Corvallis, OR.

T.J. Sullivan, 1990, Historical changes in surface water acid–base chemistry in response to acidic deposition, State of the Science, SOS/T 11, National Acid Precipitation Assessment Program.

T.J. Sullivan, 1991, Long-term temporal changes in surface water chemistry, in *Acid Deposition and Aquatic Ecosystems: Regional Case Studies*, D.F. Charles, Ed., Springer-Verlag, New York.

T.J. Sullivan, 1993, Whole ecosystem nitrogen effects research in Europe, *Environ. Sci. Technol.*, 27(8), 1482.

T.J. Sullivan, 1994, Progress in quantifying the role of aluminum in acidification of surface waters, *J. Ecol. Chem.*, 3, 157.

T.J. Sullivan, 1997, Ecosystem manipulation experimentation as a means of testing a biogeochemical model, *Environ. Manage.*, 21(1), 15.

T.J. Sullivan and B.J. Cosby, 1995, Testing, improvement, and confirmation of a watershed model of acid–base chemistry, *Water Air Soil Pollut.*, 85, 2607.

T.J. Sullivan and B.J. Cosby, 1998, Modeling the concentration of aluminum in surface waters, *Water Air Soil Pollut.*, 105, 643.

T.J. Sullivan and J.M. Eilers, 1994, Assessment of deposition levels of sulfur and nitrogen required to protect aquatic resources in selected sensitive regions of North America, final report prepared for Technical Resources, Inc., Rockville, MD, under contract to Environmental Protection Agency, Environmental Research Laboratory-Corvallis, Corvallis, OR.

T.J. Sullivan, J.A. Bernert, E.A. Jenne, J.M. Eilers, B.J. Cosby, D.F. Charles, and A.R. Selle, 1991, Comparison of MAGIC and diatom paleolimnological model hindcasts of lakewater acidification in the Adirondack region of New York, prepared for the Department of Energy under Contract DE-AC06-76RLO 1830, Pacific Northwest Laboratory, Richland, WA.

T.J. Sullivan, D.F. Charles, J.A. Bernert, B. McMartin, K.B. Vaché, and J. Zehr, 1999, Relationship between landscape characteristics, history, and lakewater acidification in the Adirondack Mountains, New York, *Water Air Soil Pollut.*, 112, 407.

T.J. Sullivan, D.F. Charles, J.P. Smol, B.F. Cumming, A.R. Selle, D.R. Thomas, J.A. Bernert, and S.S. Dixit, 1990a, Quantification of changes in lakewater chemistry in response to acidic deposition, *Nature*, 345, 54.

T.J. Sullivan, N. Christophersen, I.P. Muniz, H.M. Seip, and P.D. Sullivan, 1986, Aqueous aluminum chemistry response to episodic increases in discharge, *Nature*, 323, 324.

T.J. Sullivan, B.J. Cosby, J.A. Bernert, and J.M. Eilers, 1998, Model evaluation of dose/response relationships and critical loads for nitrogen and sulfur deposition to the watersheds of Lower Saddlebag and White Dome Lakes, Report No. 97-10-01 to USDA Forest Service, E&S Environmental Chemistry, Inc., Corvallis, OR.

T.J. Sullivan, B.J. Cosby, C.T. Driscoll, D.F. Charles, and H.F. Hemond, 1996a, Influence of organic acids on model projections of lake acidification, *Water Air Soil Pollut.*, 91, 271.

T.J. Sullivan, B.J. Cosby, C.T. Driscoll, H.F. Hemond, D.F. Charles, S.A. Norton, H.M. Seip, and G. Taugbøl, 1994, Confirmation of the MAGIC model using independent data: influence of organic acids on model estimates of lakewater acidification, Report No. DOE/ER/30196-4, final report prepared for Department of Energy under Agreement No. DE-FG02-92ER30196.

T.J. Sullivan, B.J. Cosby, and K.A. Tonnessen, in preparation, Calculation of critical loads of sulfur and nitrogen deposition to Loch Vale Watershed, Colorado, *Water Resour. Res.*

T.J. Sullivan, C.T. Driscoll, S.A. Gherini, R.K. Munson, R.B. Cook, D.F. Charles, and C.P. Yatsko, 1989, The influence of organic acid anions and aqueous aluminum on measurements of acid neutralizing capacity in surface waters, *Nature*, 338, 408.

T.J. Sullivan, J.M. Eilers, M.R. Church, D.J. Blick, K.N. Eshleman, D.H. Landers, and M.S. DeHaan, 1988, Atmospheric wet sulfate deposition and lakewater chemistry, *Nature*, 331, 607.

T.J. Sullivan, J.M. Eilers, B.J. Cosby, and K.B. Vaché, 1997, Increasing role of nitrogen in the acidification of surface waters in the Adirondack Mountains, New York, *Water, Air, Soil Pollut.*, 95, 313.

T.J. Sullivan, D.L. Kugler, M.J. Small, C.B. Johnson, D.H. Landers, B.J. Rosenbaum, W.S. Overton, W.A. Kretser, and J. Gallagher, 1990b, Variation in Adirondack, New York, lakewater chemistry as a function of surface area, *Water Resour. Bull.*, 26, 167.

T.J. Sullivan, B. McMartin, and D.F. Charles, 1996b, Re-examination of the role of landscape change in the acidification of lakes in the Adirondack Mountains, New York, *Sci. Total Environ.*, 183(3), 231.

T.J. Sullivan, D.L. Peterson, and C. Blanchard, in preparation, Assessment of Air Quality and Air Pollutant Impacts in Class I National Parks of California, U.S. Department of the Interior, National Park Service, Air Resources Division, Denver, CO.

T.J. Sullivan, R.S. Turner, D.F. Charles, B.F. Cumming, J.P. Smol, C.L. Schofield, C.T. Driscoll, B.J. Cosby, H.J.B. Birks, A.J. Uutala, J.C. Kingston, S.S. Dixit, J.A. Bernert, P.F. Ryan, and D.R. Marmorek, 1992, Use of historical assessment for evaluation of process-based model projections of future environmental change: lake acidification in the Adirondack Mountains, New York, U.S.A., *Environ. Pollut.*, 77, 253.

H. Sverdrup, W. de Vries, and A. Henriksen, 1990, Mapping critical loads, Nordic Council of Ministers, Copenhagen.

H. Sverdrup, P. Warfvinge, and K. Rosen, 1992, A model for the impact of soil solution Ca : Al ratio on tree base cation uptake, *Water Air Soil Pollut.*, 61, 365.

P.R. Sweets, 1992, Diatom paleolimnological evidence for lake acidification in the Trail Ridge region of Florida, *Water Air Soil Pollut.*, 65, 43.

P.R. Sweets, R.W. Binert, T.L. Cusimono, and M.W. Binford, 1990, Paleoecological investigations of recent lake acidification in northern Florida, *J. Paleolimnol.*, 4, 103.

Y.L. Tan and M. Heit, 1981, Biogenic and abiogenic polynuclear aromatic hydrocarbons in sediments from two remote Adirondack lakes, *Geochim. Cosmochim. Acta*, 45, 2267.

M.C. Taylor, H.C. Duthie, and S.M. Smith, 1988, Errors associated with diatom-inferred indices for predicting pH in Canadian Shield lakes, in *Proceedings of the 9th International Diatom Symposium*, F. Round, Ed., Bristol, UK, 273.

C.J.F. ter Braak, 1986, Canonical correspondence analysis: a new eigenvector technique for multivariate direct gradient analysis, *Ecology*, 67, 1167.

C.J.F. ter Braak, 1988, Partial canonical correspondence analysis, in *Classification and Related Methods of Data Analysis*, H.H. Bock, Ed., North-Holland, Amsterdam, 551.

C.J.F. ter Braak and N.J.M. Gremmen, 1987, Ecological amplitudes of plant species and the internal consistency of Ellenberg's indicator values for moisture, *Vegetatio*, 69, 79.

C.J.F. ter Braak and C.W.N. Looman, 1986, Weighted averaging, logistic regression and the Gaussian response model, *Vegetatio*, 65, 3.

C.J.F. ter Braak and I.C. Prentice, 1988, The theory of gradient analysis, *Advan. Ecolog. Res.*, 18, 271.

C.J.F. ter Braak and H. van Dam, 1989, Inferring pH from diatoms: a comparison of old and new calibration methods, *Hydrobiologia*, 178, 209.

M.E. Thompson, 1982, The cation denudation rate as a quantitative index of sensitivity of eastern Canadian rivers to acidic atmospheric precipitation, *Water Air Soil Pollut.*, 18, 215.

K. Thornton, D. Marmorek, and P. Ryan, 1990, Methods for projecting future changes in surface water acid–base chemistry, State of the Science, SOS/T 14, National Acid Precipitation Assessment Program.

A. Tietema, 1993, Mass loss and nitrogen dynamics in decomposing litter of five forest ecosystems in relation to increased nitrogen deposition, *Biogeochemistry*, 20, 45.

A. Tietema, 1998, Microbial carbon and nitrogen dynamics in coniferous forest floor material collected along a European nitrogen deposition gradient, *For. Ecol. Manage.*, 101, 29.

A. Tietema and C. Beier, 1995, A correlative evaluation of nitrogen cycling in the forest ecosystems of the EC projects NITREX and EXMAN, *For. Ecol. Manage.*, 71, 143.

D. Toetz and J. Windell, 1993, Phytoplankton in a high-elevation lake, Colorado Front Range: application to lake acidification, *Great Basin Nat.*, 53, 350.

K.A. Tonnessen, 1991, The Emerald Lake Watershed study: introduction and site description, *Water Resour. Res.*, 27, 1537.

P. Torssander and C.-M. Mörth, 1998, Sulfur dynamics in the roof experiment at lake Gårdsjön deduced from sulfur and oxygen isotope ratios in sulfate, in *Experimental Reversal of Acid Rain Effects. The Gårdsjön Roof Project*, H. Hultberg, and R. Skeffington, Eds., John Wiley & Sons, Chichester, UK, 185.

J.T. Turk and D.H. Campbell, 1997, Are aquatic resources of the Mt. Zirkel Wilderness Area in Colorado affected by acid deposition and what will emissions reductions at the local power plant do?, U.S. Geological Survey Fact Sheet FS-043-97.

J.T. Turk and N.E. Spahr, 1991, Rocky Mountains: controls on lake chemistry, in *Acidic Deposition and Aquatic Ecosystems: Regional Case Studies*, D.F. Charles, Ed., Springer-Verlag, New York.

J.T. Turk, D.H. Campbell, G.P. Ingersoll, and D.W. Clow, 1992, Initial findings of synoptic snowpack sampling in Colorado Rocky Mountains, U.S. Geological Survey, Open-File Report 92-645.

J.T. Turk, D.H. Campbell, and N.E. Spahr, 1993, Use of chemistry and stable sulfur isotopes to determine sources of trends in sulfate of Colorado lakes, *Water Air Soil Pollut.*, 67, 415.

R.S. Turner, R.B. Cook, H. van Miegroet, D.W. Johnson, J.W. Elwood, O.P. Bricker, S.E. Lindberg, and G.M. Hornberger, 1990, Watershed and lake processes affecting chronic surface water acid–base chemistry, State of the Science, SOS/T 10, National Acid Precipitation Assessment Program.

R.S. Turner, P.F. Ryan, D.R. Marmorek, K.W. Thornton, T.J. Sullivan, J.P. Baker, S.W. Christensen, and M.J. Sale, 1992, Sensitivity to change for low-ANC eastern US lakes and streams and brook trout populations under alternative sulfate deposition scenarios, *Environ. Pollut.*, 77, 269.

B. Ulrich, 1983, Soil acidity and its relations to acid deposition, in *Effects of Accumulation of Air Pollutants on Forest Ecosystems*, B. Ulrich and J. Pankrath, Eds., D. Reidel, Boston, 127.

B. Ulrich, R. Mayer, and T.K. Khanna, 1980, Chemical changes due to acid precipitation in a loess-derived soil in central Europe, *Soil Sci.*, 130, 193.

U.S. Department of Commerce, 1990 Census.

A.J. Uutala, 1990, *Chaoborus* (Diptera: Chaoboridae) mandibles—paleolimnological indicators of the historical status of fish populations in acid-sensitive lakes, *J. Paleolimnol.*, 4, 139.

N. van Breemen and H.F.G. van Dijk, 1988, Ecosystem effects of atmospheric deposition of nitrogen in The Netherlands, *Environ. Pollut.*, 54, 249.

D. Van Dam and N. van Breemen, 1995, NIICE: a model for cycling of nitrogen isotopes in coniferous forest ecosystems, *Ecol. Model.*, 79, 255.

H. van Miegroet, 1994, The relative importance of sulfur and nitrogen compounds in the acidification of fresh water, in *Acidification of Freshwater Ecosystems: Implications for the Future*, C.E.W. Steinberg and R.F. Wright, Eds., John Wiley & Sons, New York.

H. van Miegroet, D.W. Cole, and N.W. Foster, 1992, Nitrogen distribution and cycling, in *Atmospheric Deposition and Forest Nutrient Cycling: A Synthesis of the Integrated Forest Study*, D.W. Johnson and S.E. Lindberg, Eds., Springer-Verlag, New York, 178.

J. Van Sickle and M.R. Church, 1995, Methods for estimating the relative effects of sulfur and nitrogen deposition on surface water chemistry, EPA/600/R-95/172, Environmental Protection Agency, Washington, DC.

F.A. Vertucci and P.S. Corn, in press, Declines of amphibian populations in the Rocky Mountains are not due to episodic acidification, *Ecol. Appl.*

F.A. Vertucci and P.S. Corn, 1996, Evaluation of episodic acidification and amphibian declines in the Rocky Mountains, *Ecol. Appl.*, 6, 449.

F.A. Vertucci and J.M. Eilers, 1993, Issues in monitoring wilderness lake chemistry: a case study in the Sawtooth Mountains, Idaho, *Environ. Monitor. Assess.*, 28, 277.

P.M. Vitousek, 1977, The regulation of element concentrations in mountain streams in the northeastern United States, *Ecol. Monogr.*, 47, 65.

P.M. Vitousek, J.D. Aber, R.W. Howarth, G.E. Likens, P.A. Mason, D.W. Schindler, W.H. Schlesinger, and D. Tilman, 1997, Human alteration of the global nitrogen cycle: sources and consequences, *Ecol. Appl.*, 7, 737.

R.A. Vollenweider, 1975, Input–output models with special reference to the phosphorus loading concept in limnology, *Schweiz. Z. Hydrol.*, 37, 53.

D.B. Wake, 1991, Declining amphibian populations, *Science*, 253, 860.

P.M. Walthall, 1985, Acidic deposition and the soil environment of Loch Vale Watershed in Rocky Mountain National Park, Ph.D. dissertation, Colorado State University, Fort Collins, CO.

P. Warfvinge, M. Holmberg, M. Posch, and R.F. Wright, 1992, The use of dynamic models to set target loads, *Ambio*, 21, 369.

C.J. Watras and T.M. Frost, 1989, Little Rock Lake (Wisconsin): perspectives on an experimental ecosystem approach to seepage lake acidification, *Arch. Environ. Contam. Toxicol.*, 18, 157.

J.R. Webb, B.J. Cosby, K. Eshleman, and J. Galloway, 1995, Change in the acid–base status of Appalachian Mountain catchments following forest defoliation by the gypsy moth, *Water Air Soil Pollut.*, 85, 535.

J.R. Webb, F.A. Deviney, J.N. Galloway, C.A. Rinehart, P.A Thompson, and S. Wilson, 1994, The acid–base status of native brook trout streams in the mountains of Virginia. A regional assessment based on the Virginia Trout Stream Sensitivity Study, University of Virginia, Charlottesville, VA.

K.E. Webster and J.M. Eilers, 1994, Clara and Vandercook Lakes: limnological and watershed studies for 1981–1983, Tech. Bull., Wisconsin Department of Natural Resources, Madison, WI.

K.E. Webster, P.L. Brezonik, and B.J. Holdhusen, 1993, Temporal trends in low alkalinity lakes of the Upper Midwest (1983–1989), *Water Air Soil Pollut.*, 67, 397.

K.E. Webster, A.D. Newell, L.A. Baker, and P.L. Brezonik, 1990, Climatically induced rapid acidification of a softwater seepage lake, *Nature*, 347, 374.

D.A. Wentz and W.J. Rose, 1989, Interrelationships among hydrologic-budget components of a northern Wisconsin seepage lake and implications for acid-deposition modeling, *Arch. Environ. Contam. Toxicol.*, 18, 147.

M.C. Whiting, D.R. Whitehead, R.B. Holmes, and S.A. Norton, 1989, Paleolimnological reconstruction of recent acidity changes in four Sierra Nevada lakes, *J. Paleolimnol.*, 2, 285.

P.J. Wigington, J.P. Baker, D.R. DeWalle, W.A. Kretser, P.S. Murdoch, H.A. Simonin, J. Van Sickle, M.K. McDowell, D.V. Peck, and W.R. Barchet, 1993, Episodic acidification of streams in the northeastern United States: chemical and biological results of the Episodic Response Project, EPA/600/R-93/190, Environmental Protection Agency, Washington, DC.

P.J. Wigington, Jr., T.D. Davies, M. Tranter, and K.N. Eshleman, 1990, Episodic acidification of surface waters due to acidic deposition, State of Science and Technology Report No. 12, National Acid Precipitation Assessment Program, Washington, DC.

P.J. Wigington, Jr., D.R. DeWalle, P.S. Murdoch, W.A. Kretser, H.A. Simonin, J. Van Sickle, and J.P. Baker, 1996, Episodic acidification of small streams in the northeastern United States: ionic controls of episodes, *Ecol. Appl.*, 6, 389.

M.W. Williams and J.M. Melack, 1991a, Solute chemistry of snowmelt and runoff in an alpine basin, Sierra Nevada, *Water Resour. Res.*, 27, 1575.

M.W. Williams and J.M. Melack, 1991b, Precipitation chemistry in and ionic loading to an alpine basin, Sierra Nevada, *Water Resour. Res.*, 27, 1563.

M.W. Williams and K.A. Tonnessen, in review, Water quality in Grand Teton National Park, *Park Sci.*

M.W. Williams, R.C. Bales, A.D. Brown, and J.M. Melack, 1995, Fluxes and transformations of nitrogen in a high-elevation catchment, Sierra Nevada, *Biogeochemistry*, 28, 1-31.

M.W. Williams, J.S. Baron, N. Caine, R. Sommerfeld, and R. Senford, Jr., 1996a, Nitrogen saturation in the Rocky Mountains, *Environ. Sci. Technol.*, 30, 640.

M.W. Williams, A.D. Brown, and J.M. Melack, 1993, Geochemical and hydrologic controls on the composition of surface water in a high-elevation basin, Sierra Nevada, California, *Limnol. Oceanogr.*, 38, 775.

M.W. Williams, M. Losleben, N. Caine, and D. Greenland, 1996c, Changes in climate and hydrochemical responses in a high elevation catchment in the Rocky Mountains, U.S.A., *Limnol. Oceanogr.*, 41, 939.

M.W. Williams, T. Platts-Mills, and N. Caine, 1996b, Landscape controls on surface water nitrate concentrations at catchment and regional scales in the Colorado Rocky Mountains, Chapman Conference, Nitrogen Cycling in Forested Catchments, Sun River, OR, September 16–20.

S.A. Wissinger and H.H. Whiteman, 1992, Fluctuation in a Rocky Mountain population of salamanders: anthropogenic acidification or natural variation?, *J. Herpetol.*, 26, 377.

D.F. Woodward, A.M. Farag, E.E. Little, B. Steadman, and R. Yancik, 1991, Sensitivity of greenback cutthroat trout to acidic pH and elevated aluminum, *Trans. Am. Fish. Soc.*, 120, 34.

D.F. Woodward, A.M. Farag, M.E. Mueller, E.E. Little, and F.A. Vertucci, 1989, Sensitivity of endemic Snake River cutthroat trout to acidity and elevated aluminum, *Trans. Am. Fish. Soc.*, 118, 630.

R.F. Wright, 1983, Predicting acidification of North American lakes, Acid Rain Research Report 3/1983, Norwegian Institute for Water Research, Oslo, Norway.

R.F. Wright, 1988, Acidification of lakes in the eastern United States and southern Norway: a comparison, *Environ. Sci. Technol.*, 22, 178.

R.F. Wright, 1989, Rain project: role of organic acids in moderating pH change following reduction in acid deposition, *Water Air Soil Pollut.*, 46, 251.

R.F. Wright, 1991, Acidification: whole catchment manipulation, in *Ecosystem Experiments*, H.A. Mooney, E. Medina, and D.W. Schindler, Eds., John Wiley & Sons, New York.

R.F. Wright and A. Henriksen, 1978, Chemistry of small Norwegian lakes, with special reference to acid precipitation, *Limnol. Oceanogr.*, 23, 487.

R.F. Wright and A. Tietema, 1995, Ecosystem response to 9 years of nitrogen addition at Sogndal, Norway, *For. Ecol. Manage.*, 71, 133.

R.F. Wright and N. van Breemen, 1995, The NITREX project: an introduction, *For. Ecol. Manage.*, 71, 1.

R.F. Wright, B.J. Cosby, M.B. Flaten, and J.O. Reuss, 1990, Evaluation of an acidification model with data from manipulated catchments in Norway, *Nature*, 343, 53.

R.F. Wright, E.T. Gjessing, N. Christophersen, E. Lotse, H.M. Seip, A. Semb, B. Sletaune, R. Storhaug, and K. Wedum, 1986, Project rain: changing acid deposition to whole catchments. The first year of treatment, *Water Air Soil Pollut.*, 30, 47.

R.F. Wright, E. Lotse, and A. Semb, 1988b, Reversibility of acidification shown by whole-catchment experiments, *Nature*, 334, 670.

R.F. Wright, E. Lotse, and A. Semb, 1993, RAIN Project: Results after 8 years of experimentally reduced acid deposition to a whole catchment, *Can. J. Fish. Aquat. Sci.*, 50, 258.

R.F. Wright, E. Lotse, and E. Semb, 1994, Experimental acidification of alpine catchments at Sogndal, Norway: results after 8 years, *Water Air Soil Pollut.*, 72, 297.

R.F. Wright, S.A. Norton, D.F. Brakke, and T. Frogner, 1988a, Experimental verification of episodic acidification of freshwater by sea salts, *Nature*, 334, 422.

N.D. Yan, W. Keller, K.M. Somers, T.W. Pawson, and R.E. Girard, 1996a, Recovery of crustacean zooplankton communities from acid and metal contamination: comparing manipulated and reference lakes, *Can. J. Fish. Aquat. Sci.*, 5, 1301.

N.D. Yan, P.G. Welsh, H. Lin, D.J. Taylor, and J.-M. Filion, 1996b, Demographic and genetic evidence of the long-term recovery of *Daphnia galeata mendotae* (Crustacea: Daphniidae) in Sudbury lakes following additions of base: the role of metal toxicity, *Can. J. Fish. Aquat. Sci.*, 53, 1328.

Index

A

Abies balsamea, 186, 238
Absaroka-Beartooth wilderness (ABW), 113
ABW, *see* Absaroka-Beartooth wilderness
Acid
  anions, 10
  –base chemistry, measured changes in, 116
  deposition standards, 131
  loading, 51
  neutralizing capacity (ANC), 10, 13
    Adirondack lakes, 244
    chrysophyte reconstructions of, 78
    consumption, 11\
    deacidification change in, 118
    decreases in, 22
    depression, 100
    diatom-inferred, 112, 116, 120
    effect of logging on, 53
    episodic, 148
    fall overturn, 270
    generation, watershed processes and, 38
    Gran titration, 27
    influence of aluminum on, 17
    large scale titration of, 50, 78
    projected changes in, 83
    surface water, 301
  stress, 19
Acid Deposition Control Act, 120
Acidic deposition research developments, 6
Acidification, *see also* Chronic acidification;
        Episodic acidification; Surface water
        acidification, extent and magnitude
        of
  artificial, 183
  causes of, 5, 38
  diatom-inferred, 246, 249
  DOC/TOC responses to, 47
  dynamics, models of, 196
  effects of, 60
  estimates of historical, 249
  MAGIC projections of stream water, 229
  major cause of, 56

models, 33
process(es)
  experimental field studies of, 175
  long-term, 12
  role of nitrogen in, 248
quantification of, 15
recovery, 73
S-driven, 255
sensitivity to, 10, 19
simulated historical, 103
soil, 16, 125
watershed, 81
whole-lake, 115
Acidifying agent, 132
Acidity, chronic, 135
Acid Precipitation Act of 1980, 1
Adirondack Lake Survey Corporation
        (ALSC), 39, 203, 239
Adirondack Long-Term Monitoring Program
        (ALTM), 71, 240
Adirondack Mountains, 70
Adirondack Park, NY, 237–257
  background and available data, 237–241
    ALSC, 239
    ALTM, 240–241
    DDRP, 240
    ELS-I, 239
    ELS-II, 240
    ERP, 241
    PIRLA, 240
  lake-water chemistry, 244–246
  organic acidity, 246–248
  overall assessment, 256–257
  role of landscape and disturbance in
        acidification processes, 252–256
  role of nitrogen in acidification processes,
        248–252
  watershed history, 241–244
AERP, *see* Environmental Protection Agency
        Aquatic Effects Research Program
Air pollution, 138
Air quality related values (AQRVs), 260
ALCHEMI model, 218
Alfisols, 274